Lecture Notes on the

Mathematics of ACOUSTICS

Lecture Notes on the Mathematics of ACOUSTICS

Edited by

M. C. M. WRIGHT

(University of Southampton, UK)

ICP

Imperial College Press

Published by

Imperial College Press
57 Shelton Street
Covent Garden
London WC2H 9HE

Distributed by

World Scientific Publishing Co. Pte. Ltd.
5 Toh Tuck Link, Singapore 596224
USA office: 27 Warren Street, Suite 401-402, Hackensack, NJ 07601
UK office: 57 Shelton Street, Covent Garden, London WC2H 9HE

British Library Cataloguing-in-Publication Data
A catalogue record for this book is available from the British Library.

LECTURE NOTES ON THE MATHEMATICS OF ACOUSTICS

ISBN 1-86094-496-5

Printed in Singapore by World Scientific Printers (S) Pte Ltd

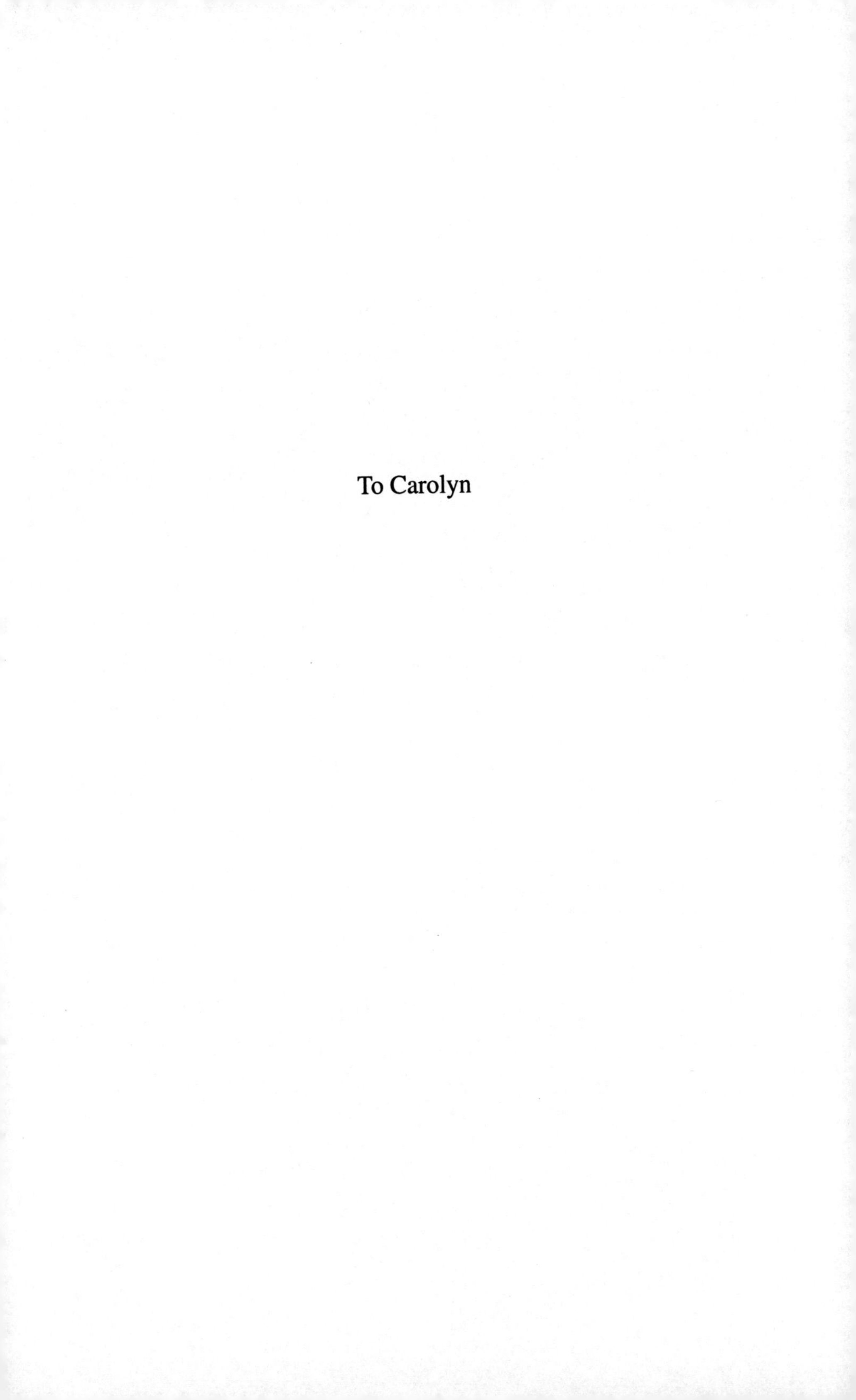

To Carolyn

Preface

The chapters in this volume are based on lectures given at a one week summer school on mathematics for acoustics, funded by the UK Engineering and Physical Sciences Research Council (EPSRC) and held at the University of Southampton from the 14th to the 18th of July, 2003. The idea of such a summer school was conceived by Keith Attenborough in response to a call from EPSRC for proposals for funding to increase the level of mathematical competency for engineering researchers in the UK, issued by Engineering Programme Manager Elizabeth Hilton, and administered by Stephen Elsby. The SMART (Support Mathematics for Acoustics Research Training) summer school was the first such summer school to be funded and run; another has since taken place and a call has been issued for more to follow.

Although it was arranged at fairly short notice the summer school proved a great success. Over forty young acoustics researchers and about twenty lecturers made their way to Southampton for an intensive week of mathematics, along with staff and research students of Southampton's School of Mathematics who provided additional tutorial support. The students lived together, ate together and spent most of the day in lectures, tutorials, seminars and one-to-one surgeries where they could raise subjects of interest with any of the staff they chose. They also had the chance to exchange ideas with one another, an opportunity that I hope will benefit those who go on to form part of the UK's future acoustics research capability. The response was such that it is planned to run such a summer school biannually with this book as a textbook, though this does not, of course, preclude the addition of other topics in future. We do not claim that the topics discussed here provide anything like complete coverage of the field of mathematical acoustics, merely that it forms a useful body of material at a level that may be of use to students of the subject.

Chapters 8 and 11 are based on seminars, which were given at the end of each day and the morning of the final day. These aimed to give an overview of a particular area without fitting into the series of lectures which had a theme for each day, similar though not identical to the parts of this book. Additional seminars on subjects such as nonlinear acoustics, quantum chaos in acoustics and atmospheric acoustics could not be included for one reason or another. The remaining chapters are based on lectures apart from Chapters 5 and 7, which didn't find a place in the final timetable.

During the editing process the material in this book has been somewhat expanded from what was presented during the summer school. It was decided at the start of our preparations for the school that total consistency was not only impractical but actually undesirable, since several notations are in common use in the literature and it would be unfair to gloss over this inevitable feature of acoustical research. That excuse is less valid for this book; nevertheless a few instances of diversity remain where the alternative would be too restrictive. Principally, in some chapters $\hat{f}$ denotes an integral transform of f whereas in others it is given by F. On the other hand exponential time factors are given as $\exp(-\mathrm{i}\omega t)$ throughout. The following people should be thanked for their help with the production of this book: Mirko Schaedlich, Chris Howls, Chris Morfey and Jeremy Astley at Southampton; Dr K. K. Phua and Ying Oi Chiew at Imperial College Press and finally the dedicatee who, in the time since our wedding a month after the summer school, has never complained about the time I've spent poring through LaTeX manuals or hunched over a laptop wrestling with this manuscript.

Matthew Wright

Contents

Part II Wave Motion

Part III Aeroacoustics

PART I
Mathematical Methods

Chapter 1

Vector Calculus

J. W. Elliott

Department of Mathematics
University of Hull

1.1 Motivation

In the absence of any mean flow, viscous dissipation and heat transfer, the propagation of sound in a stationary medium of uniform mean density ρ_0 and pressure p_0 is governed by the linearised Euler equations of motion

$$\frac{1}{c_0^2}\frac{\partial p}{\partial t} + \rho_0 \boldsymbol{\nabla} \cdot \mathbf{u} = \rho_0 q, \qquad \rho_0 \frac{\partial \mathbf{u}}{\partial t} = -\boldsymbol{\nabla} p + \mathbf{f}. \tag{1}$$

Here c_0 is the uniform speed of sound, $p = p(\mathbf{r}, t)$ is the acoustic pressure and $\mathbf{u} = \mathbf{u}(\mathbf{r}, t)$ is the fluid velocity at a point P with position vector $\mathbf{r} = \overrightarrow{OP}$ at a time t. In addition we have considered the presence of both an unsteady body force $\mathbf{f}$, which vanishes in the undisturbed state, and volume sources q.

We can eliminate the velocity $\mathbf{u}$ as follows

$$\begin{aligned}\frac{1}{c_0^2}\frac{\partial^2 p}{\partial t^2} &= \frac{\partial}{\partial t}(\rho_0 q - \rho_0 \boldsymbol{\nabla} \cdot \mathbf{u}) = \rho_0 \frac{\partial q}{\partial t} - \boldsymbol{\nabla} \cdot \left(\rho_0 \frac{\partial \mathbf{u}}{\partial t}\right) \\ &= \rho_0 \frac{\partial q}{\partial t} - \boldsymbol{\nabla} \cdot (-\boldsymbol{\nabla} p + \mathbf{f}) = \rho_0 \frac{\partial q}{\partial t} + \nabla^2 p - \boldsymbol{\nabla} \cdot \mathbf{f}. \end{aligned} \tag{2}$$

Thus the governing equation for the pressure p is

$$\left(\frac{1}{c_0^2}\frac{\partial^2}{\partial t^2} - \nabla^2\right) p(\mathbf{r}, t) = F(\mathbf{r}, t) \equiv \rho_0 \frac{\partial q}{\partial t} - \boldsymbol{\nabla} \cdot \mathbf{f}. \tag{3}$$

This is the inhomogeneous wave equation describing the production of sound by the volume source q and body force $\mathbf{f}$.

1.1.1 *Velocity potential*

In the absence of body forces, where $\mathbf{f} = \mathbf{0}$, the flow is **irrotational** since

$$\frac{\partial \boldsymbol{\omega}}{\partial t} = \frac{\partial}{\partial t}(\boldsymbol{\nabla} \times \mathbf{u}) = \boldsymbol{\nabla} \times \left(\frac{\partial \mathbf{u}}{\partial t}\right) = -\frac{1}{\rho_0}\boldsymbol{\nabla} \times (\boldsymbol{\nabla} p) \equiv 0. \tag{4}$$

Since the fluid is disturbed from an initial state of rest, we can set

$$\boldsymbol{\omega} = \boldsymbol{\nabla} \times \mathbf{u} \equiv \mathbf{0}, \qquad \text{so} \qquad \mathbf{u} = \boldsymbol{\nabla}\phi(\mathbf{r}, t), \tag{5}$$

where ϕ is the **velocity potential**. The linearised equations of motion are then

$$\frac{1}{c_0^2}\frac{\partial p}{\partial t} + \rho_0 \nabla^2 \phi = \rho_0 q, \quad \boldsymbol{\nabla}\left(p + \rho_0 \frac{\partial \phi}{\partial t}\right) = \mathbf{0}, \quad \text{so} \quad p = -\rho_0 \frac{\partial \phi}{\partial t} \tag{6}$$

where the momentum equation integrates to yield the linearised Bernoulli equation. Furthermore, it follows that

$$\left(\frac{1}{c_0^2}\frac{\partial^2}{\partial t^2} - \nabla^2\right)\phi(\mathbf{r}, t) = -q(\mathbf{r}, t). \tag{7}$$

Outside the source region, where $q = 0$, the fluctuations $\mathbf{u}$ and p all propagate as sound waves governed by the homogeneous wave equation, namely

$$\left(\frac{1}{c_0^2}\frac{\partial^2}{\partial t^2} - \nabla^2\right)\phi(\mathbf{r}, t) = 0. \tag{8}$$

The velocity $\mathbf{u}$ associated with the sound is called the **acoustic particle velocity**.

1.1.2 *Plane waves*

A plane wave is one in which

$$\mathbf{u} = u(x, t)\hat{\mathbf{i}}, \qquad p = p(x, t). \tag{9}$$

Here x measures distance in the direction of the unit vector $\hat{\mathbf{i}}$, which is (say) to the right. In the absence of body forces the linearised equations of motion are

$$\frac{1}{c_0^2}\frac{\partial p}{\partial t} + \rho_0 \frac{\partial u}{\partial x} = 0, \qquad \rho_0 \frac{\partial u}{\partial t} = -\frac{\partial p}{\partial x}. \tag{10}$$

It is easily seen that both p and u satisfy the **classical wave equation**

$$\left(\frac{1}{c_0^2}\frac{\partial^2}{\partial t^2} - \frac{\partial^2}{\partial x^2}\right)p(x, t) = 0, \qquad \left(\frac{1}{c_0^2}\frac{\partial^2}{\partial t^2} - \frac{\partial^2}{\partial x^2}\right)u(x, t) = 0. \tag{11}$$

These have a general solution

$$p(x,t) = F(x - c_0 t) + G(x + c_0 t),$$
$$u(x,t) = \frac{1}{\rho_0 c_0}\left[F(x - c_0 t) - G(x + c_0 t)\right], \tag{12}$$

where F and G are arbitrary, known as **d'Alembert's solution**. This solution represents the linear superposition of two arbitrary disturbances of invariant form both moving with a speed c, one to the right and one to the left. In particular for a wave simply travelling to the right

$$p = \rho_0 c_0 u = F(x - c_0 t), \qquad \text{and} \qquad Z = \frac{p}{u} = \rho_0 c_0, \tag{13}$$

where Z is the **acoustic impedance** and $\beta = 1/Z$ is the **acoustic admittance**.

1.1.2.1 *Planar harmonic wave*

More generally a planar harmonic wave of angular frequency ω is of the form

$$\begin{aligned} p(\mathbf{r},t) &= a\cos(k_1 x + k_2 y + k_3 z - \omega t) + b\sin(k_1 x + k_2 y + k_3 z - \omega t) \\ &= R\cos(k_1 x + k_2 y + k_3 z - \omega t + \epsilon). \end{aligned} \tag{14}$$

Taking $\tan\epsilon = b/a$, with $\mathbf{r} = (x,y,z)$ the position vector and $\mathbf{k} = (k_1, k_2, k_3)$ the **wavenumber vector**, then we can write

$$\begin{aligned} p(\mathbf{r},t) &= R\cos(\mathbf{k}\cdot\mathbf{r} - \omega t + \epsilon) \\ &= \mathrm{Re}\left\{R\mathrm{e}^{\mathrm{i}(\mathbf{k}\cdot\mathbf{r} - \omega t + \phi)}\right\} = \mathrm{Re}\left\{A\mathrm{e}^{\mathrm{i}(\mathbf{k}\cdot\mathbf{r} - \omega t)}\right\} \end{aligned} \tag{15}$$

where $A = R\mathrm{e}^{\mathrm{i}\phi}$ is the **complex amplitude**. The points of constant phase, where $\phi(\mathbf{r},t) \equiv \mathbf{k}\cdot\mathbf{r} - \omega t =$ constant define a **wave front**, and such points satisfy

$$\mathrm{d}\phi \equiv \mathbf{k}\cdot\mathrm{d}\mathbf{r} - \omega\,\mathrm{d}t = 0. \tag{16}$$

Thus at any given time t the wave-front is given by $\mathrm{d}\phi \equiv \mathbf{k}\cdot\mathrm{d}\mathbf{r} = 0$, so the wavenumber vector is perpendicular to any wave-front (crest or trough). If we move with a given crest (or trough) then we must have

$$\frac{\mathrm{d}\phi}{\mathrm{d}t} \equiv \mathbf{k}\cdot\frac{\mathrm{d}\mathbf{r}}{dt} - \omega = 0, \qquad \text{so} \qquad \mathbf{v}_\phi = \frac{\mathrm{d}\mathbf{r}}{\mathrm{d}t} = \frac{\omega}{|\mathbf{k}|^2}\mathbf{k}, \tag{17}$$

where $\mathbf{v}_\phi$ is the **phase-velocity**. Thus the wave propagates in the direction of the wavenumber vector at a **wave-speed**

$$c_0 = |\mathbf{v}_\phi| = \frac{\omega}{|k|} = \frac{\omega}{\sqrt{k_1^2 + k_2^2 + k_3^2}} \tag{18}$$

and the **wavelength**, the distance between two adjacent crests, is $\lambda = 2\pi/|\mathbf{k}|$.

1.1.3 *Spherical waves*

A spherical wave is one in which

$$\mathbf{u} = u(r,t)\hat{\mathbf{r}}, \qquad p = p(r,t), \tag{19}$$

where r measures distance in the radial direction $\hat{\mathbf{r}}$. In the absence of body forces the linearised equations of motion are

$$\frac{1}{c_0^2}\frac{\partial p}{\partial t} + \rho_0 \frac{1}{r^2}\frac{\partial}{\partial r}\left(r^2 u\right) = 0, \qquad \rho_0 \frac{\partial u}{\partial t} = -\frac{\partial p}{\partial r}. \tag{20}$$

It is easily seen that p and u satisfy the (spherical) wave equation

$$\left(\frac{1}{c_0^2}\frac{\partial^2}{\partial t^2} - \frac{\partial^2}{\partial r^2}\right)(rp) = 0, \qquad \left(\frac{1}{c_0^2}\frac{\partial^2}{\partial t^2} - \frac{\partial^2}{\partial r^2}\right)(ru) = 0, \tag{21}$$

since here

$$\nabla^2 p(r,t) \equiv \frac{1}{r^2}\frac{\partial}{\partial r}\left[r^2 \frac{\partial p}{\partial r}\right] \equiv \frac{\partial^2 p}{\partial r^2} + \frac{2}{r}\frac{\partial p}{\partial r} = \frac{1}{r}\frac{\partial^2}{\partial r^2}(rp). \tag{22}$$

Thus we have a general solution

$$p(r,t) = \frac{1}{r}\left[F(r - c_0 t) + G(r + c_0 t)\right] \tag{23}$$

where F and G correspond to waves radiating outward to infinity from the origin, and inward to the origin from infinity respectively.

1.1.3.1 *Causality*

Usually when sound is generated in open space we insist that the sound must not anticipate its cause, the so-called **causality condition**. Thus $G(r + c_0 t) = 0$ since it must have existed for an infinitely long time in the past. This condition is indistinguishable from Sommerfeld's *radiation condition*, namely

$$\lim_{r\to\infty} r\left(\frac{\partial p}{\partial t} + c_0 \frac{\partial p}{\partial r}\right) = 0. \tag{24}$$

1.1.3.2 *Harmonic spherical waves*

A harmonic outward propagating spherical wave of frequency ω is of the form

$$p = \mathrm{Re}\left[p'(r,t)\right], \qquad \text{where } p'(r,t) = \frac{A}{r}\mathrm{e}^{-\mathrm{i}\omega(t - r/c_0)} = \frac{A}{r}\mathrm{e}^{\mathrm{i}(\kappa r - \omega t)}, \tag{25}$$

where $\kappa = \omega/c_0$ is and A is a complex amplitude. Now if $u = \mathrm{Re}\,[u'(r,t)]$ then

$$\frac{\partial u'}{\partial t} = -\frac{\partial p'}{\partial r} = \frac{A}{r^2}\left(1 - \mathrm{i}\frac{\omega r}{c_0}\right)\mathrm{e}^{-\mathrm{i}\omega(t-r/c_0)} \tag{26}$$

so

$$u' = \frac{A}{\rho_0 c_0 r}\left(1 + \mathrm{i}\frac{c_0}{\omega r}\right)\mathrm{e}^{-\mathrm{i}\omega(t-r/c_0)}, \tag{27}$$

which we can write as

$$\begin{aligned} u' &= \frac{A}{\rho_0 c_0 r}\sqrt{1 + \left(\frac{c_0}{\omega r}\right)^2}\,\mathrm{e}^{\mathrm{i}\phi}\mathrm{e}^{-\mathrm{i}\omega(t-r/c_0)} \\ &= \frac{1}{\rho_0 c_0}\sqrt{1 + \left(\frac{c_0}{\omega r}\right)^2}\,p'(r, t - \frac{\phi}{\omega}) \end{aligned} \tag{28}$$

where $\phi = \tan^{-1}(c_0/\omega r) = \pi/2 - \tan^{-1}(\omega r/c_0)$ is the **phase angle**. Thus, in contrast to the planar case, the velocity and pressure are *not* in phase. Indeed as $r \to 0$, $\phi \to \pi/2$ and $p' \sim -\mathrm{i}\omega\rho_0 r u'$. We also see the **radiation impedance**

$$Z = \left|\frac{p'}{u'}\right| = \frac{\rho_0 c_0}{\sqrt{1 + (c_0/\omega r)^2}} \leq \rho_0 c_0. \tag{29}$$

As $r \to \infty$, $\phi \to 0$ the pressure and velocity are very nearly in phase, with the spherical wave acting like a plane wave with $p' \sim \rho_0 c_0 u'$.

1.2 Scalars and Vectors

Any entity that can be represented by a single real number, $\lambda \in \mathbb{R}$, is called a **scalar**. A **vector**, $\mathbf{a} \in \mathbb{R}^3$, is usually defined to be a quantity that has both a magnitude and a direction.

1.2.1 *Rectangular Cartesian coordinates*

In 3-D Euclidean $\mathbb{R}^3$-space we draw, through a fixed point O, the origin, three fixed, mutually perpendicular, lines Ox, Oy and Oz called the x-axis, y-axis and z-axis respectively, and collectively known as rectangular Cartesian axes $Oxyz$, as shown in Fig. 1.1. Place the thumb, index finger and middle finger of the right hand, at right angles in the most natural way. If the index finger points along Ox and the middle finger point along Oy then, for a **right-handed set**, the thumb points along Oz. This statement exhibits **cyclic symmetry** in x, y, z since it remains true when we replace x by y, y by z, and z by x.

The position of a general point P may be specified by drawing a line from O to P, of length $r = OP$. If α, β, γ denote the angles OP makes with Ox, Oy Oz,

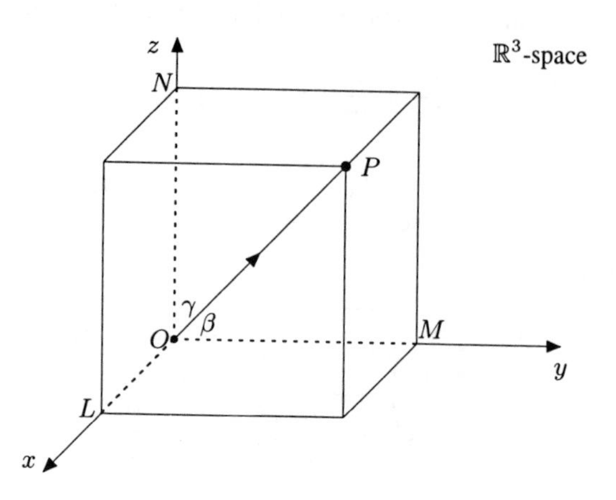

Fig. 1.1 Rectangular Cartesian axes.

then we define the Cartesian coordinates of P to be the numbers

$$x = r\cos\alpha, \qquad y = r\cos\beta, \qquad z = r\cos\gamma. \tag{30}$$

By considering the right-angled triangles OPL, OPM, OPN where L, M, N are the feet of the perpendiculars from P to the x, y and z-axes, we see that

$$x = \pm\,\text{length of } OL, \qquad y = \pm\,\text{length of } OM, \qquad z = \pm\,\text{length of } ON.$$

We refer to P as the point $P\colon (x, y, z)$. By Pythagoras

$$r = OP = \sqrt{x^2 + y^2 + z^2}. \tag{31}$$

1.2.2 *Geometric vectors*

A *geometrical* representation of a vector is the directed line segment $\mathbf{a} = \overrightarrow{OA}$, which has both a magnitude (the length OA) and a direction (indicated by an arrow in Fig. 1.2).

Let the vectors $\mathbf{a} = \overrightarrow{OA}$ and $\mathbf{b} = \overrightarrow{OB}$ represent adjacent sides of a parallelogram. This representation is *not* unique, since we also have $\mathbf{a} = \overrightarrow{BC}$. The **triangle** (or **parallelogram**) law states that $\mathbf{c} = \mathbf{a} + \mathbf{b}$ is represented by the diagonal $\overrightarrow{OC}$.

A **unit vector** is a vector of unit magnitude, often denoted by a circumflex. We denote the **modulus**, or magnitude, of a vector $\mathbf{a}$ by the scalar $a = |\mathbf{a}|$, so $\hat{\mathbf{a}} = \mathbf{a}/|\mathbf{a}|$ is a unit vector lying in the direction of $\mathbf{a}$. The zero vector $\mathbf{0}$, represented by the point O, is the only vector without an associated direction.

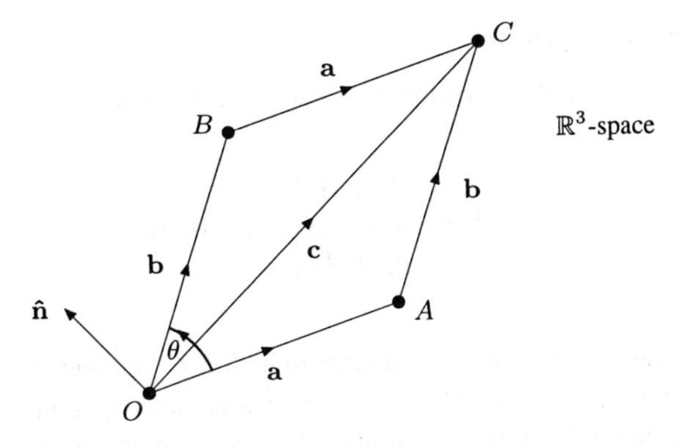

Fig. 1.2 A vector in the plane.

1.2.3 *Vector algebra*

Let $\hat{\mathbf{i}}$, $\hat{\mathbf{j}}$ and $\hat{\mathbf{k}}$ denote unit vectors in the directions of the x-axis, y-axis and z-axis respectively, then for any vector $\mathbf{a} \in \mathbb{R}^3$, we can uniquely write

$$\mathbf{a} = a_1\hat{\mathbf{i}} + a_2\hat{\mathbf{j}} + a_3\hat{\mathbf{k}} \qquad \text{or} \qquad \mathbf{a} = (a_1, a_2, a_3), \tag{32}$$

where $a_1, a_2, a_3 \in \mathbb{R}$. For example $\hat{\mathbf{i}} = (1, 0, 0)$, $\hat{\mathbf{j}} = (0, 1, 0)$, $\hat{\mathbf{k}} = (0, 0, 1)$. Then the modulus or magnitude of a vector $\mathbf{a} = (a_1, a_2, a_3)$ is the scalar

$$a = |\mathbf{a}| = \sqrt{a_1^2 + a_2^2 + a_3^2}. \tag{33}$$

Given vectors $\mathbf{a} = (a_1, a_2, a_3)$, $\mathbf{b} = (b_1, b_2, b_3)$ and a scalar $\lambda \in \mathbb{R}$, then

$$\mathbf{a} + \mathbf{b} = (a_1 + b_1, a_2 + b_2, a_3 + b_3), \qquad \lambda\mathbf{a} = (\lambda a_1, \lambda a_2, \lambda b_3). \tag{34}$$

1.2.4 *The scalar product*

The scalar product of two vectors $\mathbf{a} = (a_1, a_2, a_3)$, $\mathbf{b} = (b_1, b_2, b_3)$ is the scalar

$$\mathbf{a} \cdot \mathbf{b} = a_1b_1 + a_2b_2 + a_3b_3 = |\mathbf{a}|\,|\mathbf{b}| \cos\theta \tag{35}$$

where if $\mathbf{a} = \overrightarrow{OA}$ and $\mathbf{b} = \overrightarrow{OB}$ then θ is the angle between OA and OB. Consequently

$$\mathbf{a} \cdot \mathbf{a} = a_1^2 + a_2^2 + a_3^2 = |\mathbf{a}|^2 = a^2. \tag{36}$$

Two non-zero vectors $\mathbf{a}$ and $\mathbf{b}$ are **orthogonal** to each other if $\mathbf{a} \cdot \mathbf{b} = 0$.

1.2.5 *The vector product*

The vector product of two vectors $\mathbf{a} = (a_1, a_2, a_3)$ and $\mathbf{b} = (b_1, b_2, b_3)$ then

$$\mathbf{a} \times \mathbf{b} = \begin{vmatrix} \hat{\mathbf{i}} & \hat{\mathbf{j}} & \hat{\mathbf{k}} \\ a_1 & a_2 & a_3 \\ b_1 & b_2 & b_3 \end{vmatrix} = |\mathbf{a}||\mathbf{b}| \sin\theta \, \hat{\mathbf{n}}, \tag{37}$$

where the unit vector $\hat{\mathbf{n}}$ is perpendicular to both $\mathbf{a} = \overrightarrow{OA}$ and $\mathbf{b} = \overrightarrow{OB}$. Here θ, the angle between $\overrightarrow{OA}$ and $\overrightarrow{OB}$, is such that when looking along $\overrightarrow{ON}$, the sense of θ increasing is clockwise. Two non-zero vectors $\mathbf{a}$ and $\mathbf{b}$ are parallel to each other if $\mathbf{a} \times \mathbf{b} = \mathbf{0}$. Also if $\mathbf{a} = \overrightarrow{OA}$, $\mathbf{b} = \overrightarrow{OB}$ represent two adjacent sides of a parallelogram, then $|\mathbf{a} \times \mathbf{b}| = |\mathbf{a}|\,|\mathbf{b}| \sin\theta$ represents the area of the parallelogram.

1.2.6 *The triple scalar product*

For three vectors $\mathbf{a} = (a_1, a_2, a_3)$, $\mathbf{b} = (b_1, b_2, b_3)$ and $\mathbf{c} = (c_1, c_2, c_3)$ the triple scalar product, sometimes denoted by $[\mathbf{a}, \mathbf{b}, \mathbf{c}]$, is defined to be the scalar

$$\mathbf{a} \cdot (\mathbf{b} \times \mathbf{c}) = \mathbf{b} \cdot (\mathbf{c} \times \mathbf{a}) = \mathbf{c} \cdot (\mathbf{a} \times \mathbf{b}) = \begin{vmatrix} a_1 & a_2 & a_3 \\ b_1 & b_2 & b_3 \\ c_1 & c_2 & c_3 \end{vmatrix}. \tag{38}$$

If $\mathbf{a} = \overrightarrow{OA}$, $\mathbf{b} = \overrightarrow{OB}$ and $\mathbf{c} = \overrightarrow{OC}$ represent adjacent edges of a parallelepiped, then $|\mathbf{a} \cdot (\mathbf{b} \times \mathbf{c})\,|$ represents the volume of the parallelepiped. Consequently three non-zero vectors $\mathbf{a}$, $\mathbf{b}$ and $\mathbf{c}$ are **co-planar** if and only if $\mathbf{a} \cdot (\mathbf{b} \times \mathbf{c}) = 0$.

1.2.7 *The triple vector product*

The triple vector product of three vectors $\mathbf{a} = (a_1, a_2, a_3)$, $\mathbf{b} = (b_1, b_2, b_3)$ and $\mathbf{c} = (c_1, c_2, c_3)$ is the vector defined by

$$(\mathbf{a} \times \mathbf{b}) \times \mathbf{c} = (\mathbf{a} \cdot \mathbf{c})\mathbf{b} - (\mathbf{b} \cdot \mathbf{c})\mathbf{a}. \tag{39}$$

From which it follows that

$$(\mathbf{a} \times \mathbf{b}) \cdot (\mathbf{c} \times \mathbf{d}) = (\mathbf{a} \cdot \mathbf{c})(\mathbf{b} \cdot \mathbf{d}) - (\mathbf{a} \cdot \mathbf{d})(\mathbf{b} \cdot \mathbf{c}). \tag{40}$$

1.2.8 *The standard basis*

The set of vectors $\{\hat{\mathbf{i}}, \hat{\mathbf{j}}, \hat{\mathbf{k}}\}$, also denoted by $\{\hat{\mathbf{e}}_1, \hat{\mathbf{e}}_2, \hat{\mathbf{e}}_3\}$ and $\{\hat{\mathbf{i}}_1, \hat{\mathbf{i}}_2, \hat{\mathbf{i}}_3\}$, is known as the standard basis. It is a right-handed, orthonormal basis set satisfying

$$\hat{\mathbf{i}}\cdot\hat{\mathbf{i}} = \hat{\mathbf{j}}\cdot\hat{\mathbf{j}} = \hat{\mathbf{k}}\cdot\hat{\mathbf{k}} = 1, \qquad \hat{\mathbf{i}}\cdot\hat{\mathbf{j}} = \hat{\mathbf{j}}\cdot\hat{\mathbf{k}} = \hat{\mathbf{k}}\cdot\hat{\mathbf{i}} = 0, \tag{41}$$

$$\hat{\mathbf{i}}\times\hat{\mathbf{i}} = \hat{\mathbf{j}}\times\hat{\mathbf{j}} = \hat{\mathbf{k}}\times\hat{\mathbf{k}} = \mathbf{0}, \tag{42}$$

$$\hat{\mathbf{i}}\times\hat{\mathbf{j}} = \hat{\mathbf{k}}, \qquad \hat{\mathbf{j}}\times\hat{\mathbf{k}} = \hat{\mathbf{i}}, \qquad \hat{\mathbf{k}}\times\hat{\mathbf{i}} = \hat{\mathbf{j}}. \tag{43}$$

1.2.9 *The position vector*

The position vector of a general point P with Cartesian coordinates (x, y, z) relative to an origin O is the directed line segment $\mathbf{r} = \overrightarrow{OP}$, which we write as

$$\mathbf{r} = x\hat{\mathbf{i}} + y\hat{\mathbf{j}} + z\hat{\mathbf{k}}, \qquad \text{or} \qquad \mathbf{r} = (x, y, z)\,. \tag{44}$$

The modulus $r = OP = |\mathbf{r}|$, in agreement with Eq. (31) is given by

$$r^2 = |\mathbf{r}|^2 = \mathbf{r}\cdot\mathbf{r} = x^2 + y^2 + z^2. \tag{45}$$

An alternative notation to the above is to write $\mathbf{x} = \overrightarrow{OP}$ as the position vector of the point $P\colon (x_1, x_2, x_3)$, and correspondingly write

$$\mathbf{x} = x_1\hat{\mathbf{e}}_1 + x_2\hat{\mathbf{e}}_2 + x_3\hat{\mathbf{e}}_3, \qquad \text{or} \qquad \mathbf{x} = (x_1, x_2, x_3)\,. \tag{46}$$

1.3 Vector and Scalar Functions

1.3.1 *Vector-valued functions of a real variable*

A vector-valued function $\mathbf{F}\colon \mathbb{R} \mapsto \mathbb{R}^3$ of a real variable t assigns to each scalar $t \in I = [a, b] \subset \mathbb{R}$ a unique vector $\mathbf{F} \in \mathbb{R}^3$. In Cartesian coordinates we write

$$\mathbf{F} = \mathbf{F}(t) = u(t)\hat{\mathbf{i}} + v(t)\hat{\mathbf{j}} + w(t)\hat{\mathbf{k}} \qquad \text{for } a \le t \le b, \tag{47}$$

where $u, v, w\colon \mathbb{R} \mapsto \mathbb{R}$ are scalar functions of t. As $\{\hat{\mathbf{i}}, \hat{\mathbf{j}}, \hat{\mathbf{k}}\}$ are **uniform**, we have

$$\mathbf{F}'(t) \equiv \frac{\mathrm{d}\mathbf{F}}{\mathrm{d}t} = \frac{\mathrm{d}u}{\mathrm{d}t}\hat{\mathbf{i}} + \frac{\mathrm{d}v}{\mathrm{d}t}\hat{\mathbf{j}} + \frac{\mathrm{d}w}{\mathrm{d}t}\hat{\mathbf{k}} = \left(\frac{\mathrm{d}u}{\mathrm{d}t}, \frac{\mathrm{d}v}{\mathrm{d}t}, \frac{\mathrm{d}w}{\mathrm{d}t}\right). \tag{48}$$

The **product rules** of differentiation for scalar functions f and Ω, are

$$\frac{\mathrm{d}}{\mathrm{d}t}(f\Omega) = f\frac{\mathrm{d}\Omega}{\mathrm{d}t} + \frac{\mathrm{d}f}{\mathrm{d}t}\Omega, \qquad \frac{\mathrm{d}}{\mathrm{d}t}[\Omega\mathbf{F}] = \Omega\frac{\mathrm{d}\mathbf{F}}{\mathrm{d}t} + \frac{\mathrm{d}\Omega}{\mathrm{d}t}\mathbf{F}, \tag{49}$$

$$\frac{\mathrm{d}}{\mathrm{d}t}[\mathbf{F}\cdot\mathbf{G}] = \mathbf{F}\cdot\frac{\mathrm{d}\mathbf{G}}{\mathrm{d}t} + \frac{\mathrm{d}\mathbf{F}}{\mathrm{d}t}\cdot\mathbf{G}, \tag{50}$$

$$\frac{\mathrm{d}}{\mathrm{d}t}[\mathbf{F}\times\mathbf{G}] = \mathbf{F}\times\frac{\mathrm{d}\mathbf{G}}{\mathrm{d}t} + \frac{\mathrm{d}\mathbf{F}}{\mathrm{d}t}\times\mathbf{G}. \tag{51}$$

1.3.2 *Vector and scalar fields*

Let $\mathbf{r} = \overrightarrow{OP} = x\hat{\mathbf{i}} + y\hat{\mathbf{j}} + z\hat{\mathbf{k}}$. A scalar function of position, or **scalar field**, $\Omega\colon \mathbb{R}^3 \to \mathbb{R}$ assigns to every point $P \in \mathbb{R}^3$, a unique scalar $\Omega = \Omega(\mathbf{r}) = \Omega(x, y, z)$. A vector-valued function of position, or **vector field**, $\mathbf{F}\colon \mathbb{R}^3 \to \mathbb{R}^3$ assigns to every point $P \in \mathbb{R}^3$, a unique vector $\mathbf{F} = \mathbf{F}(\mathbf{r}) = \mathbf{F}(x, y, z)$. If we write

$$\mathbf{F} = (u, v, w) = u(x, y, z)\hat{\mathbf{i}} + v(x, y, z)\hat{\mathbf{j}} + w(x, y, z)\hat{\mathbf{k}}, \tag{52}$$

then $u, v, w\colon \mathbb{R}^3 \mapsto \mathbb{R}$ are the scalar functions of position. Furthermore

$$\Omega_x \equiv \frac{\partial\Omega}{\partial x}, \qquad \mathbf{F}_x \equiv \frac{\partial\mathbf{F}}{\partial x} = \frac{\partial u}{\partial x}\hat{\mathbf{i}} + \frac{\partial v}{\partial x}\hat{\mathbf{j}} + \frac{\partial w}{\partial x}\hat{\mathbf{k}}. \tag{53}$$

Similarly for $\Omega_y = \partial\Omega/\partial y$, $\Omega_z = \partial\Omega/\partial z$, $\mathbf{F}_y = \partial\mathbf{F}/\partial y$ and $\mathbf{F}_z = \partial\mathbf{F}/\partial z$.

1.3.3 *Differential operators*

The **gradient** of a scalar field $\Omega = \Omega(\mathbf{r})$ is a vector field defined by

$$\operatorname{grad}\Omega \equiv \boldsymbol{\nabla}\Omega = \frac{\partial\Omega}{\partial x}\hat{\mathbf{i}} + \frac{\partial\Omega}{\partial y}\hat{\mathbf{j}} + \frac{\partial\Omega}{\partial z}\hat{\mathbf{k}}. \tag{54}$$

The **divergence** of a vector field $\mathbf{F} = (u, v, w)$ is a scalar field defined by

$$\operatorname{div}\mathbf{F} \equiv \boldsymbol{\nabla}\cdot\mathbf{F} = \frac{\partial u}{\partial x} + \frac{\partial v}{\partial y} + \frac{\partial w}{\partial z}. \tag{55}$$

The **curl** of a vector field $\mathbf{F} = (u, v, w)$ is another vector field defined by

$$\operatorname{curl}\mathbf{F} \equiv \boldsymbol{\nabla}\times\mathbf{F} = \begin{vmatrix} \hat{\mathbf{i}} & \hat{\mathbf{j}} & \hat{\mathbf{k}} \\ \partial/\partial x & \partial/\partial y & \partial/\partial z \\ u & v & w \end{vmatrix}. \tag{56}$$

The **directional derivative** of a scalar field $\Omega = \Omega(\mathbf{r})$ is defined to be

$$\frac{\partial \Omega}{\partial n} = \nabla \Omega \cdot \hat{\mathbf{n}}, \tag{57}$$

where n measures distance in the direction of some unit vector $\hat{\mathbf{n}}$.

1.3.4 *The chain rules*

Let $\Omega = \Omega(x, y, z)$ and $x = x(t)$, $y = y(t)$ and $z = z(t)$, then $\Omega = \Omega(t)$ and

$$\frac{\mathrm{d}\Omega}{\mathrm{d}t} = \frac{\partial \Omega}{\partial x}\frac{\mathrm{d}x}{\mathrm{d}t} + \frac{\partial \Omega}{\partial y}\frac{\mathrm{d}y}{\mathrm{d}t} + \frac{\partial \Omega}{\partial z}\frac{\mathrm{d}z}{\mathrm{d}t} = \nabla \Omega \cdot \frac{\mathrm{d}\mathbf{r}}{\mathrm{d}t}. \tag{58}$$

Let $\Omega = \Omega(x, y, z)$ and $x = x(\xi, \eta, \zeta)$, $y = y(\xi, \eta, \zeta)$ and $z = z(\xi, \eta, \zeta)$ then $\Omega = \Omega(\xi, \eta, \zeta)$ and the chain rule gives

$$\frac{\partial \Omega}{\partial \xi} = \frac{\partial \Omega}{\partial x}\frac{\partial x}{\partial \xi} + \frac{\partial \Omega}{\partial y}\frac{\partial y}{\partial \xi} + \frac{\partial \Omega}{\partial z}\frac{\partial z}{\partial \xi} = \nabla \Omega \cdot \frac{\partial \mathbf{r}}{\partial \xi}, \tag{59}$$

with similar expressions for $\partial\Omega/\partial\eta$ and $\partial\Omega/\partial\zeta$.

1.3.5 *The del operator*

We define the (vector) del operator, ∇, also called simply *del* or *nabla*, by

$$\nabla \equiv \frac{\partial}{\partial x}\hat{\mathbf{i}} + \frac{\partial}{\partial y}\hat{\mathbf{j}} + \frac{\partial}{\partial z}\hat{\mathbf{k}}. \tag{60}$$

We treat ∇ as a *symbolic*, but *non-commutative*, vector, with $(\mathbf{F} \cdot \nabla)\,\Omega = \mathbf{F} \cdot (\nabla \Omega)$ and $(\mathbf{F} \times \nabla)\,\Omega = \mathbf{F} \times (\nabla \Omega)$ but $\mathbf{F} \cdot \nabla \neq \nabla \cdot \mathbf{F}$ and $\mathbf{F} \times \nabla \neq \nabla \times \mathbf{F}$. In addition

$$\operatorname{div}(\operatorname{grad}\Omega) \equiv \nabla^2 \Omega = \frac{\partial^2 \Omega}{\partial x^2} + \frac{\partial^2 \Omega}{\partial y^2} + \frac{\partial^2 \Omega}{\partial z^2}, \tag{61}$$

where $\nabla^2 = \nabla \cdot \nabla$ is the **Laplacian**, or (scalar) *del-squared* operator.

1.3.6 *Vector identities*

Given scalar fields λ, Ω and vector fields $\mathbf{F}$, $\mathbf{G}$ we have the product rules

$$\nabla(\lambda\Omega) = \lambda\nabla\Omega + \Omega\nabla\lambda, \tag{62}$$

$$\nabla \cdot (\Omega\mathbf{F}) = \Omega\nabla \cdot \mathbf{F} + \nabla\Omega \cdot \mathbf{F}, \tag{63}$$

$$\nabla \times (\Omega\mathbf{F}) = \Omega\nabla \times \mathbf{F} + \nabla\Omega \times \mathbf{F}, \tag{64}$$

$$\nabla(\mathbf{F}\cdot\mathbf{G}) = (\mathbf{F}\cdot\nabla)\mathbf{G} + (\mathbf{G}\cdot\nabla)\mathbf{F} + \mathbf{G}\times(\nabla\times\mathbf{F}) + \mathbf{F}\times(\nabla\times\mathbf{G}), \quad (65)$$

$$\nabla\cdot(\mathbf{F}\times\mathbf{G}) = \mathbf{G}\cdot(\nabla\times\mathbf{F}) - \mathbf{F}\cdot(\nabla\times\mathbf{G}), \quad (66)$$

$$\nabla\times(\mathbf{F}\times\mathbf{G}) = (\nabla\cdot\mathbf{G})\mathbf{F} - (\nabla\cdot\mathbf{F})\mathbf{G} + (\mathbf{G}\cdot\nabla)\mathbf{F} - (\mathbf{F}\cdot\nabla)\mathbf{G}, \quad (67)$$

$$\nabla\cdot[\nabla\times\mathbf{F}] = 0, \qquad \nabla\times[\nabla\Omega] = \mathbf{0}. \quad (68)$$

1.3.7 *Taylor's theorem*

For a sufficiently differentiable scalar field, Ω, Taylor's theorem states that

$$\begin{aligned}\Omega(x+h, y+k, z+m) =& \Omega(x,y,z) + \left(h\frac{\partial}{\partial x} + k\frac{\partial}{\partial y} + m\frac{\partial}{\partial z}\right)\Omega(x,y,z) \\ &+ \frac{1}{2!}\left(h\frac{\partial}{\partial x} + k\frac{\partial}{\partial y} + m\frac{\partial}{\partial z}\right)^2\Omega \\ &+ \frac{1}{3!}\left(h\frac{\partial}{\partial x} + k\frac{\partial}{\partial y} + m\frac{\partial}{\partial z}\right)^3\Omega + \cdots. \end{aligned} \quad (69)$$

Alternatively Taylor's theorem can be written more compactly as

$$\delta\Omega \equiv \Omega(\mathbf{r}+\delta\mathbf{r}) - \Omega(\mathbf{r}) = \nabla\Omega\cdot\delta\mathbf{r} + O\left(|\delta\mathbf{r}|^2\right), \quad (70)$$

where to first order we have the **linear approximation** $\delta\Omega = \nabla\Omega\cdot\delta\mathbf{r}$.

1.4 Curves and Surfaces in 3-D Space

1.4.1 *Curves*

A general point P on a curve $C \subset \mathbb{R}^3$ has a position vector $\mathbf{r} = \overrightarrow{OP}$ given by

$$\mathbf{r} = \mathbf{r}(t) = x(t)\hat{\mathbf{i}} + y(t)\hat{\mathbf{j}} + z(t)\hat{\mathbf{k}}, \qquad \text{for } a \le t \le b. \quad (71)$$

We regard Eq. (71) as a mapping $\mathbf{r}\colon \mathbb{R} \mapsto \mathbb{R}^3$, from $I = [a,b]$ onto C. Here Eq. (71) is *not* a unique parametric representation since any transformation $t = t(\eta)$ yields a new representation $\mathbf{r} = \mathbf{r}(\eta)$, where $\mathrm{d}t/\mathrm{d}\eta \ge 0$ preserves the sense of direction.

The **unit tangent vector** $\hat{\mathbf{T}}$ to the curve C is given by

$$\hat{\mathbf{T}} = \frac{\mathrm{d}\mathbf{r}}{\mathrm{d}t}\Big/\left|\frac{\mathrm{d}\mathbf{r}}{\mathrm{d}t}\right| = \frac{\mathrm{d}\mathbf{r}}{\mathrm{d}s}, \qquad \text{where } \frac{\mathrm{d}s}{\mathrm{d}t} = \left|\frac{\mathrm{d}\mathbf{r}}{\mathrm{d}t}\right|, \qquad \text{so} \qquad \frac{\mathrm{d}\mathbf{r}}{\mathrm{d}t} = \frac{\mathrm{d}s}{\mathrm{d}t}\hat{\mathbf{T}}, \quad (72)$$

where s measures the **arc length** along the curve C. As $\mathrm{d}s/\mathrm{d}t > 0$, $s = s(t)$ has a *unique* inverse $t = t(s)$, yielding the **intrinsic equation** of C,

$$\mathbf{r} = \mathbf{r}(s) = x(s)\hat{\mathbf{\imath}} + y(s)\hat{\mathbf{\jmath}} + z(s)\hat{\mathbf{k}}, \qquad 0 \le s \le L. \tag{73}$$

1.4.2 *Surfaces*

1.4.2.1 *Cartesian representations*

The *implicit* and *explicit* representations of a surface $S \subset \mathbb{R}^3$ are given by

$$\Omega(x, y, z) = 0, \qquad \text{and} \qquad z = f(x, y), \qquad \text{for } (x, y) \in D, \tag{74}$$

where D is a region of the xy-plane. One possible choice for Ω is $\Omega \equiv z - f(x, y)$. The **unit normal** $\hat{\mathbf{n}}$ to a surface $\Omega \equiv z - f(x, y) = 0$ is given by

$$\hat{\mathbf{n}} = \pm \frac{\nabla\Omega}{|\nabla\Omega|} = \pm \frac{1}{\sqrt{1 + f_x^2 + f_y^2}} \left(-f_x, -f_y, 1\right), \tag{75}$$

since there are two sides to each surface.

1.4.2.2 *Parametric representation*

A general point P on a surface $S \subset \mathbb{R}^3$ has a position vector $\mathbf{r} = \overrightarrow{OP}$ given by

$$\mathbf{r} = \mathbf{r}(u, v) = x(u, v)\hat{\mathbf{\imath}} + y(u, v)\hat{\mathbf{\jmath}} + z(u, v)\hat{\mathbf{k}}, \qquad \text{for } (u, v) \in D^\star. \tag{76}$$

Here Eq. (76) is a mapping $\mathbf{r}\colon \mathbb{R}^2 \mapsto \mathbb{R}^3$ from $D^\star$, a region of the uv-plane, onto S. Now $\mathbf{r} = \mathbf{r}(u, v_0)$, in which $v = v_0$ is fixed, but u varies, is the equation of one of the **u-coordinate curves**. Similarly $\mathbf{r} = \mathbf{r}(u_0, v)$ is the equation of one of the v-coordinate curves. From Eq. (72) the tangents to the coordinate curves are

$$\mathbf{r}_u = \frac{\partial \mathbf{r}}{\partial u} = \frac{\partial x}{\partial u}\hat{\mathbf{\imath}} + \frac{\partial y}{\partial u}\hat{\mathbf{\jmath}} + \frac{\partial z}{\partial u}\hat{\mathbf{k}}, \qquad \mathbf{r}_v = \frac{\partial \mathbf{r}}{\partial v} = \frac{\partial x}{\partial v}\hat{\mathbf{\imath}} + \frac{\partial y}{\partial v}\hat{\mathbf{\jmath}} + \frac{\partial z}{\partial v}\hat{\mathbf{k}}. \tag{77}$$

Consequently the unit normal $\hat{\mathbf{n}}$ to S must be of the form

$$\hat{\mathbf{n}} = \pm \frac{\mathbf{r}_u \times \mathbf{r}_v}{|\mathbf{r}_u \times \mathbf{r}_v|}. \tag{78}$$

For a surface S given by $z = f(x, y)$, we see that

$$\mathbf{r} = \mathbf{r}(x, y) = \mathbf{r}\left[x, y, f(x, y)\right] = x\hat{\mathbf{\imath}} + y\hat{\mathbf{\jmath}} + f(x, y)\hat{\mathbf{k}} \tag{79}$$

is the equation of S parameterised by x and y. In this case

$$\mathbf{r}_x \times \mathbf{r}_y = \left(-f_x, -f_y, 1\right), \quad \text{so} \quad \hat{\mathbf{n}} = \pm \frac{1}{\sqrt{1 + f_x^2 + f_y^2}} \left(-f_x, -f_y, 1\right). \tag{80}$$

1.5 Curvilinear Coordinate Systems

1.5.1 *Cartesian coordinates*

For every point P we can associate a right-handed orthonormal basis set of uniform vectors $\{\hat{\mathbf{i}}, \hat{\mathbf{j}}, \hat{\mathbf{k}}\}$, where $\mathbf{r} = \overrightarrow{OP} = x\hat{\mathbf{i}} + y\hat{\mathbf{j}} + z\hat{\mathbf{k}}$. Let Q be a neighbouring point to P, with $\overrightarrow{PQ} = \delta\mathbf{r} = (\delta x, \delta y, \delta z)$, then the elementary (cuboidal) volume element at P, of volume δV, with $\delta x\hat{\mathbf{i}}$, $\delta y\hat{\mathbf{j}}$ and $\delta z\hat{\mathbf{k}}$ representing adjacent sides, has

$$\delta s^2 = |\delta\mathbf{r}|^2 = \delta x^2 + \delta y^2 + \delta z^2, \qquad \delta V = \delta x\, \delta y\, \delta z. \tag{81}$$

1.5.2 *Curvilinear coordinates*

A new coordinate system is defined by a continuously differentiable mapping

$$x = x(u, v, w), \qquad y = y(u, v, w), \qquad z = z(u, v, w). \tag{82}$$

Thus $\mathbf{r} = \overrightarrow{OP} = x\hat{\mathbf{i}} + y\hat{\mathbf{j}} + z\hat{\mathbf{k}}$, is given by the mapping $\mathbf{r}\colon \mathbb{R}^3 \mapsto \mathbb{R}^3$, where

$$\mathbf{r} = \mathbf{r}(u, v, w) = x(u, v, w)\hat{\mathbf{i}} + y(u, v, w)\hat{\mathbf{j}} + z(u, v, w)\hat{\mathbf{k}}. \tag{83}$$

Assuming a one-to-one mapping then there must exist a unique inverse mapping, such that, except at isolated points, the **Jacobian** J of the mapping

$$J = \left|\frac{\partial\mathbf{r}}{\partial u} \cdot \left(\frac{\partial\mathbf{r}}{\partial v} \times \frac{\partial\mathbf{r}}{\partial w}\right)\right| = \frac{\partial(x, y, z)}{\partial(u, v, w)} = \begin{vmatrix} x_u & y_u & z_u \\ x_v & y_v & z_v \\ x_w & y_w & z_w \end{vmatrix} \neq 0. \tag{84}$$

Suppose the point P is specified uniquely by $u = u_0$, $v = v_0$ and $w = w_0$, then the vector equations of the u-, v- and w-**coordinate lines** through P are

$$\mathbf{r} = \mathbf{r}(u, v_0, w_0), \qquad \mathbf{r} = \mathbf{r}(u_0, v, w_0), \qquad \mathbf{r} = \mathbf{r}(u_0, v_0, w). \tag{85}$$

Given **scale factors** $h_u = |\mathbf{r}_u|$, $h_v = |\mathbf{r}_v|$ and $h_w = |\mathbf{r}_w|$, the unit tangents at P to coordinate lines are given by

$$\hat{\mathbf{e}}_u = \frac{1}{h_u}\frac{\partial\mathbf{r}}{\partial u}, \qquad \hat{\mathbf{e}}_v = \frac{1}{h_v}\frac{\partial\mathbf{r}}{\partial v}, \qquad \hat{\mathbf{e}}_w = \frac{1}{h_w}\frac{\partial\mathbf{r}}{\partial w}. \tag{86}$$

For a neighbouring point Q, where

$$\delta\mathbf{r} = \overrightarrow{PQ} = \mathbf{r}_u\, \delta u + \mathbf{r}_v\, \delta v + \mathbf{r}_w\, \delta w = h_u\, \delta u\, \hat{\mathbf{e}}_u + h_u\, \delta v\, \hat{\mathbf{e}}_v + h_w\, \delta w\, \hat{\mathbf{e}}_w \tag{87}$$

the arc length $\delta s = |\delta \mathbf{r}|$ is given by

$$\delta s^2 = |\delta \mathbf{r}|^2 = h_u^2\,\delta u^2 + h_v^2\,\delta v^2 + h_w^2\,\delta w^2 \\ + 2\left[h_u h_v\,\delta u\,\delta v\,(\hat{\mathbf{e}}_u \cdot \hat{\mathbf{e}}_v) + h_v h_w\,\delta v\,\delta w\,(\hat{\mathbf{e}}_v \cdot \hat{\mathbf{e}}_w) + h_u h_w\,\delta u\,\delta w\,(\hat{\mathbf{e}}_u \cdot \hat{\mathbf{e}}_w)\right]. \tag{88}$$

A (cuboidal) volume element in uvw-space, with $\delta u\,\hat{\mathbf{e}}_u$, $\delta v\,\hat{\mathbf{e}}_v$, $\delta w\,\hat{\mathbf{e}}_w$ representing adjacent sides, corresponds to a curvilinear volume element in xyz-space, with volume

$$\delta V = J\delta u\,\delta v\,\delta w = h_u h_v h_w |\hat{\mathbf{e}}_u \cdot (\hat{\mathbf{e}}_v \times \hat{\mathbf{e}}_w)\,|. \tag{89}$$

Finally the equations for the **coordinate surfaces** through P are given by

$$\mathbf{r} = \mathbf{r}(u_0, v, w), \qquad \mathbf{r} = \mathbf{r}(u, v_0, w), \qquad \mathbf{r} = \mathbf{r}(u, v, w_0). \tag{90}$$

Thus the coordinate surface, $w =$ const, has a unit normal $\hat{\mathbf{n}}$ with an elementary surface element of area δA, given by

$$\hat{\mathbf{n}} = \pm\frac{\mathbf{r}_u \times \mathbf{r}_v}{|\mathbf{r}_u \times \mathbf{r}_v|}, \qquad \delta A = |\mathbf{r}_u \times \mathbf{r}_v|\,\delta u\,\delta v = h_u h_v |\hat{\mathbf{e}}_u \times \hat{\mathbf{e}}_v|\,\delta u\,\delta v. \tag{91}$$

1.5.3 *Orthogonal curvilinear coordinates*

For *orthogonal* curvilinear coordinates we require

$$\frac{\partial \mathbf{r}}{\partial u} \cdot \frac{\partial \mathbf{r}}{\partial v} = 0, \qquad \frac{\partial \mathbf{r}}{\partial u} \cdot \frac{\partial \mathbf{r}}{\partial w} = 0, \qquad \frac{\partial \mathbf{r}}{\partial v} \cdot \frac{\partial \mathbf{r}}{\partial w} = 0, \tag{92}$$

with u, v and w chosen so that $\{\hat{\mathbf{e}}_u, \hat{\mathbf{e}}_v, \hat{\mathbf{e}}_w\}$ is a **right-handed orthonormal** basis set satisfying

$$\hat{\mathbf{e}}_u \cdot \hat{\mathbf{e}}_u = \hat{\mathbf{e}}_v \cdot \hat{\mathbf{e}}_v = \hat{\mathbf{e}}_w \cdot \hat{\mathbf{e}}_w = 1, \qquad \hat{\mathbf{e}}_u \cdot \hat{\mathbf{e}}_v = \hat{\mathbf{e}}_u \cdot \hat{\mathbf{e}}_w = \hat{\mathbf{e}}_v \cdot \hat{\mathbf{e}}_w = 0, \tag{93}$$

$$\hat{\mathbf{e}}_u \times \hat{\mathbf{e}}_v = \hat{\mathbf{e}}_w, \qquad \hat{\mathbf{e}}_v \times \hat{\mathbf{e}}_w = \hat{\mathbf{e}}_u, \qquad \hat{\mathbf{e}}_w \times \hat{\mathbf{e}}_u = \hat{\mathbf{e}}_v. \tag{94}$$

We can show that the arc length $\delta s = |\delta \mathbf{r}|$ is given by

$$\delta s^2 = |\delta \mathbf{r}|^2 = h_u^2\,\delta u^2 + h_v^2\,\delta v^2 + h_w^2 \delta w^2. \tag{95}$$

Similarly the elementary volume element has a volume

$$\delta V = J\,\delta u\,\delta v\,\delta w = h_u h_v h_w\,\delta u\,\delta v\,\delta w. \tag{96}$$

and for the coordinate surface, $w =$ const, with unit normal $\hat{\mathbf{e}}_w$, we have

$$\delta A = |\mathbf{r}_u \times \mathbf{r}_v|\,\delta u\,\delta v = |\mathbf{r}_u|\,|\mathbf{r}_v|\,\delta u\,\delta v = h_u h_v\,\delta u\,\delta v. \tag{97}$$

1.5.3.1 *Vector fields*

A vector field $\mathbf{F} = (F_1, F_2, F_3)$ can also be written in the form

$$\mathbf{F} = \mathbf{F}(u,v,w) = F_u(u,v,w)\hat{\mathbf{e}}_u + F_v(u,v,w)\hat{\mathbf{e}}_v + F_w(u,v,w)\hat{\mathbf{e}}_w, \tag{98}$$

where F_u, F_v and F_w are the components of $\mathbf{F}$ along the coordinate lines. For orthogonal curvilinear coordinates $F_u = \mathbf{F}\cdot\hat{\mathbf{e}}_u$, $F_v = \mathbf{F}\cdot\hat{\mathbf{e}}_v$ and $F_w = \mathbf{F}\cdot\hat{\mathbf{e}}_w$ so that

$$\mathbf{F} = \frac{1}{h_u}\left(\mathbf{F}\cdot\frac{\partial\mathbf{r}}{\partial u}\right)\hat{\mathbf{e}}_u + \frac{1}{h_v}\left(\mathbf{F}\cdot\frac{\partial\mathbf{r}}{\partial v}\right)\hat{\mathbf{e}}_v + \frac{1}{h_w}\left(\mathbf{F}\cdot\frac{\partial\mathbf{r}}{\partial w}\right)\hat{\mathbf{e}}_w. \tag{99}$$

For example, as regards the position vector $\mathbf{r} = (x, y, z)$, we have

$$\mathbf{r} = \frac{1}{h_u}\frac{\partial}{\partial u}\left(\frac{1}{2}r^2\right)\hat{\mathbf{e}}_u + \frac{1}{h_v}\frac{\partial}{\partial v}\left(\frac{1}{2}r^2\right)\hat{\mathbf{e}}_v + \frac{1}{h_w}\frac{\partial}{\partial w}\left(\frac{1}{2}r^2\right)\hat{\mathbf{e}}_w. \tag{100}$$

In general the basis set of vectors $\{\hat{\mathbf{e}}_u, \hat{\mathbf{e}}_v, \hat{\mathbf{e}}_w\}$ are *not* uniform. Indeed

$$\begin{aligned}\frac{\partial\mathbf{F}}{\partial u} &= \left[\frac{\partial F_u}{\partial u} + \frac{F_v}{h_v}\frac{\partial h_u}{\partial v} + \frac{F_w}{h_w}\frac{\partial h_u}{\partial w}\right]\hat{\mathbf{e}}_u \\ &\quad + \left[\frac{\partial F_v}{\partial u} - \frac{F_u}{h_v}\frac{\partial h_u}{\partial v}\right]\hat{\mathbf{e}}_v + \left[\frac{\partial F_w}{\partial u} - \frac{F_u}{h_w}\frac{\partial h_u}{\partial w}\right]\hat{\mathbf{e}}_w,\end{aligned} \tag{101}$$

with similar formulae for $\partial\mathbf{F}/\partial v$ and $\partial\mathbf{F}/\partial w$ obtained by cyclic permutation.

1.5.3.2 *The differential operators*

Given a scalar field $\Omega(\mathbf{r})$ and a vector field $\mathbf{F}(\mathbf{r})$ we can show that

$$\boldsymbol{\nabla}\Omega = \frac{1}{h_u}\frac{\partial\Omega}{\partial u}\hat{\mathbf{e}}_u + \frac{1}{h_v}\frac{\partial\Omega}{\partial v}\hat{\mathbf{e}}_v + \frac{1}{h_w}\frac{\partial\Omega}{\partial w}\hat{\mathbf{e}}_w, \tag{102}$$

$$\boldsymbol{\nabla}\cdot\mathbf{F} = \frac{1}{h_u h_v h_w}\left\{\frac{\partial}{\partial u}(h_v h_w F_u) + \frac{\partial}{\partial v}(h_u h_w F_v) + \frac{\partial}{\partial w}(h_u h_v F_w)\right\}, \tag{103}$$

$$\boldsymbol{\nabla}\times\mathbf{F} = \frac{1}{h_u h_v h_w}\begin{vmatrix} h_u\hat{\mathbf{e}}_u & h_v\hat{\mathbf{e}}_v & h_w\hat{\mathbf{e}}_w \\ \partial/\partial u & \partial/\partial v & \partial/\partial w \\ h_u F_u & h_v F_v & h_w F_w \end{vmatrix}, \tag{104}$$

$$\nabla^2\Omega = \frac{1}{h_u h_v h_w}\left\{\frac{\partial}{\partial u}\left(\frac{h_v h_w}{h_u}\frac{\partial\Omega}{\partial u}\right) + \frac{\partial}{\partial v}\left(\frac{h_u h_w}{h_v}\frac{\partial\Omega}{\partial v}\right) + \frac{\partial}{\partial w}\left(\frac{h_u h_v}{h_w}\frac{\partial\Omega}{\partial w}\right)\right\}. \tag{105}$$

1.5.4 *Cylindrical polar coordinates:* (R, ϕ, z)

The general point P with position vector $\mathbf{r} = \overrightarrow{OP} = x\hat{\mathbf{i}} + y\hat{\mathbf{j}} + z\hat{\mathbf{k}}$ is given by

$$\mathbf{r} = R\,(\cos\phi\hat{\mathbf{i}} + \sin\phi\hat{\mathbf{j}}) + z\hat{\mathbf{k}}, \qquad \text{so} \qquad r^2 = |\mathbf{r}|^2 = R^2 + z^2. \tag{106}$$

In terms of the right-handed orthonormal vectors $\{\,\hat{\mathbf{R}}, \hat{\boldsymbol{\phi}}, \hat{\mathbf{k}}\,\}$ we have

$$\mathbf{r} = R\hat{\mathbf{R}} + z\hat{\mathbf{k}}, \quad \text{so} \quad \delta\mathbf{r} = (\delta R\,\hat{\mathbf{R}} + R\,\delta\hat{\mathbf{R}}) + \delta z\,\hat{\mathbf{k}} = \delta R\,\hat{\mathbf{R}} + R\,\delta\phi\,\hat{\boldsymbol{\phi}} + \delta z\,\hat{\mathbf{k}}, \tag{107}$$

yielding scale factors $h_R = 1$, $h_\phi = R$ and $h_z = 1$. Hence we have

$$\delta s^2 = |\delta\mathbf{r}|^2 = \delta R^2 + R^2\,\delta\phi^2 + \delta z^2, \qquad \delta V = R\,\delta R\,\delta\phi\,\delta z. \tag{108}$$

1.5.4.1 *Scalar and vector operators*

If $\Omega = \Omega(\mathbf{r})$ and $\mathbf{F} = \mathbf{F}(\mathbf{r}) = F_R\hat{\mathbf{R}} + F_\phi\hat{\boldsymbol{\phi}} + F_z\hat{\mathbf{k}}$ then

$$\boldsymbol{\nabla}\Omega = \frac{\partial\Omega}{\partial R}\hat{\mathbf{R}} + \frac{1}{R}\frac{\partial\Omega}{\partial\phi}\hat{\boldsymbol{\phi}} + \frac{\partial\Omega}{\partial z}\hat{\mathbf{k}}, \tag{109}$$

$$\boldsymbol{\nabla}\cdot\mathbf{F} = \frac{1}{R}\frac{\partial}{\partial R}(RF_R) + \frac{1}{R}\frac{\partial F_\phi}{\partial\phi} + \frac{\partial F_z}{\partial z}, \tag{110}$$

$$\boldsymbol{\nabla}\times\mathbf{F} = \frac{1}{R}\begin{vmatrix} \hat{\mathbf{R}} & R\hat{\boldsymbol{\phi}} & \hat{\mathbf{k}} \\ \partial/\partial R & \partial/\partial\phi & \partial/\partial z \\ F_R & RF_\phi & F_z \end{vmatrix}, \tag{111}$$

$$\nabla^2\Omega = \frac{1}{R}\frac{\partial}{\partial R}\left(R\frac{\partial\Omega}{\partial R}\right) + \frac{1}{R^2}\frac{\partial^2\Omega}{\partial\phi^2} + \frac{\partial^2\Omega}{\partial z^2}. \tag{112}$$

1.5.5 *Spherical polar coordinates:* (r, θ, ϕ)

The general point P with position vector $\mathbf{r} = \overrightarrow{OP} = x\hat{\mathbf{i}} + y\hat{\mathbf{j}} + z\hat{\mathbf{k}}$ is given by

$$\mathbf{r} = r\sin\theta\,(\cos\phi\hat{\mathbf{i}} + \sin\phi\hat{\mathbf{j}}) + r\cos\theta\hat{\mathbf{k}}, \qquad \text{so} \qquad |\mathbf{r}|^2 = r^2. \tag{113}$$

In terms of the right-handed orthonormal vectors $\{\hat{\mathbf{r}}, \hat{\boldsymbol{\theta}}, \hat{\boldsymbol{\phi}}\}$ we have

$$\mathbf{r} = r\hat{\mathbf{r}}, \qquad \text{so} \qquad \delta\mathbf{r} = (\delta r\,\hat{\mathbf{r}} + r\,\delta\hat{\mathbf{r}}) = \delta r\,\hat{\mathbf{r}} + r\,\delta\theta\,\hat{\boldsymbol{\theta}} + r\sin\theta\,\delta\phi\,\hat{\boldsymbol{\phi}}, \tag{114}$$

yielding scale factors $h_r = 1$, $h_\theta = r$, $h_\phi = r\sin\theta$. Hence we have

$$\delta s^2 = |\delta\mathbf{r}|^2 = \delta r^2 + r^2\,\delta\theta^2 + r^2\sin^2\theta\,\delta\phi^2, \quad \delta V = r^2\sin\theta\,\delta r\,\delta\theta\,\delta\phi. \tag{115}$$

1.5.5.1 *Scalar and vector operators*

If $\Omega = \Omega(\mathbf{r})$ and $\mathbf{F} = \mathbf{F}(\mathbf{r}) = F_r\hat{\mathbf{r}} + F_\theta\hat{\boldsymbol{\theta}} + F_\phi\hat{\boldsymbol{\phi}}$ then

$$\nabla\Omega = \frac{\partial\Omega}{\partial r}\hat{\mathbf{r}} + \frac{1}{r}\frac{\partial\Omega}{\partial\theta}\hat{\boldsymbol{\theta}} + \frac{1}{r\sin\theta}\frac{\partial\Omega}{\partial\phi}\hat{\boldsymbol{\phi}}, \tag{116}$$

$$\nabla\cdot\mathbf{F} = \frac{1}{r^2}\frac{\partial}{\partial r}(r^2F_r) + \frac{1}{r\sin\theta}\frac{\partial}{\partial\theta}(\sin\theta\, F_\theta) + \frac{1}{r\sin\theta}\frac{\partial F_\phi}{\partial\phi}, \tag{117}$$

$$\nabla\times\mathbf{F} = \frac{1}{r^2\sin\theta}\begin{vmatrix} \hat{\mathbf{r}} & r\hat{\boldsymbol{\theta}} & r\sin\theta\hat{\boldsymbol{\phi}} \\ \partial/\partial r & \partial/\partial\theta & \partial/\partial\phi \\ F_r & rF_\theta & r\sin\theta F_\phi \end{vmatrix}, \tag{118}$$

$$\nabla^2\Omega = \frac{1}{r^2}\frac{\partial}{\partial r}\left(r^2\frac{\partial\Omega}{\partial r}\right) + \frac{1}{r^2\sin\theta}\frac{\partial}{\partial\theta}\left(\sin\theta\frac{\partial\Omega}{\partial\theta}\right) + \frac{1}{r^2\sin^2\theta}\frac{\partial^2\Omega}{\partial\phi^2}. \tag{119}$$

1.6 Integrals

1.6.1 *Line integrals*

Suppose the position vector of a general point P on a curve $C\colon A\to B$ is

$$\mathbf{r} = \mathbf{r}(t) = x(t)\hat{\mathbf{\imath}} + y(t)\hat{\mathbf{\jmath}} + z(t)\hat{\mathbf{k}}, \qquad \text{for } t_0 \le t \le t_1. \tag{120}$$

Then, on C, the scalar field $\Omega = \Omega(t)$, and the line integral of Ω along C from A to B reduces to a standard (Riemann) integral of the form

$$\int_{C\colon A\to B}\Omega(\mathbf{r})\,\mathrm{d}s = \int_{t=t_0}^{t=t_1}\Omega\left[x(t),y(t),z(t)\right]\frac{\mathrm{d}s}{\mathrm{d}t}\,\mathrm{d}t = \int_{t=t_0}^{t=t_1}\Omega(t)\left|\frac{\mathrm{d}\mathbf{r}}{\mathrm{d}t}\right|\,\mathrm{d}t. \tag{121}$$

We also write $\oint_C \Omega(\mathbf{r})\,\mathrm{d}s$ to denote a line integral around a *closed* curve C.

1.6.2 *Surface integrals*

Suppose the position vector of a general point P on a surface S is

$$\mathbf{r} = \mathbf{r}(u,v) = x(u,v)\hat{\mathbf{\imath}} + y(u,v)\hat{\mathbf{\jmath}} + z(u,v)\hat{\mathbf{k}}, \qquad \text{for } (u,v)\in D. \tag{122}$$

Then on S, the scalar field $\Omega = \Omega(u,v)$, and the surface integral of Ω over S reduces to a double integral over the region D of the *flat* uv-plane, of the form

$$\iint_S \Omega(\mathbf{r})\,\mathrm{d}A(\mathbf{r}) = \iint_D \Omega\left[x(u,v),y(u,v),z(u,v)\right]\left|\mathbf{r}_u\times\mathbf{r}_v\right|\mathrm{d}u\,\mathrm{d}v, \tag{123}$$

where we regard $\iint_S$ as a single symbol.

1.6.3 *Multiple integrals*

We evaluate a double or triple integral by expressing it as a repeated integral. For example the triple, or volume, integral of a scalar field $\Omega = \Omega(\mathbf{r})$ over a region $\Sigma \subset \mathbb{R}^3$ is given by

$$I = \iiint_\Sigma \Omega(\mathbf{r})\,\mathrm{d}V(\mathbf{r}) = \iiint_\Sigma \Omega(x, y, z)\,\mathrm{d}x\,\mathrm{d}y\,\mathrm{d}z, \tag{124}$$

where $\mathrm{d}V = \mathrm{d}x\,\mathrm{d}y\,\mathrm{d}z$. If we integrate first with respect to z, keeping both x and y fixed, then the result is a double integral of the form

$$I = \iint_D f(x, y)\,\mathrm{d}x\,\mathrm{d}y = \iint_D \left(\int_{z=g(x,y)}^{z=h(x,y)} \Omega(x, y, z)\,\mathrm{d}z \right) \mathrm{d}x\,\mathrm{d}y \tag{125}$$

where D is a domain of the xy-plane. This double integral is, in turn, evaluated by first integrating with respect to x (or y), keeping y (or x) fixed, and then integrating with respect to y (or x). Note that for a *separable* integrand, over a rectangular domain, a multiple integral reduces to a product of standard (Riemann) integrals.

Moreover if the position vector of a general point P in Σ is given by

$$\mathbf{r} = \mathbf{r}(u, v, w) = x(u, v, w)\hat{\mathbf{\imath}} + y(u, v, w)\hat{\mathbf{\jmath}} + z(u, v, w)\hat{\mathbf{k}}, \quad \text{for } (u, v, w) \in \Sigma^\star, \tag{126}$$

where $\Sigma^\star$ is the corresponding region in uvw-space, then $\Omega = \Omega(u, v, w)$ and we can represent Eq. (124) as the alternative triple integral

$$I = \iiint_{\Sigma^\star} \Omega\left[x(u, v, w), y(u, v, w), z(u, v, w)\right] |J|\,\mathrm{d}u\,\mathrm{d}v\,\mathrm{d}w, \tag{127}$$

where J is the Jacobian of the mapping.

Example 1.1 Using cylindrical polar coordinates, namely

$$x = R\cos\phi, \qquad y = R\sin\phi, \qquad z = z, \tag{128}$$

find the curved surface area A of the cylinder $S\colon x^2 + y^2 = a^2, 0 \le z \le H$ and the volume V of the region $\Sigma\colon x^2 + y^2 \le a^2, 0 \le z \le H$ within the cylinder.

Solution The position vector $\mathbf{r} = \overrightarrow{OP}$ of a general point on the curved surface of S is

$$\mathbf{r} = \mathbf{r}(\phi, z) = (a\cos\phi, a\sin\phi, z). \tag{129}$$

Here θ and z parameterise S with $D\colon 0 \le \theta \le 2\pi$, $0 \le z \le H$ mapping to S. Here

$$|\mathbf{r}_\phi \times \mathbf{r}_z| = \left|a\left(-\sin\phi\hat{\mathbf{i}} + \cos\phi\hat{\mathbf{j}}\right) \times \hat{\mathbf{k}}\right| = a\left|\cos\phi\hat{\mathbf{i}} + \sin\phi\hat{\mathbf{j}}\right| = a, \tag{130}$$

or $\delta A = h_\phi h_z\,\delta\phi\,\delta z = a\,\delta\phi\,\delta z$. Thus the curved surface area of the cylinder is given by

$$A = \iint_S \mathrm{d}A = \iint_D a\,\mathrm{d}\phi\,\mathrm{d}z = a\left(\int_{\phi=0}^{\phi=2\pi}\mathrm{d}\phi\right)\left(\int_{z=0}^{z=h}\mathrm{d}z\right) = 2\pi a h. \tag{131}$$

Here $\delta V = J\delta R\,\delta\phi\,\delta z$ where the Jacobian $J = h_R h_\phi h_z = R$. Thus if $\Sigma^\star$ is the region $0 \le R \le a$, $0 \le \phi \le 2\pi$, $0 \le z \le H$ which maps to Σ, then

$$\begin{aligned} V &= \iiint_\Sigma \mathrm{d}V = \iiint_{\Sigma^\star} R\,\mathrm{d}R\,\mathrm{d}\phi\,\mathrm{d}z \\ &= \left(\int_{R=0}^{R=a} R\,\mathrm{d}R\right)\left(\int_{\phi=0}^{\phi=2\pi}\mathrm{d}\phi\right)\left(\int_{z=0}^{z=h}\mathrm{d}z\right) = \pi a^2 H. \end{aligned} \tag{132}$$

Example 1.2 Using spherical polar coordinates, namely

$$x = r\sin\theta\cos\phi, \qquad y = r\sin\theta\sin\phi, \qquad z = r\cos\theta, \tag{133}$$

find the surface area A of the sphere $S\colon x^2 + y^2 + z^2 = a^2$ and the volume V of the region $\Sigma\colon x^2 + y^2 + z^2 \le a^2$ within the sphere.

Solution The position vector $\mathbf{r} = \overrightarrow{OP}$ of a general point on the surface S is

$$\mathbf{r} = \mathbf{r}(\theta,\phi) = a(\sin\theta\cos\phi, \sin\theta\sin\phi, \cos\theta). \tag{134}$$

Here θ and ϕ parameterise S with $D\colon 0 \le \theta \le \pi$, $0 \le \phi \le 2\pi$ mapping to S. Also

$$\mathbf{r}_\theta = a\left[\cos\theta\left(\cos\phi\hat{\mathbf{i}} + \sin\phi\hat{\mathbf{j}}\right) - \sin\theta\hat{\mathbf{k}}\right], \tag{135}$$

$$\mathbf{r}_\phi = a\sin\theta\left(-\sin\phi\hat{\mathbf{i}} + \cos\phi\hat{\mathbf{j}}\right), \tag{136}$$

$$|\mathbf{r}_\theta \times \mathbf{r}_\phi| = a^2\sin\theta\left|\sin\theta\left(\cos\phi\hat{\mathbf{i}} + \sin\phi\hat{\mathbf{j}}\right) + \cos\theta\hat{\mathbf{k}}\right| = a^2\sin\theta, \tag{137}$$

or $\delta A = h_\theta h_\phi\,\delta\theta\,\delta\phi = a^2\sin\theta\,\delta\theta\,\delta\phi$. Thus the surface area of the sphere is given by

$$\begin{aligned} A &= \iint_S \mathrm{d}A = \iint_D a^2\sin\theta\,\mathrm{d}\theta\,\mathrm{d}\phi \\ &= a^2\left(\int_{\theta=0}^{\theta=\pi}\sin\theta\,\mathrm{d}\theta\right)\left(\int_{\phi=0}^{\phi=2\pi}\mathrm{d}\phi\right) = 4\pi a^2. \end{aligned} \tag{138}$$

Here $\delta V = J\,\delta r\,\delta\theta\,\delta\phi$ where the Jacobian $J = h_r h_\theta h_\phi = r^2 \sin\theta$. Thus if $\Sigma^\star$ is the region $0 \le r \le a$, $0 \le \theta \le \pi$, $0 \le \phi \le 2\pi$ which maps to Σ, then

$$V = \iiint_\Sigma \mathrm{d}V = \iiint_{\Sigma^\star} r^2 \sin\theta\,\mathrm{d}r\,\mathrm{d}\theta\,\mathrm{d}\phi$$
$$= \left(\int_{r=0}^{r=a} r^2\,\mathrm{d}r\right)\left(\int_{\theta=0}^{\theta=\pi} \sin\theta\,\mathrm{d}\theta\right)\left(\int_{\phi=0}^{\phi=2\pi} \mathrm{d}\phi\right) = \tfrac{4}{3}\pi a^3. \tag{139}$$

1.6.3.1 *Alternative notation*

When we take $\mathbf{x} = (x, y, z) = \overrightarrow{OP}$ to denote the position vector of the point P, the triple integral Eq. (124) may be written as

$$I = \iiint_\Sigma \Omega(\mathbf{x})\,\mathrm{d}^3\mathbf{x} = \iiint_\Sigma \Omega(x, y, z)\,\mathrm{d}x\,\mathrm{d}y\,\mathrm{d}z,$$
$$= \iiint_{\Sigma^\star} \Omega\left[x(u,v,w), y(u,v,w), z(u,v,w)\right] |J|\,\mathrm{d}u\,\mathrm{d}v\,\mathrm{d}w. \tag{140}$$

Moreover, when the integrand is a function of several variables, then

$$I(\mathbf{x}) = \iiint_\Sigma f(\mathbf{x}, \mathbf{y})\,\mathrm{d}V(\mathbf{y}) = \iiint_\Sigma f(\mathbf{x}, \mathbf{y})\,\mathrm{d}^3\mathbf{y}. \tag{141}$$

Both notations indicate that here $\mathbf{y}$ is the **dummy** vector of integration, its value ranging over all the points of the region Σ.

1.7 Integral Theorems

1.7.1 *Green's lemma*

Lemma 1.1 *Consider a curve C in the xy-plane enclosing a convex simply-connected region D in a* positive *anticlockwise sense, then*

$$\iint_D \left(\frac{\partial Q}{\partial x} - \frac{\partial P}{\partial y}\right) \mathrm{d}x\,\mathrm{d}y = \oint_C \left[P(x,y)\,\mathrm{d}x + Q(x,y)\,\mathrm{d}y\right], \tag{142}$$

where $\oint_C \Omega(\mathbf{r})\,\mathrm{d}s$, denotes a line integral around a closed *curve C.*

This result is also known as **Stokes's theorem in the plane**.

1.7.2 *Gauss's divergence theorem*

Theorem 1.1 *Let Σ be a finite region bounded by a simple closed surface S, which has an outward-drawn unit normal $\hat{\mathbf{n}}$. If $\mathbf{F}(\mathbf{r}) = (F_1, F_2 F_3)$ is a vector* *field defined and continually differentiable throughout* Σ *and on* S*, then*

$$\iiint_{\Sigma} \operatorname{div}\mathbf{F}\,\mathrm{d}V = \iint_{S} \mathbf{F}\cdot\hat{\mathbf{n}}\,\mathrm{d}A. \tag{143}$$

Corollary 1.1 *If both* $\Omega(\mathbf{r})$ *and* $\mathbf{F}(\mathbf{r})$ *are continuously differentiable in* Σ *and on* S*, then*

$$\iiint_{\Sigma} \operatorname{grad}\Omega\,\mathrm{d}V = \iint_{S} \Omega\hat{\mathbf{n}}\,\mathrm{d}A, \tag{144}$$

$$\iiint_{\Sigma} \operatorname{curl}\mathbf{F}\,\mathrm{d}V = \iint_{S} \hat{\mathbf{n}}\times\mathbf{F}\,\mathrm{d}A. \tag{145}$$

1.7.3 *Stokes's theorem*

Theorem 1.2 *Let* S *be an* open *surface with a unit normal* $\hat{\mathbf{n}}$ *and* C *be a simple closed boundary curve to* S *with a unit normal* $\hat{\mathbf{N}}$ *and unit tangent vector* $\hat{\mathbf{T}}$ *taken in the positive anticlockwise sense* $(\hat{\mathbf{N}}\times\hat{\mathbf{T}} = \hat{\mathbf{n}})$*. If* $\mathbf{F}(\mathbf{r})$ *is a vector field then*

$$\iint_{S} \operatorname{curl}\mathbf{F}\cdot\hat{\mathbf{n}}\,\mathrm{d}A = \oint_{C} \mathbf{F}\cdot\mathrm{d}\mathbf{r} = \oint_{C} \mathbf{F}\cdot\hat{\mathbf{T}}\,\mathrm{d}s, \tag{146}$$

where $\mathrm{d}s$ *is an element of arc length along the curve* C*.*

By the triple scalar product identity Eq. (146) can be re-stated as

$$\iint_{S} (\hat{\mathbf{n}}\times\boldsymbol{\nabla})\cdot\mathbf{F}\,\mathrm{d}A = \oint_{C} \mathrm{d}\mathbf{r}\cdot\mathbf{F}. \tag{147}$$

Corollary 1.2 *If both* $\Omega(\mathbf{r})$ *and* $\mathbf{F}(\mathbf{r})$ *are continuously differentiable in* S *and on* C*, then*

$$\iint_{S} \hat{\mathbf{n}}\times\boldsymbol{\nabla}\Omega\,\mathrm{d}A = \oint_{C} \Omega\,\mathrm{d}\mathbf{r}, \tag{148}$$

$$\iint_{S} (\hat{\mathbf{n}}\times\boldsymbol{\nabla})\times\mathbf{F}\,\mathrm{d}A = \oint_{C} \mathrm{d}\mathbf{r}\times\mathbf{F}. \tag{149}$$

1.7.4 *Green's theorems*

Following from the divergence theorem (143) we have

Theorem 1.3 ***Green's first theorem*** *(or identity)*

$$\iint_S \phi \frac{\partial \psi}{\partial n}\, \mathrm{d}A = \iiint_\Sigma \mathrm{div}(\phi \boldsymbol{\nabla} \psi)\, \mathrm{d}V = \iiint_\Sigma \left[\phi \nabla^2 \psi + \boldsymbol{\nabla}\phi \cdot \boldsymbol{\nabla}\psi\right] \mathrm{d}V. \tag{150}$$

Theorem 1.4 ***Green's second theorem*** *(or identity)*

$$\iint_S \left(\phi \frac{\partial \psi}{\partial n} - \psi \frac{\partial \phi}{\partial n}\right) \mathrm{d}A = \iiint_\Sigma \left[\phi \nabla^2 \psi - \psi \nabla^2 \phi\right] \mathrm{d}V. \tag{151}$$

1.7.5 *Fundamental solutions*

The fundamental solution to the inhomogeneous **Helmholtz** equation

$$\left(\nabla^2 + \kappa^2\right) \phi(\mathbf{r}) = \rho(\mathbf{r}), \qquad \text{for all } \mathbf{r} \in \Sigma \subset \mathbb{R}^3, \tag{152}$$

where $\Sigma \subset \mathbb{R}^3$ is a finite region of 3-D space, is any solution $G = G(\mathbf{r}; \mathbf{y})$ to

$$\left(\nabla^2 + \kappa^2\right) G(\mathbf{r}; \mathbf{y}) = \delta(\mathbf{r} - \mathbf{y}) = \begin{cases} 0 & \text{if } \mathbf{r} \neq \mathbf{y}, \\ `\infty' & \text{if } \mathbf{r} = \mathbf{y}. \end{cases} \tag{153}$$

Here $\delta(\mathbf{r})$ is the 3-D Dirac delta function, which has the property

$$\iiint_\Sigma \delta(\mathbf{r} - \mathbf{a}) f(\mathbf{r})\, \mathrm{d}V(\mathbf{r}) = \begin{cases} f(\mathbf{a}) & \text{if } \mathbf{a} \in \Sigma, \\ 0 & \text{if } \mathbf{a} \notin \Sigma. \end{cases} \tag{154}$$

Thus $G(\mathbf{r}; \mathbf{y})$ is the solution at the point $\mathbf{r}$ due to a **point source** placed at the point $\mathbf{y}$. By the principle of superposition we have

$$\phi(\mathbf{r}) = \iiint_\Sigma \rho(\mathbf{y}) G(\mathbf{r}; \mathbf{y})\, \mathrm{d}V(\mathbf{y}), \tag{155}$$

where $\mathbf{y}$ is now the dummy variable of integration. For then

$$\begin{aligned} \left(\nabla^2 + \kappa^2\right) \phi &= \iiint_\Sigma \rho(\mathbf{y}) \left(\nabla^2 + \kappa^2\right) G(\mathbf{r}; \mathbf{y})\, \mathrm{d}V(\mathbf{y}) \\ &= \iiint_\Sigma \rho(\mathbf{y}) \delta(\mathbf{r} - \mathbf{y})\, \mathrm{d}V(\mathbf{y}) = \rho(\mathbf{r}). \end{aligned} \tag{156}$$

In the absence of boundaries, where $\Sigma \equiv \mathbb{R}^3$, we have the **free-space** solution

$$G(\mathbf{r}; \mathbf{y}) = G(\mathbf{y}; \mathbf{r}) = G(|\mathbf{r} - \mathbf{y}|) = -\frac{1}{4\pi} \frac{1}{|\mathbf{r} - \mathbf{y}|} \mathrm{e}^{\pm \mathrm{i}\kappa |\mathbf{r} - \mathbf{y}|}. \tag{157}$$

Here $\mathrm{i} = \sqrt{(-1)}$ is the **imaginary unit**, and the sign is fixed by the causality **Sommerfeld radiation condition**. This, in turn, yields

$$\phi(\mathbf{r}) = -\frac{1}{4\pi}\iiint_{\mathbb{R}^3} \frac{\rho(\mathbf{y})}{|\mathbf{r}-\mathbf{y}|} \mathrm{e}^{\pm \mathrm{i}\kappa|\mathbf{r}-\mathbf{y}|}\,\mathrm{d}V(\mathbf{y}), \tag{158}$$

with the case $\kappa = 0$ corresponding to Laplace's equation, $\nabla^2\phi = 0$. Indeed near $\mathbf{r} = \mathbf{y}$, Helmholtz's equation effectively reduces to Laplace's equation, having the same singular behaviour. Moreover for any $\Sigma \subset \mathbb{R}$ it can be shown that

$$G(\mathbf{r};\mathbf{y}) = -\frac{1}{4\pi}\frac{1}{|\mathbf{r}-\mathbf{y}|}\mathrm{e}^{\pm \mathrm{i}\kappa|\mathbf{r}-\mathbf{y}|} + H(\mathbf{r};\mathbf{y}) \tag{159}$$

where H satisfies the homogeneous Helmholtz equation

$$\left(\nabla^2 + \kappa^2\right) H(\mathbf{r};\mathbf{y}) = 0, \qquad \text{for all } \mathbf{r} \in \Sigma \subset \mathbb{R}^3. \tag{160}$$

1.7.6 *The Green's function*

Let $\Sigma \subset \mathbb{R}^3$ be a finite region of 3-D space bounded by a surface S, with an outward-drawn unit normal $\hat{\mathbf{n}}$. Suppose we wish to find $\phi = \phi(\mathbf{r})$ such that

$$\left(\nabla^2 + \kappa^2\right)\phi(\mathbf{r}) = \rho(\mathbf{r}), \qquad \text{for all } \mathbf{r} \in \Sigma, \tag{161}$$

subject to the boundary condition

$$\phi = f(\mathbf{r}), \qquad \text{for all } \mathbf{r} \in S, \tag{162}$$

where f or g are given scalar fields on S. From Green's 2nd identity Eq. (151) we have

$$\begin{aligned}\iint_S \left(\phi\frac{\partial G}{\partial n} - G\frac{\partial \phi}{\partial n}\right)\mathrm{d}A(\mathbf{r}) &= \iiint_\Sigma \left[\phi\nabla^2 G - G\nabla^2\phi\right]\mathrm{d}V(\mathbf{r}) \\ &= \iiint_\Sigma \left[\phi\left(\nabla^2+\kappa^2\right)G - G\left(\nabla^2+\kappa^2\right)\phi\right]\mathrm{d}V(\mathbf{r}). \end{aligned} \tag{163}$$

Now if ϕ satisfies Eq. (161) and G satisfies Eq. (153), then

$$\iint_S \left(\phi\frac{\partial G}{\partial n} - G\frac{\partial \phi}{\partial n}\right)\mathrm{d}A(\mathbf{r}) = \phi(\mathbf{y}) - \iiint_\Sigma G(\mathbf{r},\mathbf{y})\rho(\mathbf{r})\,\mathrm{d}V(\mathbf{r}). \tag{164}$$

Moreover if the **Green's function** G also satisfies the boundary condition

$$G = 0, \qquad \text{for all } \mathbf{r} \in S, \tag{165}$$

then the solution at the point $\mathbf{y}$

$$\phi(\mathbf{y}) = \iiint_\Sigma G(\mathbf{r};\mathbf{y})\rho(\mathbf{r})\,\mathrm{d}V(\mathbf{r}) + \iint_S f(\mathbf{r})\frac{\partial G}{\partial n}(\mathbf{r};\mathbf{y})\,\mathrm{d}A(\mathbf{r}). \tag{166}$$

Similarly in terms of $\mathbf{r}$ with $\mathbf{y}$ now the dummy variable of integration

$$\phi(\mathbf{r}) = \iiint_\Sigma G(\mathbf{y};\mathbf{r})\rho(\mathbf{y})\,\mathrm{d}V(\mathbf{y}) + \iint_S f(\mathbf{y})\frac{\partial G}{\partial n'}(\mathbf{y};\mathbf{r})\,\mathrm{d}A(\mathbf{y}). \tag{167}$$

1.8 Suffix Notation

In suffix notation the equation $\mathbf{c} = \mathbf{a} + \mathbf{b}$ is simply written as

$$c_i = a_i + b_i, \tag{168}$$

where $\mathbf{a} = (a_1, a_2, a_3)$ and $\mathbf{b} = (b_1, b_2, b_3)$. It is understood that this equation holds for all values of the free suffix i. The **summation convention** states that when a suffix is repeated in a single term, we sum over all values of that suffix. Thus

$$\mathbf{a}\cdot\mathbf{b} = a_j b_j, \qquad \text{so} \qquad \mathbf{a}\cdot\mathbf{b} = a_1b_1 + a_2b_2 + a_3b_3. \tag{169}$$

To avoid ambiguity, never use any suffix more than *twice* in any one term. Thus

$$(\mathbf{a}\cdot\mathbf{b})\,(\mathbf{c}\cdot\mathbf{d}) = a_j b_j c_k d_k = a_r b_r c_s d_s \tag{170}$$

where we use two dummy variables to indicate that it is $\mathbf{b}$ dotted with $\mathbf{a}$.

1.8.1 *The Kronecker delta*

The Kronecker delta δ_{ij} consists of 9 quantities defined by

$$\delta_{ij} = [I]_{ij} = \begin{cases} 1 & \text{for } i = j, \\ 0 & \text{for } i \neq j, \end{cases} \qquad \text{where } I = \begin{pmatrix} 1\,0\,0 \\ 0\,1\,0 \\ 0\,0\,1 \end{pmatrix}. \tag{171}$$

We see that the **trace** of the matrix I

$$\mathrm{Tr}(I) = \delta_{ii} = \delta_{11} + \delta_{22} + \delta_{33} = 3. \tag{172}$$

For any vector $\mathbf{u} = (u, u_2, u_3)$ we have

$$\delta_{ij}u_j = \delta_{i1}u_1 + \delta_{i2}u_2 + \delta_{i3}u_3, \qquad \text{so} \qquad \delta_{ij}u_j = u_i, \tag{173}$$

since as j runs over all values $\delta_{ij} \neq 0$ only for $j = i$. Similarly

$$\delta_{ij}\delta_{jk} = \sum_{j=1}^{3} \delta_{ij}\delta_{jk} = \delta_{i1}\delta_{1k} + \delta_{i2}\delta_{2k} + \delta_{i3}\delta_{3k} = \delta_{ik}. \tag{174}$$

1.8.2 *The alternating tensor*

The **alternating tensor** ϵ_{ijk} is a set of 27 quantities defined by

$$\epsilon_{ijk} = \begin{cases} +1 & \text{if } (i,j,k) = (1,2,3) \text{ or } (2,3,1) \text{ or } (3,1,2), \\ -1 & \text{if } (i,j,k) = (1,3,2) \text{ or } (2,1,3) \text{ or } (3,2,1), \\ 0 & \text{if any } i,j,k \text{ equal.} \end{cases} \tag{175}$$

There are only 6 non-zero elements. A **cyclic permutation** of suffixes yields

$$\epsilon_{ijk} = \epsilon_{jki} = \epsilon_{kij}, \qquad \text{but} \qquad \epsilon_{ijk} = -\epsilon_{jik}, \tag{176}$$

since the sign of ϵ_{ijk} is changed if any two suffixes are interchanged.

1.8.2.1 *The vector product*

The vector product of $\mathbf{a} = (a_1, a_2, a_3)$ and $\mathbf{b} = (b_1, b_2, b_3)$ has an ith component

$$[\mathbf{a} \times \mathbf{b}]_i = \epsilon_{ijk} a_j b_k \tag{177}$$

for $i = 1, 2$ and 3. Thus the triple scalar product is given by

$$\mathbf{a} \cdot (\mathbf{b} \times \mathbf{c}) = a_i [\mathbf{b} \times \mathbf{c}]_i = a_i \epsilon_{ijk} b_j c_k = \epsilon_{ijk} a_i b_j c_k. \tag{178}$$

1.8.2.2 *The ϵ–δ relation*

A very useful result is that

$$\epsilon_{ijk}\epsilon_{imn} = \delta_{jm}\delta_{kn} - \delta_{jn}\delta_{km}. \tag{179}$$

from which it follows that

$$\delta_{ij}\epsilon_{ijk} = \epsilon_{iik} = 0, \tag{180}$$

$$\epsilon_{ijk}\epsilon_{ijk} = \delta_{jj}\delta_{kk} - \delta_{jk}\delta_{kj} = (\delta_{kk})^2 - \delta_{kk} = 9 - 3 = 6, \tag{181}$$

$$\epsilon_{ijk}\epsilon_{ijn} = \delta_{jj}\delta_{kn} - \delta_{jn}\delta_{kj} = 3\delta_{kn} - \delta_{kn} = 2\delta_{kn}. \tag{182}$$

1.8.3 *Cartesian coordinates*

When using suffix notation we use Cartesian axes $Ox_1x_2x_3$, and we can state that the standard basis $\{\hat{\mathbf{e}}_1, \hat{\mathbf{e}}_2, \hat{\mathbf{e}}_3\}$ is both orthonormal and right-handed by writing

$$\hat{\mathbf{e}}_i \cdot \hat{\mathbf{e}}_j = \delta_{ij}, \qquad \hat{\mathbf{e}}_i \times \hat{\mathbf{e}}_j = \epsilon_{ijk}\hat{\mathbf{e}}_k. \tag{183}$$

If $\mathbf{r} = \overrightarrow{OP}$ is the position vector of the general point $P\colon (x_1, x_2, x_3)$, then

$$\mathbf{r} = x_i\hat{\mathbf{e}}_i \qquad \text{is equivalent to} \qquad \mathbf{r} = x_1\hat{\mathbf{e}}_1 + x_2\hat{\mathbf{e}}_2 + x_3\hat{\mathbf{e}}_3. \tag{184}$$

Note that $r^2 = |\mathbf{r}|^2 = x_i x_i$, but that $r^2 \neq x_i^2$.

1.8.4 *Differential operators*

For a scalar field $\Omega(\mathbf{r})\colon \mathbb{R}^3 \mapsto \mathbb{R}$ the ith component of the vector field $\boldsymbol{\nabla}\Omega(\mathbf{r})$ is

$$[\boldsymbol{\nabla}\Omega]_i = \frac{\partial \Omega}{\partial x_i}, \qquad \text{where } \boldsymbol{\nabla} = \hat{\mathbf{e}}_j \frac{\partial}{\partial x_j}. \tag{185}$$

Thus for a vector field $\mathbf{F} = (F_1, F_2, F_3)$ we have

$$\boldsymbol{\nabla} \cdot \mathbf{F} = \frac{\partial F_i}{\partial x_i}, \qquad [\boldsymbol{\nabla} \times \mathbf{F}]_i = \epsilon_{ijk}\frac{\partial F_k}{\partial x_j}. \tag{186}$$

Note that, in suffix notation the chain rule for a mapping $\mathbf{r} = \mathbf{r}(u, v, w)$, is simply

$$\frac{\partial \Omega}{\partial u} = \frac{\partial \Omega}{\partial x_k}\frac{\partial x_k}{\partial u}, \qquad \frac{\partial \Omega}{\partial v} = \frac{\partial \Omega}{\partial x_k}\frac{\partial x_k}{\partial v}, \qquad \frac{\partial \Omega}{\partial w} = \frac{\partial \Omega}{\partial x_k}\frac{\partial x_k}{\partial w}. \tag{187}$$

1.8.4.1 *Second order operators*

Here div(grad Ω), also known as the **Laplacian** of a scalar field $\Omega(\mathbf{r})$, is given by

$$\nabla^2\Omega \equiv \boldsymbol{\nabla} \cdot (\boldsymbol{\nabla}\Omega) = \frac{\partial}{\partial x_i}\left([\boldsymbol{\nabla}\Omega]_i\right) = \frac{\partial}{\partial x_i}\left(\frac{\partial \Omega}{\partial x_i}\right) = \frac{\partial^2 \Omega}{\partial x_i \partial x_i}, \tag{188}$$

Also the ith component of the Laplacian of a vector field $\mathbf{F}(\mathbf{r})$ is

$$\left[\nabla^2\mathbf{F}\right]_i = \frac{\partial^2 F_i}{\partial x_1^2} + \frac{\partial^2 F_i}{\partial x_2^2} + \frac{\partial^2 F_i}{\partial x_3^2} = \frac{\partial^2 F_i}{\partial x_j \partial x_j}. \tag{189}$$

Similarly we have grad(div $\mathbf{F}$), of which the ith component is given by

$$[\boldsymbol{\nabla}(\boldsymbol{\nabla} \cdot \mathbf{F})]_i = \frac{\partial}{\partial x_i}\left(\frac{\partial F_j}{\partial x_j}\right) = \frac{\partial^2 F_j}{\partial x_i \partial x_j}. \tag{190}$$

Problems

Exercise 1.1 Let $\mathbf{r} = \overrightarrow{OP} = (x, y, z)$ be the position vector of the general point P and $G = G(\mathbf{r})$ be the scalar field

$$G(\mathbf{r}) = -\frac{1}{4\pi}\frac{e^{\lambda|\mathbf{r}-\mathbf{y}|}}{|\mathbf{r}-\mathbf{y}|} = -\frac{1}{4\pi R}e^{\lambda R}, \qquad \text{defined for } R = |\mathbf{r}-\mathbf{y}| \neq 0,$$

where $\mathbf{y} = (x_s, y_s, z_s)$ is some given uniform vector. Obtain ∇G and show that G is a solution to

$$\left(\nabla^2 - \lambda^2\right) G = 0, \qquad \text{for all } R \neq 0.$$

Show that for the case $\lambda = \pm i\kappa$, where $\kappa > 0$, then the condition of an outward propagating wave $Ge^{-i\kappa c_0 t}$, is equivalent to the Sommerfeld condition

$$r\left(\frac{\partial G}{\partial r} - i\kappa G\right) \to 0, \qquad \text{as } r = |\mathbf{r}| \to \infty.$$

Exercise 1.2

(i) Show that one corollary of the divergence theorem is that

$$\iiint_\Sigma \nabla^2 G \, dV = \iint_S \frac{\partial G}{\partial n} \, dA,$$

where Σ is a finite region Σ of 3-D space enclosed by the surface S, and in which n measures distances in the direction of the outward-drawn unit normal, $\hat{\mathbf{n}}$, to S.

(ii) By considering the scalar field G of Ex. 1.1, and a sphere $S\colon |\mathbf{r}-\mathbf{y}| = r_0$, show that

$$\iint_S \frac{\partial G}{\partial n} \, dA = (1 - \lambda r_0)\, e^{\lambda r_0}.$$

Hence confirm the above corollary for this G by direct numerical evaluation for the case where Σ is the region enclosed by spheres $|\mathbf{r}-\mathbf{y}| = b$ and $|\mathbf{r}-\mathbf{y}| = \epsilon$ where $0 < \epsilon < b$. To evaluate the volume integral you may wish to use spherical polars

$$x = x_s + r\cos\phi\sin\theta, \quad y = y_s + r\sin\phi\sin\theta, \quad z = z_s + r\cos\theta.$$

Confirm that this corollary does *not* hold for the case where Σ is the region inside a single sphere $S\colon |\mathbf{r}-\mathbf{y}| = r_0$ and explain this result.

Exercise 1.3 Use suffix notation and the summation convention to prove the identities

(i) $\boldsymbol{\nabla} \cdot (\mathbf{F} \times \mathbf{G}) = \mathbf{G} \cdot (\boldsymbol{\nabla} \times \mathbf{F}) - \mathbf{F} \cdot (\boldsymbol{\nabla} \times \mathbf{G})$,

(ii) $\boldsymbol{\nabla} \times (\mathbf{F} \times \mathbf{G}) = (\boldsymbol{\nabla} \cdot \mathbf{G})\mathbf{F} - (\boldsymbol{\nabla} \cdot \mathbf{F})\mathbf{G} + (\mathbf{G} \cdot \boldsymbol{\nabla})\mathbf{F} - (\mathbf{F} \cdot \boldsymbol{\nabla})\mathbf{G}$,

(iii) $\boldsymbol{\nabla}(\mathbf{F} \cdot \mathbf{G}) = (\mathbf{F} \cdot \boldsymbol{\nabla})\mathbf{G} + (\mathbf{G} \cdot \boldsymbol{\nabla})\mathbf{F} + \mathbf{G} \times (\boldsymbol{\nabla} \times \mathbf{F}) + \mathbf{F} \times (\boldsymbol{\nabla} \times \mathbf{G})$.

Chapter 2

Functions of a Complex Variable

J. W. Elliott
Department of Mathematics
University of Hull

2.1 Complex Numbers

The **Cartesian representation** of a complex number, $z \in \mathbb{C}$, is given by

$$z = x + \mathrm{i}y, \qquad \text{where } \mathrm{i} = \sqrt{(-1)}. \tag{1}$$

We call i the **imaginary unit**, $x = \mathrm{Re}(z) \in \mathbb{R}$ the **real part** and $y = \mathrm{Im}(z) \in \mathbb{R}$ the **imaginary part**. We call z **purely real** if $y = 0$ and **purely imaginary** if $x = 0$. The **zero** element $0 = 0 + 0\mathrm{i}$.

2.1.1 *Algebraic operations*

If $z_1 = a_1 + \mathrm{i}b_1$ and $z_2 = a_2 + \mathrm{i}b_2$ then we define

$$z_1 \pm z_2 = (a_1 + \mathrm{i}b_1) \pm (a_2 + \mathrm{i}b_2) = (a_1 \pm a_2) + \mathrm{i}\,(b_1 \pm b_2)\,, \tag{2}$$

$$z_1 z_2 = (a_1 + \mathrm{i}b_1)\,(a_2 + \mathrm{i}b_2) = (a_1 a_2 - b_1 b_2) + \mathrm{i}\,(a_1 b_2 + a_2 b_1) = z_2 z_1, \tag{3}$$

$$\frac{z_1}{z_2} = \frac{a_1 + \mathrm{i}b_1}{a_2 + \mathrm{i}b_2} = \frac{(a_1 + \mathrm{i}b_1)\,(a_2 - \mathrm{i}b_2)}{(a_2 + \mathrm{i}b_2)\,(a_2 - \mathrm{i}b_2)} = \frac{(a_1 a_2 + b_1 b_2) + \mathrm{i}\,(a_2 b_1 - a_1 b_2)}{a_2^2 + b_2^2}. \tag{4}$$

2.1.2 *Complex conjugate and modulus*

For any complex number $z = x + \mathrm{i}y$ we define its **complex conjugate** z^* by

$$z^* = x - \mathrm{i}y. \tag{5}$$

We see then that

$$x = \mathrm{Re}\, z = \tfrac{1}{2}\left(z + z^*\right), \qquad y = \mathrm{Im}\, z = -\tfrac{1}{2}\mathrm{i}\left(z - z^*\right). \tag{6}$$

It can be shown that

$$(z^*)^* = z, \quad (z_1 + z_2)^* = z_1^* + z_2^*, \quad (z_1 z_2)^* = z_1^*\, z_2^*, \quad (z_1/z_2)^* = z_1^*/z_2^*. \tag{7}$$

We define its **modulus**, or **magnitude**, $|z|$, by

$$|z| = \sqrt{x^2 + y^2}. \tag{8}$$

It can be shown that

$$|z^*| = |z|, \qquad |z_1 z_2| = |z_1||z_2|, \qquad |z_1/z_2| = |z_1|/|z_2|. \tag{9}$$

We note that since

$$|z|^2 = zz* = (x + \mathrm{i}y)(x - \mathrm{i}y) = x^2 + y^2, \tag{10}$$

then we can define the reciprocal by

$$z^{-1} = \frac{1}{z} = \frac{z^*}{|z|^2}, \qquad \text{and} \qquad \frac{z_1}{z_2} = z_1 z_2^{-1} = \frac{z_1 z_2^*}{|z_2|^2}. \tag{11}$$

Finally we have the triangle inequalities

$$|z_1 + z_2| \leq |z_1| + |z_2|, \qquad |z_1 - z_2| \geq |z_1| - |z_2|. \tag{12}$$

2.1.3 *The Argand diagram*

A geometrical representation of a complex number is the **Argand diagram** shown in Fig. 2.1, where $z = x + \mathrm{i}y$ is the point P in the complex z-plane with Cartesian coordinates (x, y). We split up the plane into four quadrants, I–IV. Here z^* is the reflection of z in the real axis and $|z|$ is the length of the vector $\overrightarrow{OP}$. We represent the sum $z_1 + z_2$ *geometrically* as the point P, where $\overrightarrow{OP} = \overrightarrow{OP}_1 + \overrightarrow{OP}_2$, and P_1 and P_2 are the points representing z_1 and z_2.

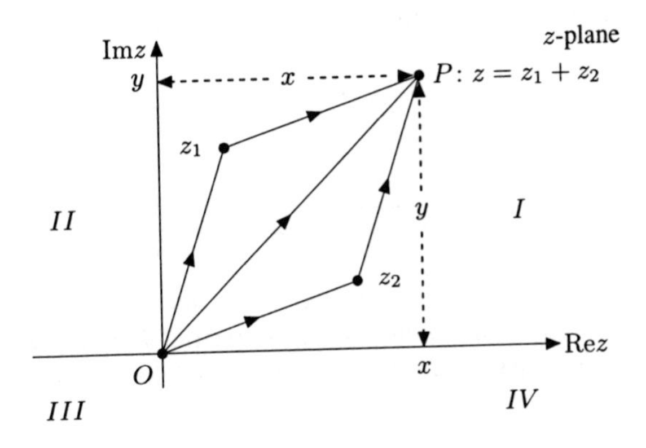

Fig. 2.1 The complex plane.

2.1.4 *Polar coordinates*

Every complex number $z \in \mathbb{C}$ has a **polar representation**

$$z = r\cos\theta + \mathrm{i}r\sin\theta, \tag{13}$$

where $r = |z|$ is the magnitude of z and $\theta = \arg z$, the **argument** of z, is the angle between the real axis and the line OP, measured in an anticlockwise sense, as shown in Fig. 2.2.

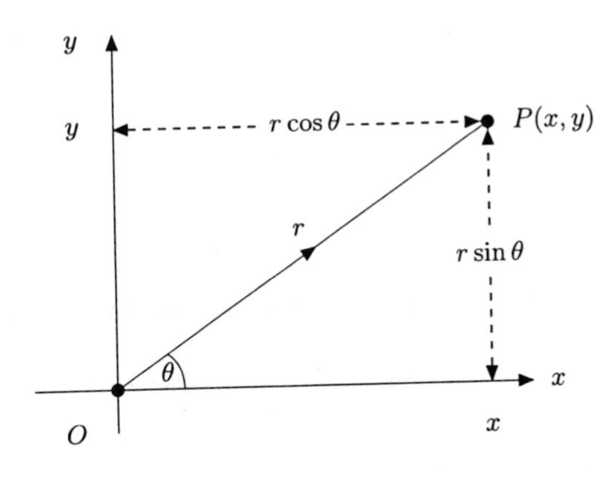

Fig. 2.2 Polar coordinates.

In terms of the Cartesian form $z = x + \mathrm{i}y$ we see that

$$x = r\cos\theta, \qquad y = r\sin\theta \tag{14}$$

or equivalently

$$r = |z| = \sqrt{x^2 + y^2}, \qquad \theta = \arg z = \tan^{-1}(y/x). \tag{15}$$

If we increase θ to $\theta + 2k\pi$, $k = 0, \pm 1, \pm 2, \ldots$ then we are at the same point in the z-plane. To avoid this ambiguity we can restrict θ to the range $-\pi < \theta \leq \pi$, known as the **principal value** of the argument, and denoted by $\mathrm{Arg}\, z$.

2.1.5 *The complex exponential form*

Euler's identity gives us that

$$\mathrm{e}^{\mathrm{i}\theta} = \cos\theta + \mathrm{i}\sin\theta, \qquad \mathrm{e}^{-\mathrm{i}\theta} = \cos\theta - \mathrm{i}\sin\theta. \tag{16}$$

The complex number $\mathrm{e}^{\mathrm{i}\theta}$, and its complex conjugate $\mathrm{e}^{-\mathrm{i}\theta} = \mathrm{e}^{\mathrm{i}(-\theta)}$, both represent points lying on the unit circle, $|z| = 1$, in the complex plane, since

$$|\mathrm{e}^{\pm\mathrm{i}\theta}| = \sqrt{\cos^2\theta + \sin^2\theta} = 1. \tag{17}$$

Thus we can write the polar form Eq. (13) simply as

$$z = r\,(\cos\theta + \mathrm{i}\sin\theta) = r\mathrm{e}^{\mathrm{i}\theta} = |z|\mathrm{e}^{\mathrm{i}\arg z}, \tag{18}$$

avoiding the need to use $\cos\theta$ and $\sin\theta$. This is known as the **complex exponential form**. Note that if

$$z = r\mathrm{e}^{\mathrm{i}\theta}, \qquad \text{then} \qquad z^* = r\mathrm{e}^{-\mathrm{i}\theta}, \qquad \text{so} \qquad \frac{1}{z} = \frac{z^*}{|z|^2} = \frac{1}{r}\mathrm{e}^{-\mathrm{i}\theta}. \tag{19}$$

2.1.5.1 *Multiplication*

If $z_1 = r_1\mathrm{e}^{\mathrm{i}\theta_1}$ and $z_2 = r_2\mathrm{e}^{\mathrm{i}\theta_2}$ then for multiplication and division we have

$$z_1 z_2 = (r_1 r_2)\,\mathrm{e}^{\mathrm{i}(\theta_1+\theta_2)}, \qquad \frac{z_1}{z_2} = \frac{r_1}{r_2}\mathrm{e}^{\mathrm{i}(\theta_1-\theta_2)}. \tag{20}$$

Further, since $|\mathrm{e}^{\mathrm{i}\theta}| = 1$ we recover the results

$$|z_1 z_2| = r_1 r_2 = |z_1||z_2|, \qquad \left|\frac{z_1}{z_2}\right| = \frac{r_1}{r_2} = \frac{|z_1|}{|z_2|}. \tag{21}$$

2.1.5.2 *De Moivre's theorem*

If $z = r\mathrm{e}^{\mathrm{i}\theta}$ then repeated application of Eq. (20) leads to the power law form

$$z^n = \left(r\mathrm{e}^{\mathrm{i}\theta}\right)^n = r^n \mathrm{e}^{\mathrm{i}n\theta}, \tag{22}$$

for any integer $n \in \mathbb{Z}$. If $z = \mathrm{e}^{\mathrm{i}\theta}$, then the power law Eq. (22) yields **de Moivre's theorem**, which is usually expressed as

$$(\cos\theta + \mathrm{i}\sin\theta)^n = \left(\mathrm{e}^{\mathrm{i}\theta}\right)^n = \mathrm{e}^{\mathrm{i}n\theta} = \cos(n\theta) + \mathrm{i}\sin(n\theta) \tag{23}$$

for any integer $n \in \mathbb{Z}$.

2.1.5.3 *Rotation*

Multiplying any complex number z by $\mathrm{e}^{\mathrm{i}\alpha}$ is equivalent to an anticlockwise rotation of the point z through an angle α, since

$$z = r\mathrm{e}^{\mathrm{i}\theta}, \qquad \text{so} \qquad z\mathrm{e}^{\mathrm{i}\alpha} = r\mathrm{e}^{\mathrm{i}(\alpha+\theta)}. \tag{24}$$

For example multiplication of z by $\mathrm{i} = \mathrm{e}^{\mathrm{i}\pi/2}$ is equivalent to an anticlockwise rotation of z by $\pi/2$. Indeed if $z = x + \mathrm{i}y$ then

$$\mathrm{i}z = -y + \mathrm{i}x, \qquad \mathrm{i}^2 z = -x - \mathrm{i}y, \qquad \mathrm{i}^3 z = y - \mathrm{i}x. \tag{25}$$

2.1.6 *Waves*

If x measures distance and t is the time then a plane wave of **amplitude** A_m, **wavenumber** k and **angular velocity** ω is of the form

$$u(x,t) = A_m \cos(kx - \omega t + \phi), \tag{26}$$

where ϕ is the **initial phase**. From Euler's identity

$$\mathrm{e}^{\mathrm{i}\theta} = \cos\theta + \mathrm{i}\sin\theta \qquad \text{so} \qquad \cos\theta = \mathrm{Re}\left(\mathrm{e}^{\mathrm{i}\theta}\right), \qquad \sin\theta = \mathrm{Im}\left(\mathrm{e}^{\mathrm{i}\theta}\right). \tag{27}$$

Consequently we see that we can write the wave as

$$u(t) = \mathrm{Re}\left\{A_m \mathrm{e}^{\mathrm{i}(kx-\omega t+\phi)}\right\} = \mathrm{Re}\left\{A\mathrm{e}^{\mathrm{i}(kx-\omega t)}\right\}, \tag{28}$$

where $A = A_m \mathrm{e}^{\mathrm{i}\phi}$ is the **complex magnitude** incorporating the initial phase.

2.1.7 *Complex polynomials*

A complex polynomial equation, of degree n, is of the form

$$P_n(z) \equiv a_n z^n + a_{n-1} z^{n-1} + \cdots + a_1 z + a_0 = 0, \tag{29}$$

where the constants $a_0, a_1, \ldots, a_n \in \mathbb{C}$, $a_n \neq 0$. The **fundamental theorem of algebra** states that there are n complex **roots** $z = z_1, z_2, \ldots, z_n$ to this equation. That is we can always factorise $P_n(z)$ as

$$P_n(z) \equiv a_n(z - z_1)(z - z_2)\cdots(z - z_n). \tag{30}$$

If the coefficients a_j for $j = 1, 2, \ldots, n$ are in fact *real*, then the roots are either real or appear as complex conjugate pairs, recalling that if $z_k = x_k + \mathrm{i}y_k$

$$(z - z_k)(z - z_k^*) = z^2 - (z_k + z_k^*) + z_k z_k^* = z^2 - 2x_k z + (x_k^2 + y_k^2). \tag{31}$$

2.1.8 *Square and* n*th roots*

There are two roots to a complex quadratic of the form

$$a_2 z^2 + a_1 z + a_0 = a_2(z - z_1)(z - z_2) = 0. \tag{32}$$

Finding the square root of the complex number $z_0 = r_0 \mathrm{e}^{\mathrm{i}\theta_0}$ is equivalent to solving the quadratic $z^2 - z_0 = 0$. Since $\mathrm{e}^{\mathrm{i}2k\pi} = 1$ for $k = 0, \pm 1, \pm 2, \ldots$ we see that

$$z^2 = z_0 = r_0 \mathrm{e}^{\mathrm{i}\theta_0} = r_0 \mathrm{e}^{\mathrm{i}(\theta_0 + 2k\pi)}, \qquad \text{for } k = 0, \pm 1, \pm 2, \ldots. \tag{33}$$

Squaring a complex number entails squaring its modulus and doubling its argument. Thus the inverse requires taking the square root of the modulus and halving the argument. This yields the two *distinct* square roots $z = z_1$ and $z = z_2 = -z_1$

$$z = \sqrt{z_0} = \sqrt{r_0}\,\mathrm{e}^{\mathrm{i}(\theta_0 + 2k\pi)/2} = \begin{cases} z_1 = \sqrt{r_0}\,\mathrm{e}^{\frac{1}{2}\mathrm{i}\theta_0} & \text{for } k = 0, \pm 2, \pm 4, \ldots, \\ z_2 = -\sqrt{r_0}\,\mathrm{e}^{\frac{1}{2}\mathrm{i}\theta_0} & \text{for } k = \pm 1, \pm 3, \pm 5, \ldots. \end{cases} \tag{34}$$

Thus the nth roots of a complex number z_0 are those n complex numbers z_r for $r = 1, 2, \ldots, n$ that satisfy the equation $z^n - z_0 = 0$. In polar form we have

$$z^n = z_0 = r_0 \mathrm{e}^{\mathrm{i}\theta_0} = r_0 \mathrm{e}^{\mathrm{i}[\theta_0 + 2(r-1)\pi]} = r_0 \mathrm{e}^{\mathrm{i}\theta_r} \qquad \text{for } r = 1, 2, \ldots, \tag{35}$$

where $\theta_r = [\theta_0 + 2(r-1)\pi]$. Therefore we have n distinct complex solutions

$$z_r = \sqrt[n]{r_0}\mathrm{e}^{\mathrm{i}\theta_r/n} = \sqrt[n]{r_0}\,[\cos(\theta_r/n) + \mathrm{i}\sin(\theta_r/n)] \quad \text{for } r = 1, 2, \ldots, n. \tag{36}$$

Geometrically the n roots are equally spaced out around the circle $|z| = \sqrt[n]{r_0}$.

2.2 Functions of a Complex Variable

A **function** $f\colon \mathbb{C} \mapsto \mathbb{C}$ assigns to each $z \in D \subset \mathbb{C}$, a *unique* complex number w. Writing $z = x + \mathrm{i}y$ the **Cartesian representation** is given by

$$w = f(z) = u(x, y) + \mathrm{i}v(x, y), \tag{37}$$

where $u = \operatorname{Re} f$, $v = \operatorname{Im} f$. Writing $z = r\mathrm{e}^{\mathrm{i}\theta}$ the **polar representation** is given by

$$w = f(z) = \rho(r, \theta)\mathrm{e}^{\mathrm{i}\phi(r,\theta)} = u(r, \theta) + \mathrm{i}v(r, \theta), \tag{38}$$

where $\rho = |f|$, $\phi = \arg f$. For example

$$w = f(z) = z^2 = \begin{cases} (x + \mathrm{i}y)^2 = (x^2 - y^2) + \mathrm{i}2xy, \\ r^2\mathrm{e}^{2\mathrm{i}\theta} = r^2\left[\cos(2\theta) + \mathrm{i}\sin(2\theta)\right], \end{cases} \tag{39}$$

so that

$$u(x, y) = x^2 - y^2 = r^2\cos(2\theta), \qquad v(x, y) = 2xy = r^2\sin(2\theta). \tag{40}$$

2.2.1 *Mappings*

To display graphically $w = f(z)$, where both w and z are complex, we must introduce two separate complex planes, with $\mathbb{C}_z$ the z-plane and $\mathbb{C}_w$ the w-plane, as shown in Fig. 2.3.

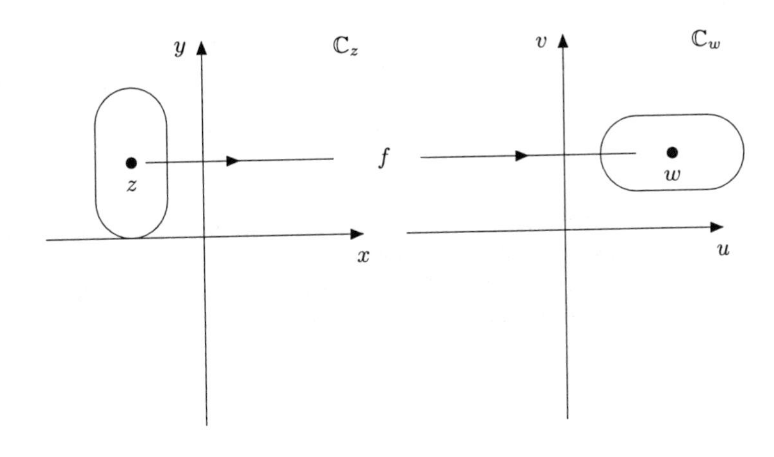

Fig. 2.3 Complex mappings.

By allowing z to vary, we see that f maps the region D, called the **domain** (of definition) of f, into a region R, called the **range** of f.

2.2.2 *The complex exponential function*

We define the exponential function by

$$f(z) = \mathrm{e}^z = \mathrm{e}^{x+\mathrm{i}y} = \mathrm{e}^x \mathrm{e}^{\mathrm{i}y} = \mathrm{e}^x \left[\cos y + \mathrm{i} \sin y\right] \tag{41}$$

so that $|\mathrm{e}^z| = \mathrm{e}^x$ and $\arg(\mathrm{e}^z) = y$. We can then define the hyperbolic functions by

$$\cosh z = \frac{1}{2}\left(\mathrm{e}^z + \mathrm{e}^{-z}\right), \qquad \sinh z = \frac{1}{2}\left(\mathrm{e}^z - \mathrm{e}^{-z}\right), \tag{42}$$

and the trigonometric functions by

$$\cos z = \frac{1}{2}\left(\mathrm{e}^{\mathrm{i}z} + \mathrm{e}^{-\mathrm{i}z}\right), \qquad \sin z = \frac{1}{2\mathrm{i}}\left(\mathrm{e}^{\mathrm{i}z} - \mathrm{e}^{-\mathrm{i}z}\right). \tag{43}$$

2.2.3 *Multiple-valued functions*

A multiple-valued function f assigns more than one value to a point $z \in D$. To avoid confusion we construct a **branch** of a multiple-valued function f, which is a single-valued function F that in some domain D takes only one of the values of $f(z)$ for $z \in D$. Often all points z on a given line, a **branch cut**, are removed from the z-plane in order to define the single-valued branch.

For example if $z = r\mathrm{e}^{\mathrm{i}\theta}$ where $-\pi < \theta \leq \pi$, then the square-root function has two possible values, namely $\pm\sqrt{r}\,\mathrm{e}^{\mathrm{i}\theta/2}$. Choosing only the positive value, taking

$$F_0(z) = \begin{cases} \sqrt{r}\,\mathrm{e}^{\mathrm{i}\theta/2} & \text{for } r > 0,\ -\pi < \theta < \pi, \\ 0 & \text{for } z = 0, \end{cases} \tag{44}$$

then this (single-valued) function F_0 is well-defined on the entire complex plane *except* along the **ray** $\theta = \pi$, the negative real axis.

Any point that is common to all branch cuts of f is called a **branch point**. In general $z = z_0$ is a branch point of f then for some $r_0 > 0$

$$f\left(z_0 + r_0\mathrm{e}^{\mathrm{i}\theta_0}\right) \neq f\left(z_0 + r_0\mathrm{e}^{\mathrm{i}(\theta_0+2\pi)}\right) \tag{45}$$

2.2.3.1 *The nth root function*

For integer $n > 0$ consider the n-valued function

$$w = f(z) = z^{1/n} = r^{1/n}\mathrm{e}^{\mathrm{i}\theta/n}, \tag{46}$$

where $z = re^{i\theta}$. Here $z = 0$ is a branch point since

$$f\left[re^{i(\theta+2\pi)}\right] = r^{1/n}e^{i(\theta+2\pi)/n} = e^{i2\pi/n}f(re^{i\theta}) \neq f(re^{i\theta}). \quad (47)$$

To make $w = z^{1/n}$ single-valued we **cut** the z-plane, between the branch point $z = 0$ along some **ray** $\theta = \alpha$, and forbid z to cross this barrier, effectively restricting $\theta = \arg z$ to the range $\alpha - 2\pi < \theta < \alpha$. For $\alpha = \pi$ we have n single-valued functions

$$w = F_k(z) = r^{1/n} \exp\left[i\left(\frac{\theta + 2k\pi}{n}\right)\right] \qquad (r > 0, -\pi < \theta < \pi), \quad (48)$$

for $k = 0, 1, 2, \ldots, n-1$. These branches of $z^{1/n}$ represent one-to-one mappings from the z-plane, cut along the negative real axis, to wedge-shaped regions D_k of the w-plane, where

$$D_k: \qquad (2k-1)\frac{\pi}{n} < \arg w < (2k+1)\frac{\pi}{n} \qquad k = 0, 1, 2, \ldots, n-1. \quad (49)$$

The case $w = F_0(z)$ is called the **principal branch**, mapping the cut z-plane to the region D_0 where $|\arg w| < \pi/n$.

For example the principal branch of the square-root function, $w = z^{1/2}$, given by

$$w = F_0(z) = r^{1/2}e^{i\theta/2}, \qquad \text{where } r > 0, -\pi < \theta < \pi, \quad (50)$$

maps the z-plane, cut along the negative real axis, onto the right-hand half of the w-plane. Similarly the branch

$$w = F_1(z) = r^{1/2}e^{i(\theta+2\pi)/2} = -r^{1/2}e^{i\theta/2} = -F_0(z) \quad (51)$$

maps the cut z-plane onto the left-hand half of the w-plane.

2.2.4 *The complex logarithmic function*

For $z = x + iy = re^{i\theta} = |z|e^{i\arg z}$, we define the logarithm function by

$$f(z) = \log z = \ln|z| + i\arg z. \quad (52)$$

Again $\log z$ is the inverse function to the complex exponential since

$$e^{\log z} = e^{[\ln|z| + i\arg z]} = e^{\ln|z|}e^{i\arg z} = |z|e^{i\arg z} = z, \quad (53)$$

$$\log(e^z) = \ln|e^z| + i\arg(e^z) = \ln(e^x) + iy = x + iy = z. \quad (54)$$

The usual properties still hold, namely

$$\log(z_1 z_2) = \log z_1 + \log z_2, \qquad \log(z_1/z_2) = \log z_1 - \log z_2. \quad (55)$$

Because of $\arg z$, $\log z$ is a many-valued function, where clearly the origin is a branch point. We define the principal value of $\log z$, denoted by $\operatorname{Log} z$, to be the single-valued function

$$w = \operatorname{Log} z = \ln r + \mathrm{i}\theta = \ln|z| + \mathrm{i}\operatorname{Arg} z, \quad \text{where } r > 0,\ -\pi < \theta < \pi. \tag{56}$$

Here the z-plane is cut from the origin along negative real axis. There is a one-to-one mapping between the cut z-plane and the strip $-\pi < v = \operatorname{Im}(w) < \pi$ in the w-plane.

This, in turn allows us to define the **complex exponent** by

$$z^c = \mathrm{e}^{c \log z} \qquad \text{for } z \neq 0,\ c \in \mathbb{C}. \tag{57}$$

In particular this yields

$$z^{c_1+c_2} = \mathrm{e}^{(c_1+c_2)\log z} = \mathrm{e}^{c_1 \log z}\mathrm{e}^{c_2 \log z} = z^{c_1} z^{c_2}. \tag{58}$$

2.2.5 *Analytic functions*

We define the derivative of the complex function $w = f(z)$, to be the limit

$$f'(z) = \lim_{s \to 0} \frac{f(z+s) - f(z)}{s}. \tag{59}$$

If the limit both *exists* and is *independent* of the way $s \to 0$ we say that f is **analytic** at the point z. Otherwise we say that f is **singular** at z.

Suppose f is analytic, then if $w = f(z)$ and $z = x + \mathrm{i}y$ we can write

$$f(x + \mathrm{i}y) = u(x, y) + \mathrm{i}v(x, y). \tag{60}$$

Hence if $s = h + \mathrm{i}k$ then, by definition, the derivative is given by

$$f'(z) = \lim_{s \to 0} \frac{[u(x+h, y+k) - u(x, y)] + \mathrm{i}\,[v(x+h, y+k) - v(x, y)]}{s}. \tag{61}$$

Setting $s = h + \mathrm{i}0$ and letting $h \to 0$, then gives

$$f'(z) = \lim_{h \to 0} \frac{u(x+h, y) - u(x, y)}{h} + \mathrm{i} \lim_{h \to 0} \frac{v(x+h, y) - v(x, y)}{h} = \frac{\partial u}{\partial x} + \mathrm{i}\frac{\partial v}{\partial x}. \tag{62}$$

Alternatively setting $s = 0 + \mathrm{i}k$ and letting $k \to 0$, gives

$$f'(z) = \lim_{k \to 0} \frac{u(x, y+k) - u(x, y)}{\mathrm{i}k} + \mathrm{i} \lim_{k \to 0} \frac{v(x, y+k) - v(x, y)}{\mathrm{i}k} = -\mathrm{i}\frac{\partial u}{\partial y} + \frac{\partial v}{\partial y}. \tag{63}$$

Equating the two relations yields the **Cauchy–Riemann equations**

$$\frac{\partial u}{\partial x} = \frac{\partial v}{\partial y}, \qquad \frac{\partial u}{\partial y} = -\frac{\partial v}{\partial x}, \tag{64}$$

which form a necessary condition for f to be analytic. Further if f is analytic then both u and v are **harmonic** since they satisfy **Laplace's equation**

$$\frac{\partial^2 u}{\partial x^2} + \frac{\partial^2 u}{\partial y^2} = 0, \qquad \frac{\partial^2 v}{\partial x^2} + \frac{\partial^2 v}{\partial y^2} = 0. \tag{65}$$

2.3 Contour Integration

The **contour integral** of a function f of the complex variable $z = x + \mathrm{i}y$ along a curve (contour) C, lying in the complex z-plane, between the points A and B is

$$I = \int_{C:\, A\to B} f(z)\,\mathrm{d}z = \int_{C:\, A\to B} (u\,\mathrm{d}x - v\,\mathrm{d}y) + \mathrm{i}\int_{C:\, A\to B} (v\,\mathrm{d}x + u\,\mathrm{d}y), \tag{66}$$

since $f = u + \mathrm{i}v$ and $\mathrm{d}z = \mathrm{d}x + \mathrm{i}\mathrm{d}y$. Thus contour integration is equivalent to combining two line integrals, along the same curve C, to obtain a complex value. Provided the function is analytic within the region of consideration

$$\int_{C:\, A\to B} f(z)\,\mathrm{d}z = [F(z)]_A^B \qquad \text{where } F'(z) = f(z). \tag{67}$$

Example 2.1 Evaluate the integral

$$\begin{aligned} I &= \int_{C:\, A\to B} z^2\,\mathrm{d}z \\ &= \int_C \left[(x^2 - y^2)\,\mathrm{d}x - 2xy\,\mathrm{d}y\right] + \mathrm{i}\int_C \left[2xy\,\mathrm{d}x - (x^2 - y^2)\,\mathrm{d}y\right], \end{aligned} \tag{68}$$

where C is the straight line AB from $-1 + \mathrm{i}$ to $5 + 3\mathrm{i}$.

Solution On AB we can parameterise the curve

$$z = -1 + \mathrm{i} + 2(3 + \mathrm{i})t, \qquad \mathrm{d}z = 2(3 + \mathrm{i})\mathrm{d}t, \qquad 0 \le t \le 1, \tag{69}$$

so that

$$\begin{aligned} I &= \int_{AB} z^2\,\mathrm{d}z = \int_0^1 \left[-2\mathrm{i} + 2(-4 + 2\mathrm{i})t + 4(8 + 6\mathrm{i})t^2\right] 2(3 + \mathrm{i})\mathrm{d}t \\ &= 2(3 + \mathrm{i})\left[-2\mathrm{i}t + (-4 + 2\mathrm{i})t^2 + \tfrac{4}{3}(8 + 6\mathrm{i})t^3\right]_{-1}^{5} = -4 + \tfrac{196}{3}\mathrm{i}. \end{aligned} \tag{70}$$

Alternatively

$$I = \int_{C:\,A\to B} z^2\,\mathrm{d}z = \left[\tfrac{1}{3}z^3\right]_{-1+\mathrm{i}}^{5+3\mathrm{i}} = \tfrac{1}{3}\left[(5+3\mathrm{i})^3 - (-1+\mathrm{i})^3\right]$$
$$= \tfrac{1}{3}\left[(-10+198\mathrm{i}) - (2+2\mathrm{i})^3\right] = -4 + \tfrac{196}{3}\mathrm{i}. \tag{71}$$

2.3.1 *Properties*

If we **partition** a curve C into two smooth arcs, $C = C_1 + C_2$, it follows

$$\int_C f(z)\,\mathrm{d}z = \int_{C_1} f(z)\,\mathrm{d}z + \int_{C_2} f(z)\,\mathrm{d}z. \tag{72}$$

Linearity implies that for any functions f and g and constants $\alpha, \beta \in \mathbb{C}$, we have

$$\int_C [\alpha f(z) + \beta g(z)]\,\mathrm{d}z = \alpha \int_C f(z)\,\mathrm{d}z + \beta \int_C g(z)\,\mathrm{d}z. \tag{73}$$

The **magnitude inequality** states that if $|f(z)| \leq M$ at all points of a contour C, for some positive constant M, then

$$\left|\int_C f(z)\,\mathrm{d}z\right| \leq \int_C |f(z)||\mathrm{d}z| \leq ML, \tag{74}$$

where L is the length of the contour C.

2.3.2 *Closed contours*

For a **closed** contour C we write

$$I = \oint_C f(z)\,\mathrm{d}z.$$

We say C is **positively orientated** if it is taken in an anticlockwise sense, and that C is a **simple** closed curve if there are no self-intersections.

Example 2.2 If C is a circle, centred about $z = z_0$, of radius r, taken in an anticlockwise sense, evaluate

$$\text{(i)}\quad I_1 = \oint_C (z - z_0)^n\,\mathrm{d}z, \qquad (n \neq -1), \qquad \text{(ii)}\quad I_2 = \oint_C \frac{1}{z - z_0}\,\mathrm{d}z. \tag{75}$$

Solution On C, $z = z_0 + re^{i\theta}$, $dz = ire^{i\theta}d\theta$ for $0 \le \theta \le 2\pi$. Thus

(i)
$$I_1 = \int_0^{2\pi} (re^{i\theta})^n ire^{i\theta} d\theta = \int_0^{2\pi} ir^{n+1}e^{(n+1)i\theta} d\theta$$
$$= \left[\frac{1}{(n+1)} r^{n+1} e^{(n+1)i\theta}\right]_0^{2\pi} = 0. \quad (76)$$

(ii)
$$I_2 = \int_0^{2\pi} \frac{1}{re^{i\theta}} ire^{i\theta} d\theta = \int_0^{2\pi} i\, d\theta = [i\theta]_0^{2\pi} = 2\pi i. \quad (77)$$

2.3.3 *Cauchy's theorem*

Theorem 2.1 *Let C be a simple, positively orientated closed curve and suppose that $f(z)$ is analytic both inside and on C, then* ***Cauchy's theorem*** *states that*

$$\oint_C f(z)\, dz = 0. \quad (78)$$

Theorem 2.2 *If $f(z)$ is a continuous function defined in a simply connected domain D, such that for every simple closed curve C in D*

$$\oint_C f(z)\, dz = 0, \quad (79)$$

then ***Morera's theorem*** *states that $f(z)$ is analytic in D.*

2.3.4 *Cauchy's integral theorem*

Theorem 2.3 *Let C be a simple, positively orientated closed curve and suppose that $f(z)$ is analytic both inside and on C, then* ***Cauchy's integral theorem*** *states that if z_0 is any point inside C, then*

$$f^{(n)}(z_0) = \frac{n!}{2\pi} \oint_C \frac{f(z)}{(z - z_0)^{n+1}}\, dz. \quad (80)$$

2.4 Power Series Expansions

2.4.1 *The Taylor series*

For a function $f(z)$, which is analytic in the neighbourhood $|z - z_0| < R$, we define its Taylor series about a point $z = z_0$ to be the expansion

$$f(z) = f(z_0) + (z - z_0)f'(z_0) + \frac{1}{2!}(z - z_0)f''(z_0) + \cdots . \tag{81}$$

The series Eq. (81) converges only for $|z - z_0| < R$, where R, the **radius of convergence** equals the distance from $z = z_0$ to the nearest singular point. For example the Taylor series of $(1 - z)^{-1}$, about $z = 0$, is the geometric series

$$\frac{1}{1-z} = \sum_{k=0}^{\infty} z^k = 1 + z + z^2 + z^3 + \cdots \qquad \text{for } |z| < 1. \tag{82}$$

The radius of convergence $R = 1$ is the distance from $z = 0$ to the singularity at $z = 1$. To obtain the expansion of $(1 - z)^{-1}$, about any point $z = z_0$, we can write

$$\begin{aligned}\frac{1}{1-z} &= \frac{1}{1 - z_0 - (z - z_0)} = \frac{1}{(1 - z_0)\left[1 - \frac{z-z_0}{1-z_0}\right]} = \frac{1}{1 - z_0}\sum_{k=0}^{\infty}\left(\frac{z - z_0}{1 - z_0}\right)^k \\ &= \frac{1}{1 - z_0}\left\{1 + \left(\frac{z - z_0}{1 - z_0}\right) + \left(\frac{z - z_0}{1 - z_0}\right)^2 + \cdots\right\} \quad \text{for } \left|\frac{z - z_0}{1 - z_0}\right| < 1.\end{aligned} \tag{83}$$

Again the radius of convergence is equal to the distance from $z = z_0$ to $z = 1$.

2.4.2 *The Laurent series*

For a function f, which is analytic in the annular domain $0 \leq r < |z - z_0| < R$, we define its **Laurent series**, about a point $z = z_0$ (which is *not* necessarily analytic), to be the expansion

$$f(z) = \cdots + \frac{a_{-2}}{(z - z_0)^2} + \frac{a_{-1}}{(z - z_0)} + a_0 + a_1(z - z_0) + a_2(z - z_0)^2 + \cdots , \tag{84}$$

where, if γ is any anticlockwise circular contour $|z - z_0| = \rho$, for $0 < \rho < R$, then

$$a_n = \frac{1}{2\pi}\oint_\gamma \frac{f(z)}{(z - z_0)^{n+1}}\,\mathrm{d}z \tag{85}$$

From Cauchy's integral formula, Eq. (80), we see that Eq. (84) is a generalization of Taylor's series. We can write Eq. (84) as the sum of two infinite series

$$f(z) = \sum_{n=-\infty}^{n=\infty} a_n(z - z_0)^n = \sum_{n=1}^{n=\infty} \frac{a_{-n}}{(z - z_0)^n} + \sum_{n=0}^{n=\infty} a_n(z - z_0)^n. \tag{86}$$

We call the first series the **principal part**. If $f(z)$ is *regular* (analytic) at the point $z = z_0$ then the principal part is zero. If the principal part is *infinite* we say that $f(z)$ has an **essential singularity** at $z = z_0$. For example the function

$$\mathrm{e}^{\frac{1}{z}} = \sum_{k=0}^{\infty} \frac{1}{k!z^n} = 1 + \frac{1}{z} + \frac{1}{2!z^2} + \frac{1}{3!z^3} + \cdots \tag{87}$$

has an essential singularity at $z = 0$. If the principal part is *finite*, with $a_k \equiv 0$ for all $k < -m$, then f has **pole** singularity of order m, with

$$f(z) = \frac{a_{-m}}{(z - z_0)^m} + \cdots + \frac{a_{-1}}{(z - z_0)} + a_0 + a_1(z - z_0) + \cdots \tag{88}$$

Indeed, in this case, we can see that

$$\lim_{z \to z_0} (z - z_0)^m f(z) = a_{-m} \neq 0. \tag{89}$$

A pole of order one is called a **simple**. For a pole of order m the coefficient a_{-1} is called the **residue** of the pole at $z = z_0$, and is given by

$$a_{-1} = \frac{1}{(m-1)!} \lim_{z \to z_0} \frac{\mathrm{d}^{m-1}}{\mathrm{d}z^{m-1}} \left[(z - z_0)^m f(z)\right]. \tag{90}$$

Example 2.3 Obtain the residues of any pole singularities for the function

$$f(z) = \frac{1}{z^2(z+1)} = \frac{1}{z^2} - \frac{1}{z} + \frac{1}{(z+1)}. \tag{91}$$

Solution The function has a pole of order two at $z = 0$ and a simple pole at $z = -1$. Expanding about $z = 0$ we have the Laurent series

$$f(z) = \frac{1}{z^2(1+z)} = \frac{1}{z^2}\left\{1 - z + z^2 - z^3 + \cdots\right\} = \frac{1}{z^2} - \frac{1}{z} + 1 - z + z^2 + \cdots, \tag{92}$$

with a residue $r_1 = -1$. Expanding about $z = -1$ we have the Laurent series

$$\begin{aligned} f(z) &= \frac{1}{(z+1)} \frac{1}{[1-(z+1)]^2} = \frac{1}{z+1}\left\{1 + 2(z+1) + 3(z+1)^2 + \cdots\right\} \\ &= \frac{1}{(z+1)} + 2 + 3(z+1) + 4(z+1)^2 - \cdots, \end{aligned} \tag{93}$$

with residue $r_2 = 1$.

Example 2.4 Obtain the residues of any pole singularities for the function

$$f(z) = \frac{z-1}{(z+2)(z-3)^2} = \frac{2}{5}\frac{1}{(z-3)^2} + \frac{3}{25}\frac{1}{(z-3)} - \frac{3}{25}\frac{1}{(z+2)}. \tag{94}$$

Solution Here $f(z)$ has a simple pole at $z = -2$ and at pole of order two at $z = 3$. To obtain the Laurent series at $z = 3$ we write

$$\begin{aligned} f(z) &= \frac{1}{(z-3)^2}\frac{z-1}{(z+2)} = \frac{1}{(z-3)^2}\frac{2+(z-3)}{[5+(z-3)]} = \frac{1}{(z-3)^2}\frac{2+(z-3)}{5[1+\frac{1}{5}(z-3)]} \\ &= \frac{1}{(z-3)^2}\frac{1}{5}[2+(z-3)]\left\{1-\tfrac{1}{5}(z-3)-\tfrac{1}{25}(z-3)^2+\cdots\right\} \\ &= \frac{1}{(z-3)^2}\frac{1}{5}\left\{2+\tfrac{3}{5}(z-3)-\tfrac{3}{25}(z-3)^2+\cdots\right\}. \end{aligned} \tag{95}$$

Thus we have

$$f(z) = \frac{2}{5}\frac{1}{(z-3)^2} + \frac{3}{25}\frac{1}{(z-3)} - \frac{3}{125} + \cdots, \tag{96}$$

with residue $\frac{3}{25}$. Alternatively, from Eq. (90), the residue is given by

$$\begin{aligned} \lim_{z\to 3}\frac{\mathrm{d}}{\mathrm{d}z}\left[(z-3)^2 f(z)\right] &= \lim_{z\to 3}\frac{\mathrm{d}}{\mathrm{d}z}\left[\frac{z-1}{z+2}\right] = \lim_{z\to 3}\frac{\mathrm{d}}{\mathrm{d}z}\left[1-\frac{3}{z+2}\right] \\ &= \lim_{z\to 3}\frac{3}{(z+2)^2} = \frac{3}{25}. \end{aligned} \tag{97}$$

Again at $z = -2$ the residue is given by

$$\lim_{z\to -2}(z+2)f(z) = \lim_{z\to -2}\frac{z-1}{(z-3)^2} = -\frac{3}{25}. \tag{98}$$

2.5 Residue Theory

2.5.1 *Singularities*

From **Liouville's theorem**, the only function analytic everywhere, including the point at infinity, is a constant. However **entire** functions, such as polynomials and exponentials are analytic at all points of the *finite* complex plane.

A singular point z_0 is said to be **isolated** if f is analytic at every point of some **deleted neighbourhood**, $0 < |z - z_0| < \epsilon$, of z_0. We can always obtain a Laurent series expansion about an isolated singular point. Non-isolated singularities include branch points, for which the branch cut is an example of a **singular** curve, and **cluster points**. For example $z = 0$ is a cluster point of $f(z) = \cot(\pi/z)$ due to the sequence of isolated singular points at $z = 1/n$ for $n = \pm, \pm 2, \ldots$.

2.5.2 *Cauchy's residue theorem*

Theorem 2.4 *Suppose C is a simple, anticlockwise orientated, closed curve and $f(z)$ is analytic both on C and inside C, except at the isolated points $z = z_k$, $k = 1, 2, \ldots, n$ where it has pole singularities, then* ***Cauchy's residue theorem*** *states that*

$$\oint_C f(z)\,\mathrm{d}z = 2\pi\mathrm{i}\sum_{k=1}^{n} r_k, \tag{99}$$

where r_k is the residue of the pole at $z = z_k$.

Proof.

(i) Suppose that $f(z)$ has a single pole singularity, of order $m \geq 1$, at the point $z = z_1$. Let γ be a small circular anticlockwise contour, with centre z_1 and radius $\delta \ll 1$, connected to C by the straight line $L\colon A \to B$. By considering the closed contour $\Gamma = C + L - \gamma - L$, we have from Cauchy's theorem

$$\oint_\Gamma f(z)\,\mathrm{d}z = \left(\oint_C + \int_L - \oint_\gamma - \int_L\right) f(z)\,\mathrm{d}z = 0, \tag{100}$$

since $f(z)$ is analytic at all points inside the contour Γ. Consequently

$$\oint_C f(z)\,\mathrm{d}z = \oint_\gamma f(z)\,\mathrm{d}z. \tag{101}$$

In the vicinity of $z = z_1$ the function $f(z)$ has a Laurent series

$$f(z) = \sum_{n=-m}^{\infty} a_n(z-z_1)^n = \frac{a_{-m}}{(z-z_1)^m}+\cdots+\frac{a_{-1}}{(z-z_1)}+a_0+a_1(z-z_1)+\cdots, \tag{102}$$

where $r = a_{-1}$ is the residue of the pole at $z = z_1$. Consequently

$$\oint_C f(z)\,\mathrm{d}z = \oint_\gamma f(z)\,\mathrm{d}z = \sum_{n=-m}^{\infty} a_n \oint_\gamma (z-z_1)^n\,\mathrm{d}z = 2\pi\mathrm{i}r, \tag{103}$$

since from Example 2.2 we have the result

$$\oint_\gamma (z-z_1)^n \mathrm{d}z = \begin{cases} 2\pi\mathrm{i} & \text{if } n = -1, \\ 0 & \text{otherwise.} \end{cases} \tag{104}$$

(ii) For the general case we can take γ_k to be the small circular anticlockwise contour surrounding the singularity $z = z_k$, $k = 1, 2, \ldots, n$. We can similarly show that

$$\oint_C f(z)\,\mathrm{d}z = \sum_{k=1}^{n} \oint_{\gamma_k} f(z)\,\mathrm{d}z = 2\pi\mathrm{i}\sum_{k=1}^{n} r_k. \tag{105}$$

□

Example 2.5 If C is an anticlockwise simple closed curve large enough to enclose all the singularities of the integrand, evaluate

$$I = \oint_C \frac{z^3 - z^2 + z - 1}{z^3 + 4z}\,\mathrm{d}z. \tag{106}$$

Solution The singularities of the integrand are simple poles at

$$z^3 + 3z = z(z^2 + 2) = 0, \qquad \text{so} \qquad z = 0, \qquad z = \pm 2\mathrm{i}. \tag{107}$$

The residue at $z = 0$ is given by

$$r_1 = \lim_{z\to 0} z f(z) = \lim_{z\to 0} \frac{z^3 - z^2 + z - 1}{z^2 + 4} = -\tfrac{1}{4}. \tag{108}$$

The residue at $z = +2\mathrm{i}$ is given by

$$r_2 = \lim_{z\to 2\mathrm{i}} (z - 2\mathrm{i}) f(z) = \lim_{z\to 2\mathrm{i}} \frac{z^3 - z^2 + z - 1}{z(z + 2\mathrm{i})} = \frac{3 - 6\mathrm{i}}{(2\mathrm{i})(4\mathrm{i})} = -\tfrac{1}{8}(3 - 6\mathrm{i}). \tag{109}$$

The residue at $z = -2\mathrm{i}$ is given by

$$r_3 = \lim_{z\to -2\mathrm{i}} (z+2\mathrm{i}) f(z) = \lim_{z\to -2\mathrm{i}} \frac{z^3 - z^2 + z - 1}{z(z - 2\mathrm{i})} = \frac{3 + 6\mathrm{i}}{(-2\mathrm{i})(-4\mathrm{i})} = -\tfrac{1}{8}(3+6\mathrm{i}). \tag{110}$$

Thus the sum of the residues is given by

$$\sum_{k=1}^{3} r_k = -\tfrac{1}{4} - \tfrac{1}{8}(3 - 6\mathrm{i}) - \tfrac{1}{8}(3 + 6\mathrm{i}) = -1. \tag{111}$$

Consequently

$$I = \oint_C \frac{z^3 - z^2 + z - 1}{z^3 + 4z}\,\mathrm{d}z = -2\pi\mathrm{i}. \tag{112}$$

2.5.3 *Evaluation of real integrals*

Suppose we wish to evaluate the *real* integral

$$I = \int_{-\infty}^{+\infty} \frac{\mathrm{d}x}{(x^2 + a^2)(x^2 + b^2)} \qquad \text{for } b > a > 0. \tag{113}$$

We first consider the *complex* contour integral

$$\oint_C f(z)\,\mathrm{d}z = \oint_C \frac{\mathrm{d}z}{(z^2+a^2)(z^2+b^2)}, \tag{114}$$

where $C = L + \Gamma$ is a simple closed contour taken in the anticlockwise sense, shown in Fig. 2.4. Here C consists of a straight line segment L along the x-axis, $x \in [-R, R]$, and a semicircular arc C with centre $z = 0$ and of large radius $R \gg 1$.

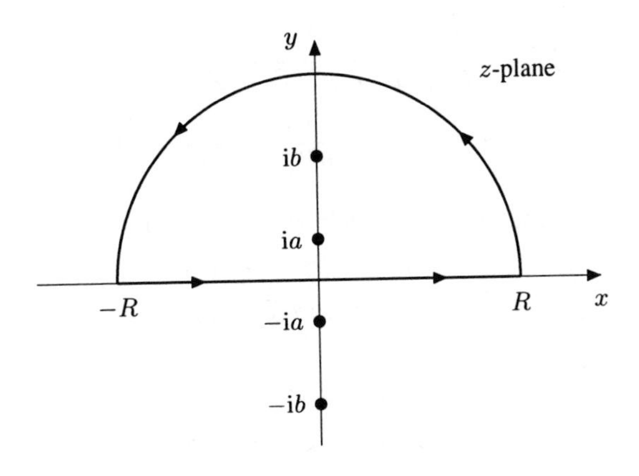

Fig. 2.4 Contour integration.

Although the integrand has *four* poles at $z = \pm \mathrm{i}a$ and at $z = \pm \mathrm{i}b$, we see that the contour encloses only the *two* poles at $z = \mathrm{i}a$, $z = \mathrm{i}b$. The residue at $z = \mathrm{i}a$ is given by

$$r_1 = \lim_{z\to a\mathrm{i}} (z - a\mathrm{i}) f(z) = \lim_{z\to a\mathrm{i}} \frac{1}{(z+a\mathrm{i})(z^2+b^2)} = \frac{1}{2\mathrm{i}a(b^2-a^2)}. \tag{115}$$

The residue at $z = \mathrm{i}b$ is given by

$$r_2 = \lim_{z\to b\mathrm{i}} (z - b\mathrm{i}) f(z) = \lim_{z\to b\mathrm{i}} \frac{1}{(z^2+a^2)(z+b\mathrm{i})} = \frac{1}{2\mathrm{i}b(a^2-b^2)}. \tag{116}$$

The sum of the residues

$$\frac{1}{2\mathrm{i}a(b^2-a^2)} + \frac{1}{2\mathrm{i}b(a^2-b^2)} = \frac{1}{2\mathrm{i}(b^2-a^2)}\left\{\frac{1}{a} - \frac{1}{b}\right\} = \frac{1}{2\mathrm{i}ab(a+b)}. \tag{117}$$

Thus

$$\oint_C f(z)\,\mathrm{d}z = 2\pi\mathrm{i}\left(\frac{1}{2\mathrm{i}ab(a+b)}\right) = \frac{\pi}{ab(a+b)}. \tag{118}$$

Now we see that since $C = L + \Gamma$ we have

$$\oint_C f(z)\,\mathrm{d}z = \int_L f(z)\,\mathrm{d}z + \int_\Gamma f(z)\,\mathrm{d}z. \tag{119}$$

On the segment L we have $z = x$, $\mathrm{d}z = \mathrm{d}x$ where $-R \le x \le R$, so that

$$\begin{aligned}\int_L f(z)\,\mathrm{d}z &= \int_{-R}^{R} \frac{\mathrm{d}x}{(x^2+a^2)(x^2+b^2)} \\ &\to \int_{-\infty}^{\infty} \frac{\mathrm{d}x}{(x^2+a^2)(x^2+b^2)} \qquad \text{as } R \to \infty.\end{aligned} \tag{120}$$

On the semicircular contour Γ, $z = R\mathrm{e}^{\mathrm{i}\theta}$, $\mathrm{d}z = \mathrm{i}R\mathrm{e}^{\mathrm{i}\theta}$, $0 \le \theta \le \pi$ so that

$$\int_\Gamma f(z)\,\mathrm{d}z = \int_0^{\pi} \frac{\mathrm{i}R\mathrm{e}^{\mathrm{i}\theta}\,\mathrm{d}\theta}{(R^2\mathrm{e}^{2\mathrm{i}\theta}+a^2)(R^2\mathrm{e}^{2\mathrm{i}\theta}+b^2)}. \tag{121}$$

Using the fact that for a real integral

$$\left|\int_0^{\pi} f(R,\theta)\,\mathrm{d}\theta\right| \le \pi R\,|f(R,\theta)|, \tag{122}$$

where πR is the length of the contour, then we see that as $R \to \infty$

$$\left|\int_0^{\pi} \frac{\mathrm{i}R\mathrm{e}^{\mathrm{i}\theta}\mathrm{d}\theta}{(R^2\mathrm{e}^{2\mathrm{i}\theta}+a^2)(R^2\mathrm{e}^{2\mathrm{i}\theta}+b^2)}\right| \le \pi R\frac{M}{R^3} = \frac{\pi M}{R^2} \to 0, \tag{123}$$

where M is some constant. Consequently

$$\int_{-\infty}^{\infty} \frac{\mathrm{d}x}{(x^2+a^2)(x^2+b^2)} = \frac{\pi}{ab(a+b)}. \tag{124}$$

2.6 Analytic Continuation

Given a function f, defined in some limited region D', we often require a representation for f which is valid in a larger region D. The process of extending the region of definition of an analytic function is called **analytic continuation**. For example suppose that f_1 and f_2 are analytic in domains D_1 and D_2 respectively, and satisfy

$$f_1(z) = f_2(z) \qquad \text{for all } z \in \Gamma = D_1 \cap D_2 \ne \emptyset, \tag{125}$$

then we call f_2 an **analytic continuation** of f_1 into D_2, and *vice versa*. Then the function $F(z)$, defined by

$$F(z) = \begin{cases} f_1(z) & \text{when } z \in D_1, \\ f_2(z) & \text{when } z \in D_2, \end{cases} \tag{126}$$

is analytic in $D = D_1 \cup D_2$. This holds even when D_1 and D_2 are disjoint but share a common boundary Γ, provided f_1 is continuous in $D_1 \cup \Gamma$ and f_2 is continuous in $D_2 \cup \Gamma$.

Consider the function f_1 defined by

$$f_1(z) = \sum_{n=0}^{\infty} z^n \qquad \text{defined for } |z| < 1. \tag{127}$$

This power series clearly diverges for $|z| \geq 1$ since the nth term does not approach zero as $n \to \infty$. We can also consider the function

$$f_2(z) = \frac{1}{1-z}, \qquad z \neq 1, \tag{128}$$

which is defined and analytic at all points except $z = 1$. Moreover

$$f_2(z) = f_1(z) \qquad \text{for all } |z| < 1, \tag{129}$$

since Eq. (127) is the Taylor series representation, about $z = 0$, of Eq. (128) inside the unit circle. In fact we can state that $f_2(z)$ is the *unique* analytic continuation of f_1 into the domain $|z| \geq 1$, $z \neq 1$. This follows from the following theorem.

Theorem 2.5 *A function that is analytic in a domain D is uniquely determined by its values in some interior domain $D' \subset D$, or along an arc Γ interior to D.*

If z_0 is a point on a curve $\Gamma \in D$ and z_n is any other point in D, then we can construct a curve C from z_0 to z_n, with intermediate points z_k, $k = 1, 2, \ldots, n-1$ such that $|z_k - z_{k-1}| < d$, where d is the smallest distance from C to the boundary of D. Since $f(z)$ is analytic in every region $|z - z_k| < d$ we can construct a Taylor series using its values at $z = z_k$ to obtain its values at $z = z_{k+1}$. By this repeated overlapping of Taylor series we can uniquely determine f at any point in D.

Example 2.6 Consider the function f_1 defined by

$$f_1(z) = \int_0^{\infty} \mathrm{e}^{-zt}\, \mathrm{d}t, \tag{130}$$

which is defined only for D_1: $\operatorname{Re} z > 0$. However we can also write

$$f_1(z) = \left[-\frac{1}{z} \mathrm{e}^{-zt} \right]_{t=0}^{t=\infty} = \frac{1}{z} \qquad \text{when } \operatorname{Re} z > 0. \tag{131}$$

Let $f_2(z)$ be defined by the geometric series

$$f_2(z) = \mathrm{i}\sum_{n=0}^{\infty}\left(\frac{z+\mathrm{i}}{\mathrm{i}}\right)^n, \qquad |z+\mathrm{i}| < 1. \tag{132}$$

Then within the circle $D_2\colon |z+\mathrm{i}| < 1$ the series converges to

$$f_2(z) = \frac{\mathrm{i}}{[1-(z+\mathrm{i})/\mathrm{i}]} = \frac{1}{z} \qquad \text{when } |z+\mathrm{i}| < 1. \tag{133}$$

Evidently $f_2(z) = f_1(z)$ for each $z \in D_1 \cap D_2$ and so f_2 is the analytic continuation of f_1 into D_2. Furthermore the function

$$F(z) = \frac{1}{z}, \qquad z \neq 0, \tag{134}$$

is the analytic continuation of both f_1 and f_2 into the entire z-plane for $z \neq 0$.

2.6.1 *Chains of analytic continuation*

Suppose we have a functions f_1, f_2 and f_3 analytic in regions D_1, D_2 and D_3 respectively. Suppose further that

$$f_1(z) = f_2(z) \quad \text{in } D_1 \cap D_2 \quad \text{and} \quad f_2(z) = f_3(z) \quad \text{in } D_2 \cap D_3, \tag{135}$$

so that f_2 is the analytic continuation of f_1 into D_2 and f_3 is the analytic continuation of f_2 into D_3. Then we *cannot* conclude that

$$f_3(z) = f_1(z) \qquad \text{for all } z \in D_1 \cap D_3 \tag{136}$$

because D_1, D_2, D_3 might enclose a branch point of a multi-valued function.

Example 2.7 Consider the branch of $z^{1/2}$

$$f_1(z) = \sqrt{r}\mathrm{e}^{\mathrm{i}\theta/2}, \qquad r > 0, \quad 0 < \theta < \pi. \tag{137}$$

An analytic continuation of f_1 into the lower half plane is given by

$$f_2(z) = \sqrt{r}\mathrm{e}^{\mathrm{i}\theta/2}, \qquad r > 0, \quad \pi/2 < \theta < 2\pi. \tag{138}$$

An analytic continuation of f_2 into the first quadrant is

$$f_3(z) = \sqrt{r}\mathrm{e}^{\mathrm{i}\theta/2}, \qquad r > 0, \quad \frac{3\pi}{2} < \theta < \frac{5\pi}{2}. \tag{139}$$

Note that $f_3(z) \neq f_1(z)$. In fact $f_3(z) = -f_1(z)$.

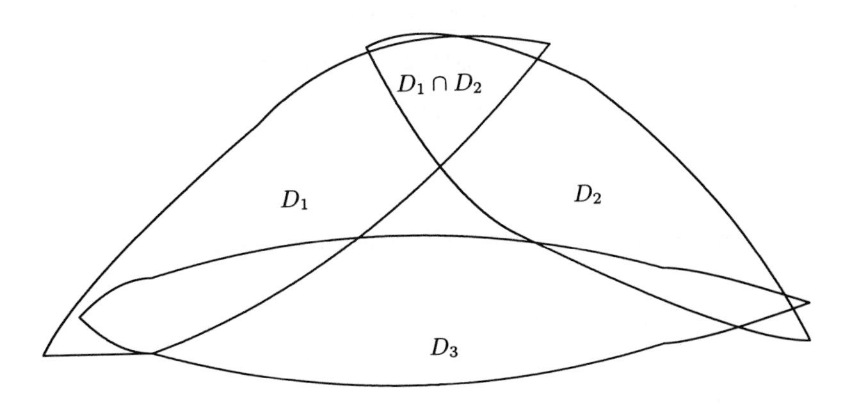

Fig. 2.5 A possible arrangement of the regions D_1, D_2 and D_3.

However suppose we analytically continue a function along two distinct curves C_1 and C_2, to a given point, so that $C_1 - C_2$ is a closed contour. If there are no singular points in the region D enclosed by $C_1 - C_2$, then the result of each analytic continuation is identical and we obtain a single-valued function for all $z \in D$.

Problems

Exercise 2.1 By using Cauchy's residue theorem, evaluate

$$I = \oint_{C:\ |z|=1} z^n \mathrm{e}^{1/z}\,\mathrm{d}z,$$

for integer $n > 0$.

Exercise 2.2 By using Cauchy's residue theorem, evaluate

$$\text{(a)} \quad I_1 = \oint_{C:\ |z|=\frac{1}{2}} \frac{z+2}{z(z+1)^2}\,\mathrm{d}z, \qquad \text{(b)} \quad I_2 = \oint_{C:\ |z|=2} \frac{z+2}{z(z+1)^2}\,\mathrm{d}z.$$

Exercise 2.3 Suppose that the rational function $f(z)$ has a simple pole at $z = 0$, such that

$$f(z) = \frac{g(z)}{h(z)} \qquad \text{where } h(z_0) = 0 \text{ and } g(z_0) \neq 0,$$

where g and h are polynomial functions. By writing

$$h(z) = (z - z_0)R(z) \qquad \text{where } R(z_0) \neq 0,$$

show that the residue of f at $z = z_0$ is given by

$$\text{Res}(f(z) : z_0) = \frac{g(z_0)}{h'(z_0)}.$$

Exercise 2.4 By using contour integral methods, evaluate the Cauchy Principal Value

$$I_P = \text{PV} \int_{-\infty}^{\infty} \frac{x^2}{x^4+1}\, dx = \lim_{R\to\infty} \int_{-R}^{R} \frac{x^2}{x^4+1}\, dx.$$

Note that this may differ in value from

$$I = \int_{-\infty}^{\infty} \frac{x^2}{x^4+1}\, dx = \lim_{L\to\infty} \int_{-L}^{\alpha} \frac{x^2}{x^4+1}\, dx + \lim_{R\to\infty} \int_{\alpha}^{R} \frac{x^2}{x^4+1}\, dx,$$

where α is some finite number. You may use the magnitude inequality, namely that

$$\left| \int_C f(z)\, dz \right| \leq \int_C |f(z)||dz| \quad \text{and} \quad |z - a| \geq |z| - |a|.$$

Chapter 3

Integral Transforms

J. W. Elliott

Department of Mathematics
University of Hull

3.1 The Laplace Transform

The **Laplace Transform** of a function $f(t)$, defined for all $t > 0$, is given by

$$\mathcal{L}\{f(t)\} = \hat{f}(s) = \int_0^\infty f(t)\mathrm{e}^{-st}\,\mathrm{d}t, \tag{1}$$

provided the integral converges. The Laplace transform converts $f(t)$ into a new function of the complex transform variable $s = \sigma + \mathrm{i}\omega$.

We say that $f(t)$ is of **exponential order** for $t \geq T$, where $T \geq 0$, if there exist real constants σ', $M > 0$ and $T > 0$, such that $|f(t)| < M\mathrm{e}^{\sigma' t}$ for all $t > T$. For a given f, σ' is *not* unique but $\sigma' > \sigma_0$, where σ_0 (the greatest lower bound for σ') is called the **abscissa of convergence**.

Theorem 3.1 *If f is piecewise-continuous and of exponential order on $t \geq T > 0$, with abscissa of convergence $\sigma_0 \geq 0$, then its Laplace transform, defined by Eq. (1), exists but converges only for* $\mathrm{Re}(s) > \sigma_0$.

Proof. The proof of existence is rather technical but from the magnitude inequality, if $s = \sigma + \mathrm{i}\omega$, then $|\mathrm{e}^{-st}| = |\mathrm{e}^{-(\sigma+\mathrm{i}\omega)t}| = |\mathrm{e}^{-\sigma t}||\mathrm{e}^{-\mathrm{i}\omega t}| = \mathrm{e}^{-\sigma t}$ and

$$|\hat{f}(s)| = \left|\int_0^\infty f(t)\mathrm{e}^{-st}\,\mathrm{d}t\right| \leq \int_0^\infty |f(t)||\mathrm{e}^{-st}|\,\mathrm{d}t \leq I(s) + M\int_T^\infty \mathrm{e}^{-(\sigma-\sigma_0)t}\,\mathrm{d}t. \tag{2}$$

If we assume that f is absolutely integrable on $[0, T]$, so that

$$I(s) = \int_0^T |f(t)| \mathrm{e}^{-\sigma t}\,\mathrm{d}t \le \int_0^T |f(t)|\,\mathrm{d}t < \infty, \tag{3}$$

then the right hand side is finite only within the **region of validity**, given by $\sigma = \mathrm{Re}(s) > \sigma_0$. □

Thus for some functions, whether it's because of their rapid growth as $t \to \infty$, for example $f(t) = \mathrm{e}^{t^2}$, or because of their integrability, for example $f(t) = 1/t^2$, their Laplace transform does not converge for *any* value of s.

Example 3.1 For integer $n \ge 0$, we have for $\mathrm{Re}(s) > 0$

$$\mathcal{L}\{t^n\} = \int_0^\infty t^n \mathrm{e}^{-st}\,\mathrm{d}t = \frac{1}{s^{n+1}} \int_0^\infty r^n \mathrm{e}^{-r}\,\mathrm{d}r = \frac{n!}{s^{n+1}}. \tag{4}$$

Example 3.2 For ν complex, we have for $\mathrm{Re}(s) > \mathrm{Re}(\nu)$

$$\mathcal{L}\{\mathrm{e}^{\nu t}\} = \int_0^\infty \mathrm{e}^{-(s-\nu)t}\,\mathrm{d}t = \left[-\frac{1}{s-\nu}\mathrm{e}^{-(s-\nu)t}\right]_0^\infty = \frac{1}{s-\nu}. \tag{5}$$

The Laplace transform is a **linear** operator, with

$$\mathcal{L}\{\alpha f(t) + \beta g(t)\} = \alpha \mathcal{L}\{f(t)\} + \beta \mathcal{L}\{g(t)\}, \tag{6}$$

for constants α and β. It follows that, for $\mathrm{Re}(s) > 0$, we have

$$\mathcal{L}\{\cos(\theta t)\} = \mathcal{L}\left\{\frac{1}{2}\left[\mathrm{e}^{\mathrm{i}\theta t} + \mathrm{e}^{-\mathrm{i}\theta t}\right]\right\} = \frac{1}{2}\left[\frac{s+\mathrm{i}\theta}{s^2+\theta^2} + \frac{s-\mathrm{i}\theta}{s^2+\theta^2}\right] = \frac{s}{s^2+\theta^2}, \tag{7}$$

$$\mathcal{L}\{\sin(\theta t)\} = \mathcal{L}\left\{\frac{1}{2}\left[\mathrm{e}^{\mathrm{i}\theta t} - \mathrm{e}^{-\mathrm{i}\theta t}\right]\right\} = \frac{1}{2}\left[\frac{s+\mathrm{i}\theta}{s^2+\theta^2} - \frac{s-\mathrm{i}\theta}{s^2+\theta^2}\right] = \frac{\theta}{s^2+\theta^2}. \tag{8}$$

Similarly for $\mathrm{Re}(s) > |k|$ we have

$$\mathcal{L}\{\cosh(kt)\} = \frac{s}{s^2-k^2}, \qquad \mathcal{L}\{\sinh(kt)\} = \frac{k}{s^2-k^2}. \tag{9}$$

3.1.1 *The Heaviside step function*

The **Heaviside unit step function**, $\mathrm{H}(t)$, is the discontinuous function

$$\mathrm{H}(t) = \begin{cases} 1 & \text{for } t > 0, \\ 0 & \text{for } t < 0, \end{cases} \qquad \mathrm{H}(-t) = 1 - \mathrm{H}(t) = \begin{cases} 0 & \text{for } t > 0, \\ 1 & \text{for } t < 0. \end{cases} \tag{10}$$

With this notation the Laplace transform Eq. (1) may be written as

$$\mathcal{L}\{f(t)\} = \hat{f}(s) = \int_{-\infty}^{\infty} \mathrm{H}(t) f(t) \mathrm{e}^{-st}\, \mathrm{d}t. \tag{11}$$

We say that $\hat{f}(s)$ is the *two-sided* Laplace transform of $\mathrm{H}(t)f(t)$.

We say that $f(t)$ is **causal** if $f(t) = 0$ for all $t < 0$. The multiplying of $f(t)$ by $\mathrm{H}(t)$ converts f into the causal function $\mathrm{H}(t)f(t)$, which, unless $f(0) = 0$, is discontinuous at $t = 0$. It follows that for $a \geq 0$

$$\mathcal{L}\{\mathrm{H}(t-a)\} = \int_a^{\infty} \mathrm{e}^{-st}\, \mathrm{d}t = \mathrm{e}^{-as} \int_0^{\infty} \mathrm{e}^{-sr}\, \mathrm{d}r = \mathrm{e}^{-as}\mathcal{L}\{1\} = \frac{1}{s}\mathrm{e}^{-as}. \tag{12}$$

3.1.2 *Properties*

3.1.2.1 *The scale rule*

$$\mathcal{L}\{f(kt)\} = \int_0^{\infty} f(kt)\mathrm{e}^{-st}\, \mathrm{d}t = \frac{1}{k}\int_0^{\infty} f(r)\mathrm{e}^{-sr/k}\, \mathrm{d}r = \frac{1}{k}\hat{f}\left(\frac{s}{k}\right). \tag{13}$$

3.1.2.2 *The 1st shifting theorem*

$$\mathcal{L}\left\{f(t)\mathrm{e}^{\alpha t}\right\} = \int_0^{\infty} f(t)\mathrm{e}^{-(s-\alpha)t}\, \mathrm{d}t = \hat{f}(s-\alpha). \tag{14}$$

3.1.2.3 *The 2nd shifting theorem*

For $a \geq 0$ we see that

$$\begin{aligned} \mathcal{L}\{f(t-a)\,\mathrm{H}(t-a)\} &= \int_a^{\infty} f(t-a)\mathrm{e}^{-st}\, \mathrm{d}t \\ &= \mathrm{e}^{-as}\int_0^{\infty} f(r)\mathrm{e}^{-sr}\, \mathrm{d}r = \mathrm{e}^{-as}\hat{f}(s). \end{aligned} \tag{15}$$

3.1.2.4 *Multiplying by t^n*

$$\begin{aligned} \mathcal{L}\{t^n f(t)\} &= \int_0^{\infty} t^n f(t)\mathrm{e}^{-st}\, \mathrm{d}t \\ &= (-1)^n \frac{\mathrm{d}^n}{\mathrm{d}s^n}\left(\int_0^{\infty} f(t)\mathrm{e}^{-st}\, \mathrm{d}t\right) = (-1)^n \frac{\mathrm{d}^n \hat{f}}{\mathrm{d}s^n}. \end{aligned} \tag{16}$$

Example 3.3 From Eqs. (14) and (16) it follows that

$$\mathcal{L}\left\{\sin(2t)\mathrm{e}^{-3t}\right\} = \frac{2}{(s+3)^2+4}, \tag{17}$$

$$\mathcal{L}\left\{t\cosh(3t)\right\} = -\frac{\mathrm{d}}{\mathrm{d}s}\left[\frac{s}{s^2-9}\right] = \frac{s^2+9}{(s^2-9)^2}, \tag{18}$$

$$\mathcal{L}\left\{t^2\cosh(3t)\right\} = \frac{\mathrm{d}^2}{\mathrm{d}s^2}\left[\frac{s}{s^2-9}\right] = -\frac{\mathrm{d}}{\mathrm{d}s}\left[\frac{s^2+9}{(s^2-9)^2}\right] = \frac{36s}{(s^2-9)^3}. \tag{19}$$

3.1.3 *Differential properties*

We have, upon using integration by parts,

$$\begin{aligned}\mathcal{L}\left\{f'(t)\right\} &= \int_0^\infty \frac{\mathrm{d}f}{\mathrm{d}t}\mathrm{e}^{-st}\,\mathrm{d}t \\ &= \left[f(t)\mathrm{e}^{-st}\right]_0^\infty + s\int_0^\infty f(t)\mathrm{e}^{-st}\,\mathrm{d}t = s\hat{f}(s) - f(0),\end{aligned} \tag{20}$$

$$\begin{aligned}\mathcal{L}\left\{f''(t)\right\} &= \int_0^\infty \frac{\mathrm{d}^2 f}{\mathrm{d}t^2}\mathrm{e}^{-st}\,\mathrm{d}t \\ &= \left[f'(t)\mathrm{e}^{-st}\right]_0^\infty + s\int_0^\infty f'(t)\mathrm{e}^{-st}\,\mathrm{d}t \\ &= s\mathcal{L}\left\{f'(t)\right\} - f(0) = s^2\hat{f}(s) - sf(0) - f'(0).\end{aligned} \tag{21}$$

Thus, by induction, we see that

$$\mathcal{L}\left\{\frac{\mathrm{d}^n f}{\mathrm{d}t^n}\right\} = s^n\hat{f}(s) - \sum_{k=1}^{n} s^{n-k} f^{(k-1)}(0). \tag{22}$$

Also for a function of more than one variable

$$\mathcal{L}\left\{\frac{\partial f}{\partial x}(x,t)\right\} = \int_0^\infty \frac{\partial f}{\partial x}\mathrm{e}^{-st}\,\mathrm{d}t = \frac{\partial}{\partial x}\left(\int_0^\infty f(t)\mathrm{e}^{-st}\,\mathrm{d}t\right) = \frac{\partial \hat{f}}{\partial x}(x,s). \tag{23}$$

3.1.4 *Convolution*

Given two piecewise-continuous functions $f(t)$ and $g(t)$, defined for all t, the **convolution** of f and g is defined by

$$f(t) * g(t) = \int_{-\infty}^{\infty} f(r)g(t-r)\,\mathrm{d}r = \int_{-\infty}^{\infty} f(t-r)g(r)\,\mathrm{d}r = g(t) * f(t). \tag{24}$$

The notation $(f * g)(t) = f(t) * g(t)$ indicates that convolution converts the two functions f and g into a third new function of t. In addition to being *commutative,* convolution is *linear* and *associative,* since for constants α and β

$$f(t) * [\alpha g(t) + \beta h(t)] = \alpha f(t) * g(t) + \beta f(t)h(t), \tag{25}$$

$$f(t) * [g(t) * h(t)] = [f(t) * g(t)] * h(t). \tag{26}$$

In particular, if both f and g are causal, so that $f(r) = g(r) = 0$ for $r < 0$ and $f(t-r) = g(t-r) = 0$ for $r > t$ then Eq. (24) reduces to

$$f(t) * g(t) = \int_0^t f(r)g(t-r)\,\mathrm{d}r = \int_0^t f(t-r)g(r)\,\mathrm{d}r = g(t) * f(t). \tag{27}$$

Henceforth, when discussing convolution in connection with Laplace transforms, we shall assume the functions involved are causal.

3.1.5 *The convolution theorem*

Theorem 3.2 *If $f(t)$ and $g(t)$ are causal and defined for $t > 0$, then*

$$\mathcal{L}\left\{\int_0^t f(r)g(t-r)\,\mathrm{d}r\right\} = \mathcal{L}\{f(t) * g(t)\} = \mathcal{L}\{f(t)\}\,\mathcal{L}\{g(t)\} = \hat{f}(s)\hat{g}(s). \tag{28}$$

For example, for any causal function $f(t)$, the convolution theorem Eq. (28) yields the integral property

$$\mathcal{L}\left\{\int_0^t f(r)\mathrm{d}r\right\} = \mathcal{L}\{f(t) * \mathrm{H}(t)\} = \mathcal{L}\{\mathrm{H}(t)\}\,\mathcal{L}\{f(t)\} = \frac{1}{s}\hat{f}(s). \tag{29}$$

3.1.6 *The inverse transform*

We take the symbol $\mathcal{L}^{-1}\left\{\hat{f}(s)\right\}$ to denote the *causal* function $f(t)$ whose Laplace transform is $\hat{f}(s)$. That is, provided f is continuous

$$\mathcal{L}\{f(t)\} = \hat{f}(s), \quad \text{so} \quad \mathcal{L}^{-1}\left\{\hat{f}(s)\right\} = \mathrm{H}(t)f(t) = \begin{cases} f(t) & \text{for } t > 0, \\ 0 & \text{for } t < 0. \end{cases} \tag{30}$$

3.1.6.1 *The 1st shifting theorem*

Suppose, for example, that after using the method of partial fractions

$$\hat{f}(s) = \frac{A}{(s-\alpha)^{n+1}} + \left[\frac{C_1(s-\beta) + C_2\gamma}{(s-\beta)^2 + \gamma^2}\right] \tag{31}$$

where $\alpha, \beta\ \gamma$ are real constants. Then, for $t > 0$, the 1st shifting theorem gives

$$f(t) = \mathcal{L}^{-1}\left\{\hat{f}(s)\right\} = \mathcal{L}^{-1}\left\{\frac{A}{s}\right\}\mathrm{e}^{\alpha t} + \mathcal{L}^{-1}\left\{\frac{C_1 s + C_2\gamma}{s^2+\gamma^2}\right\}\mathrm{e}^{\beta t}, \qquad (32)$$

From known transforms we have, for $t > 0$, the result

$$f(t) = \frac{A}{n!}t^n \mathrm{e}^{\alpha t} + \left[C_1\cos(\gamma t) + C_2\sin(\gamma t)\right]\mathrm{e}^{\beta t}. \qquad (33)$$

3.1.6.2 *Convolution*

If we take the inverse transform of the convolution theorem we obtain

$$\mathcal{L}^{-1}\left\{\mathcal{L}\left[f(t)\right]\mathcal{L}\left[g(t)\right]\right\} = \mathcal{L}^{-1}\left\{\hat{f}(s)\hat{g}(s)\right\} = f(t) * g(t). \qquad (34)$$

For example suppose

$$\hat{u}(s) = \frac{1}{(s-1)(s+2)}\hat{f}(s) = \frac{1}{3}\left[\frac{1}{(s-1)} - \frac{1}{(s+2)}\right]\hat{f}(s), \qquad (35)$$

then upon inversion, we have for $t > 0$

$$u(t) = f(t) * \tfrac{1}{3}(\mathrm{e}^t - \mathrm{e}^{-2t}) = \frac{1}{3}\int_0^t f(r)\left[\mathrm{e}^{(t-r)} - \mathrm{e}^{2(t-r)}\right]\mathrm{d}r. \qquad (36)$$

3.1.6.3 *The 2nd shifting theorem*

If we take the inverse transform of the 2nd shifting theorem we have

$$f(t) = \mathcal{L}^{-1}\left\{\hat{g}(s)\mathrm{e}^{-as}\right\} = g(t-a)\,\mathrm{H}(t-a). \qquad (37)$$

Example 3.4 Solve the partial differential equation

$$\frac{\partial u}{\partial t} + \frac{\partial u}{\partial x} + u = xt \qquad \text{in } x > 0,\ t > 0, \qquad (38)$$

subject to the conditions $u = 0$ at $x = 0$ and $t = 0$.

Solution If we take the Laplace transform of the PDE with respect to t we obtain a linear 1st order ODE for $\hat{u}(x, s)$, namely

$$\frac{\partial \hat{u}}{\partial x} + (s+1)\hat{u} = x\mathcal{L}\left\{t\right\} + u(x,0), \quad \text{so} \quad \mathrm{e}^{-(s+1)x}\frac{\partial}{\partial x}\left[\mathrm{e}^{(s+1)}\hat{u}\right] = \frac{x}{s^2}, \qquad (39)$$

where we have used an integrating factor. Using the fact that $\hat{u} = \mathcal{L}\{0\} = 0$ at $x = 0$, we have the particular solution

$$\begin{aligned}\hat{u}(x,s) &= \frac{\mathrm{e}^{-(s+1)x}}{s^2}\int_0^x r\mathrm{e}^{(s+1)r}\,\mathrm{d}r \\ &= \frac{1}{s^2(s+1)}\left\{x - \frac{1}{(s+1)}\left[1 - \mathrm{e}^{-(s+1)x}\right]\right\}. \end{aligned} \tag{40}$$

Resolving the rational terms into partial fractions gives

$$\begin{aligned}\hat{u}(x,s) =& x\left[\frac{1}{s^2} - \frac{1}{s} + \frac{1}{(s+1)}\right] \\ &- \left(1 - \mathrm{e}^{-(s+1)x}\right)\left[\frac{1}{s^2} - \frac{2}{s} + \frac{2}{(s+1)} + \frac{1}{(s+1)^2}\right]. \end{aligned} \tag{41}$$

Taking the inverse transform gives us

$$u(x,t) = x\left(t - 1 + \mathrm{e}^{-t}\right) - \left[f(t) - \mathrm{e}^{-x} f(t-x)\,\mathrm{H}(t-x)\right], \tag{42}$$

where

$$f(t) = \mathcal{L}^{-1}\left\{\frac{1}{s^2} - \frac{2}{s} + \frac{2}{(s+1)} + \frac{1}{(s+1)^2}\right\} = t - 2 + (2+t)\mathrm{e}^{-t} \tag{43}$$

and

$$\begin{aligned}&f(t) - \mathrm{e}^{-x} f(t-x)\,\mathrm{H}(t-x) \\ &\quad= \left(t - 2 + (2+t)\mathrm{e}^{-t}\right) - \mathrm{e}^{-x}\left[(t-x) - 2 + (2+t-x)\mathrm{e}^{-(t-x)}\right] \\ &\quad= x\left(t - 1 + \mathrm{e}^{-t}\right) - \left[t - 2 - (t-x-2)\mathrm{e}^{-x}\right]. \end{aligned} \tag{44}$$

Thus $u(x,t) = u(t,x)$ for $t > x$, since

$$u(x,t) = \begin{cases} x\left(t - 1 + \mathrm{e}^{-t}\right) - \left[t - 2 + (2+t)\mathrm{e}^{-t}\right] & \text{for } t < x, \\ t\left(x - 1 + \mathrm{e}^{-x}\right) - \left[x - 2 + (2+x)\mathrm{e}^{-x}\right] & \text{for } t > x. \end{cases} \tag{45}$$

3.2 The Inversion Theorem

Theorem 3.3 *The* ***inversion theorem*** *states that if*

$$\hat{f}(s) = \mathcal{L}\{f(t)\} = \int_0^\infty f(t)\mathrm{e}^{-st}\,\mathrm{d}t = \int_{-\infty}^\infty f(t)\,\mathrm{H}(t)\mathrm{e}^{-st}\,\mathrm{d}t, \tag{46}$$

where the integral converges for $\mathrm{Re}(s) > \sigma_0$*, then if* $\gamma \geq \sigma_0$

$$\mathcal{L}^{-1}\left\{\hat{f}(s)\right\} = \frac{1}{2\pi\mathrm{i}}\int_{\gamma-\mathrm{i}\infty}^{\gamma+\mathrm{i}\infty} \hat{f}(s)\mathrm{e}^{st}\,\mathrm{d}s = \begin{cases} \mathrm{H}(t)f(t), \\ \frac{1}{2}[\mathrm{H}(t+)f(t+) + \mathrm{H}(t-)f(t-)], \end{cases} \tag{47}$$

depending on whether f *is continuous at* t *or not. In particular, for* $t = 0$ *the integral converges to* $\frac{1}{2}f(0+)$*.*

3.2.1 *Evaluation of the contour integral*

Generally $\hat{f}(s)$ is an analytic function of s except at isolated poles at $s = s_k$, and at **branch points** at $s = \alpha_j$, all of which lie in $\mathrm{Re}(s) \leq \sigma_0$. In general we take branch cuts to emanate from $s = \alpha_j$, and lie along the straight lines $\arg(s - \alpha_j) = \pi$.

Typically we consider a closed contour C_R, which consists of the **Bromwich contour** $s = \gamma + \mathrm{i}r$, $-R \leq r \leq R$, which is closed by a semicircle S_R of radius R, on the right for $t < 0$, and on the left for $t > 0$. Then, by Cauchy's theorem

$$\frac{1}{2\pi\mathrm{i}}\oint_{C_R} \hat{f}(s)\mathrm{e}^{st}\,\mathrm{d}s = \frac{1}{2\pi\mathrm{i}}\left(\int_{\gamma-\mathrm{i}R}^{\gamma+\mathrm{i}R} + \int_{S_R}\right)\hat{f}(s)\mathrm{e}^{st}\,\mathrm{d}s = 0 \qquad \text{for } t < 0, \tag{48}$$

since $\hat{f}(s)\mathrm{e}^{st}$ is analytic inside C_R. For $t > 0$ we have

$$\oint_{C_R} \hat{f}(s)\mathrm{e}^{st}\,\mathrm{d}s = \left(\int_{\gamma-\mathrm{i}R}^{\gamma+\mathrm{i}R} + \int_{S_R} - \sum_j \int_{\alpha_j-R}^{(\alpha_j+)}\right)\hat{f}(s)\,^{st}\,\mathrm{d}s, \tag{49}$$

where R is sufficiently large to enclose all the poles, and the *loop integrals* wrap around each branch cut. Thus, by the *residue theorem*

$$\frac{1}{2\pi\mathrm{i}}\oint_{C_R} \hat{f}(s)\mathrm{e}^{st}\,\mathrm{d}s = \sum_k r_k \qquad \text{for } t > 0, \tag{50}$$

where r_k is the residue of the kth pole. It is standard practice then to show that the integrand is exponentially small on S_R, implying that

$$\int_{S_R} \hat{f}(s)\mathrm{e}^{st}\,\mathrm{d}s \to 0 \qquad \text{as } R \to \infty. \tag{51}$$

Here use is made of the fact that on S_R, for $R \gg 1$ we have $s = \gamma + R\,\mathrm{e}^{\mathrm{i}\theta} \approx R\,\mathrm{e}^{\mathrm{i}\theta}$ so that

$$|\mathrm{e}^{st}| = |\mathrm{e}^{Rt\mathrm{e}^{\mathrm{i}\theta}}| = \mathrm{e}^{Rt\cos\theta} = \begin{cases} \mathrm{e}^{-R|t|\cos\theta} \to 0 & \text{in } t < 0, \cos\theta > 0, \\ \mathrm{e}^{-Rt|\cos\theta|} \to 0 & \text{in } t > 0, \cos\theta < 0. \end{cases} \tag{52}$$

It then follows that

$$f(t)\,\mathrm{H}(t) = \frac{1}{2\pi\mathrm{i}} \lim_{R\to\infty} \int_{\gamma-\mathrm{i}R}^{\gamma+\mathrm{i}R} \hat{f}(s)\mathrm{e}^{st}\,\mathrm{d}s = \sum_k^n r_k + \frac{1}{2\pi\mathrm{i}} \sum_j \int_{-\infty}^{(\alpha_j+)} \hat{f}(s)\,{}^{st}\,\mathrm{d}s. \tag{53}$$

Example 3.5 Solve the wave equation

$$\frac{\partial^2 u}{\partial x^2} = \frac{1}{c_0^2}\frac{\partial^2 u}{\partial t^2} \qquad \text{in } x > 0, t > 0, \tag{54}$$

for constant $c_0 > 0$, subject to the boundary conditions

$$u = f(t) \quad \text{on } x = a, \quad \text{and} \quad u \to 0 \quad \text{at } x \to \infty, \quad \text{for all } t > 0, \tag{55}$$

and initial conditions

$$u = \partial u/\partial t = 0 \quad \text{at } t = 0, \quad \text{for all } r. \tag{56}$$

Solution Taking a Laplace transform of the differential equation we have

$$\frac{\partial^2 \hat{u}}{\partial x^2} - \frac{s^2}{c_0^2}\hat{u} = 0, \tag{57}$$

for $\hat{u}(x, s)$, subject to the boundary conditions

$$\hat{u} = \hat{f}(s), \qquad x = 0, \qquad \hat{u} \to 0, \qquad \text{as } x \to \infty. \tag{58}$$

Thus we have a general solution

$$\hat{u}(x, s) = A(s)\mathrm{e}^{sx/c_0} + B(s)\mathrm{e}^{-sx/c_0}. \tag{59}$$

Clearly $A \equiv 0$, to satisfy conditions for x large, and $B = \hat{f}(s)$ so that

$$\hat{u}(x, s) = \hat{f}(s)\mathrm{e}^{-sx/c_0}, \quad \text{so} \quad u(x, t) = \frac{1}{2\pi\mathrm{i}} \int_{\gamma-\mathrm{i}\infty}^{\gamma+\mathrm{i}\infty} \hat{f}(s)\mathrm{e}^{s(t-x/c_0)}\,\mathrm{d}s, \tag{60}$$

which, from Eq. (47), we can interpret as

$$u(x, t) = \mathrm{H}(t - x/c_0) f(t - x/c_0). \tag{61}$$

Example 3.6 Solve the partial differential equation

$$\frac{\partial^2 u}{\partial r^2} + \frac{2}{r}\frac{\partial u}{\partial r} = \frac{1}{c_0^2}\frac{\partial^2 u}{\partial t^2} \qquad \text{in} \quad 0 < r < a, \quad t > 0, \tag{62}$$

for constant $c_0 > 0$, subject to the boundary conditions

$$u = 2U_0 \text{ (const.)} \quad \text{on } r = a, \quad \text{and} \quad u \text{ is finite at } r = 0, \quad \text{for all } t > 0, \tag{63}$$

and initial conditions

$$u = U_0 \qquad \text{at } t = 0, \qquad \text{for all } r. \tag{64}$$

Solution Taking a Laplace transform of the differential equation we have

$$\frac{\partial^2 \hat{u}}{\partial r^2} + \frac{2}{r}\frac{\partial \hat{u}}{\partial r} - \frac{s^2}{c_0^2}\hat{u} \equiv \frac{1}{r}\frac{\partial^2}{\partial r^2}(r\hat{u}) - \frac{s^2}{c_0^2}\hat{u} = -\frac{sU_0}{c_0^2}, \tag{65}$$

for $\hat{u}(r, s)$, subject to the boundary conditions

$$\hat{u} \quad \text{finite at } r = 0, \qquad \hat{u} = \mathcal{L}\{2U_0\} = \frac{2}{s}U_0 \quad \text{at } r = a. \tag{66}$$

A particular integral is simply U_0/s. Thus we have a general solution

$$\hat{u}(r, s) = \frac{U_0}{s} + \frac{1}{r}\left[A(s)\cosh\left(\frac{sr}{c_0}\right) + B(s)\sinh\left(\frac{sr}{c_0}\right)\right]. \tag{67}$$

To satisfy $\hat{u}$ finite at $r = 0$ requires $A = 0$. The condition at $r = a$ yields

$$\frac{U_0}{s} + \frac{B(s)}{a}\sinh\left(\frac{sa}{c_0}\right) = \frac{2U_0}{s} \quad \text{so} \quad \hat{u}(r, s) = \frac{U_0}{s}\left[1 + \frac{a}{r}\frac{\sinh(sr/c_0)}{\sinh(sa/c_0)}\right]. \tag{68}$$

Hence by the inversion theorem (47), we have solution

$$u(r,t) = \frac{1}{2\pi \mathrm{i}}\int_{\gamma-\mathrm{i}\infty}^{\gamma+\mathrm{i}\infty} \frac{U_0}{s}\left[1 + \frac{a}{r}\frac{\sinh(sr/c_0)}{\sinh(sa/c_0)}\right] \mathrm{e}^{st}\,\mathrm{d}s. \tag{69}$$

Consider a closed contour C_R consisting of the vertical line $s = \gamma + \mathrm{i}r$, $-R \le r \le R$, closed on the left by a semicircle S_R of large radius R, as shown in Fig. 3.1. For $|s| \sim R \gg 1$ and $\mathrm{Re}(s) < 0$, the integrand is exponentially small as $R \to \infty$. Thus the contribution from S_R becomes negligible as $R \to \infty$. Thus

$$u(r,t) = \frac{1}{2\pi \mathrm{i}}\lim_{R\to\infty}\oint_{C_R} \frac{U_0}{s}\left[1 + \frac{a}{r}\frac{\sinh(sr/c_0)}{\sinh(sa/c_0)}\right] \mathrm{e}^{st}\,\mathrm{d}s. \tag{70}$$

The integrand has simple poles at $s = 0$ and also where

$$g(s) \equiv \sinh\left(\frac{sa}{c_0}\right) = 0, \quad \text{so} \quad s = \pm s_n = \pm \mathrm{i}\frac{n\pi c_0}{a} \quad n = 1, 2, \ldots. \tag{71}$$

Near $s = 0$ the integrand has the expansion

$$\frac{U_0}{s}\left\{1 + \frac{a}{r}\frac{(sr/c_0)}{(sa/c_0)}[1 + O(s)]\right\} = \frac{U_0}{s}[2 + O(s)]. \tag{72}$$

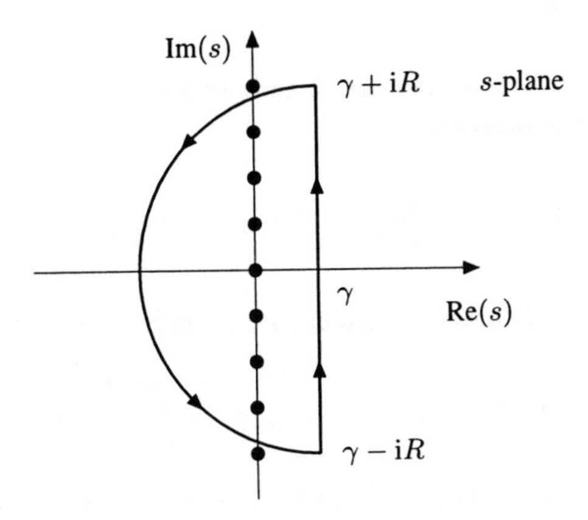

Fig. 3.1 The contour of integration for Example 3.6.

Thus the residue at $s = 0$ is $2U_0$. To consider the poles at $s = s_n$ we recall that

$$\lim_{s\to s_n} \frac{s - s_n}{g(s)} = \frac{1}{g'(s_n)}, \qquad \text{where} \qquad g'(s) = \frac{a}{c_0}\cosh\left(\frac{sa}{c_0}\right). \tag{73}$$

Thus the residue at $s = s_n$ is

$$\begin{aligned}\operatorname{Res}(s_n) &= \frac{c_0 U_0}{r s_n}\frac{\sinh(s_n r/c_0)}{\cosh(s_n a/c_0)}\mathrm{e}^{s_n t} = \frac{aU_0}{rin\pi}\frac{\sinh(in\pi r/a)}{\cosh(in\pi)}\mathrm{e}^{in\pi c_0 t/a} \\ &= \frac{aU_0}{rn\pi}\frac{\sinh(n\pi r/a)}{\cosh(n\pi)}\mathrm{e}^{in\pi c_0 t/a} = \frac{aU_0}{r\pi}\frac{(-1)^n}{n}\sin\left(\frac{n\pi r}{a}\right)\mathrm{e}^{in\pi c_0 t/a},\end{aligned} \tag{74}$$

given that $\cosh(\mathrm{i}z) = \cos z$, $\sinh(\mathrm{i}z) = \mathrm{i}\sin z$. Consequently

$$\begin{aligned}\operatorname{Res}(s_n) + \operatorname{Res}(-s_n) &= \frac{aU_0}{r\pi}\frac{(-1)^n}{n}\sin\left(\frac{n\pi r}{a}\right)\left[\mathrm{e}^{in\pi c_0 t/a} + \mathrm{e}^{-in\pi c_0 t/a}\right] \\ &= \frac{2aU_0}{r\pi}\frac{(-1)^n}{n}\sin\left(\frac{n\pi r}{a}\right)\cos\left(\frac{n\pi c_0 t}{a}\right).\end{aligned} \tag{75}$$

Hence by the residue theorem

$$u(x,t) = \sum \operatorname{Res}(s_n) = 2U_0 + \frac{2aU_0}{r\pi}\sum_{n=1}^{\infty}\frac{(-1)^n}{n}\sin\left(\frac{n\pi r}{a}\right)\cos\left(\frac{n\pi c_0 t}{a}\right). \tag{76}$$

3.3 Fourier Series

A periodic function $f(t)$, of **period** T, and **angular frequency** $\omega = 2\pi/T$ has the **Fourier series expansion**

$$f(t) = \tfrac{1}{2}a_0 + \sum_{n=1}^{\infty} \left[a_n \cos(n\omega t) + b_n \sin(n\omega t)\right]. \tag{77}$$

Here the **Fourier coefficients,** a_n and b_n, are given by **Euler's formulae**, namely

$$a_n = \frac{2}{T}\int_d^{d+T} f(t)\cos(n\omega t)\,\mathrm{d}t, \qquad b_n = \frac{2}{T}\int_d^{d+T} f(t)\sin(n\omega t)\,\mathrm{d}t, \tag{78}$$

where the choice of d is arbitrary. Here Eq. (78) follows by multiplying Eq. (77) by either $\cos(m\omega t)$ or $\sin(m\omega t)$, for some integer m, and integrating. Use is made of

$$\int_d^{d+T} \cos(m\omega t)\,\mathrm{d}t = \begin{cases} 0 & \text{for } m \neq 0, \\ T & \text{for } m = 0, \end{cases} \qquad \int_d^{d+T} \sin(m\omega t)\,\mathrm{d}t = 0, \tag{79}$$

together with the **orthogonality relations** for the trigonometric functions, namely

$$\int_d^{d+T} \cos(m\omega t)\cos(n\omega t)\,\mathrm{d}t = \begin{cases} 0 & \text{for } m \neq n, \\ T/2 & \text{for } m = n \neq 0, \end{cases} \tag{80}$$

$$\int_d^{d+T} \sin(m\omega t)\sin(n\omega t)\,\mathrm{d}t = \begin{cases} 0 & \text{for } m \neq n, \\ T/2 & \text{for } m = n \neq 0, \end{cases} \tag{81}$$

$$\int_d^{d+T} \cos(m\omega t)\sin(n\omega t)\,\mathrm{d}t = 0 \qquad \text{for all } m, n. \tag{82}$$

3.3.1 *Harmonics*

We can also write the Fourier series Eq. (77) in the form

$$f(t) = A_0 + \sum_{n=1}^{\infty} A_n \cos(n\omega t + \phi_n), \tag{83}$$

where $A_n > 0$ and ϕ_n are the **amplitude** and **phase** of the **series nth harmonic**. Comparison of Eq. (83) with Eq. (77) yields $a_n = A_n \cos\phi_n$, $b_n = -A_n \sin\phi_n$ or

$$A_0 = \tfrac{1}{2}a_0, \qquad A_n = \sqrt{a_n^2 + b_n^2}, \qquad \phi_n = \tan^{-1}\left(-b_n/a_n\right). \tag{84}$$

3.3.2 *Odd and even functions*

An even periodic function $f(t)$, in which $f(t) = f(-t)$, and an odd periodic function $g(t)$, where $g(t) = -g(-t)$, have Fourier series of the form

$$f(t) = \tfrac{1}{2}a_0 + \sum_{n=1}^{\infty} a_n \cos(n\omega t), \qquad g(t) = \sum_{n=1}^{\infty} b_n \sin(n\omega t). \tag{85}$$

3.3.3 *Convergence*

The existence of coefficients a_n and b_n is not, in itself, a guarantee that the Fourier series of f converges to the function $f(t)$. A bounded periodic function is said to satisfy **Dirichlet's conditions** if, in any one period, it has a finite number of isolated maxima and minima and points of finite discontinuity.

Theorem 3.4 *If $f(t)$ is a periodic function satisfying Dirichlet's conditions, and a_n and b_n are given by Eq. (78), then* **Fourier's theorem** *states that as $N \to \infty$*

$$\tfrac{1}{2}a_0 + \textstyle\sum_{n=1}^{N} \left[a_n \cos(n\omega t) + b_n \sin(n\omega t)\right] \to \begin{cases} f(t), \\ \tfrac{1}{2}\left[f(t+) + f(t-)\right], \end{cases} \tag{86}$$

depending upon whether of not f is continuous at t.

If f is discontinuous the convergence is slow with a_n and b_n of $O(n^{-1})$, for n large, but if $f^{(k)}$ is continuous for $k = 0, 1, \ldots, r$ and $f^{(r+1)}$ is discontinuous, then a_n and b_n are $O(n^{-r})$. Convergence is slowest near the point of discontinuity with an accompanying overshoot/undershoot at the point of discontinuity, known as the **Gibbs phenomenon**, the magnitude of which does *not* diminish as $N \to \infty$.

3.3.4 *Differentiation and integration*

If we *integrate* the Fourier series of $f(t)$ term by term, the integrated series converges to the integral of $f(t)$ *faster* than the original series. Conversely the *differentiated* series converges to $f'(t)$ only if $f'(t)$ satisfies Dirichlet's conditions, with convergence being *slower* than the original series.

3.3.5 *The complex exponential form*

Using Euler's identity

$$\cos\theta = \frac{1}{2}\left(e^{i\theta} + e^{-i\theta}\right), \qquad \sin\theta = \frac{1}{2i}\left(e^{i\theta} - e^{-i\theta}\right), \tag{87}$$

we can write the Fourier series expansion Eq. (77) in the form

$$f(t) = \tfrac{1}{2}a_0 + \sum_{n=1}^{\infty} \left[\tfrac{1}{2}(a_n - \mathrm{i}b_n)\mathrm{e}^{\mathrm{i}n\omega t} + \tfrac{1}{2}(a_n + \mathrm{i}b_n)\mathrm{e}^{-\mathrm{i}n\omega t}\right] = \sum_{n=-\infty}^{\infty} c_n \mathrm{e}^{\mathrm{i}n\omega t}, \tag{88}$$

where $c_0 = \frac{1}{2}a_0$, $c_{-n} = c_n^*$ and

$$\begin{aligned} c_n = \tfrac{1}{2}(a_n - \mathrm{i}b_n) &= \frac{1}{T}\int_d^{d+T} f(t)\left[\cos(n\omega t) - \mathrm{i}\sin(n\omega t)\right] \mathrm{d}t \\ &= \frac{1}{T}\int_d^{d+T} f(t)\mathrm{e}^{-\mathrm{i}n\omega t}\,\mathrm{d}t. \end{aligned} \tag{89}$$

Recalling the harmonic series Eq. (83) we see that

$$c_n = \tfrac{1}{2}(a_n - \mathrm{i}b_n) = A_n \mathrm{e}^{\mathrm{i}\phi_n} \qquad \text{so} \qquad |c_n| = A_n = \sqrt{a_n^2 + b_n^2}. \tag{90}$$

3.3.5.1 *Parseval's theorem*

Theorem 3.5 *If c_n and d_n are the complex Fourier coefficients of $f(t)$ and $g(t)$ then*

$$\frac{1}{T}\int_d^{d+T} f(t)g(t)\,\mathrm{d}t = \sum_{n=-\infty}^{\infty} c_n d_{-n} = \sum_{n=-\infty}^{\infty} c_n d_n^*. \tag{91}$$

Corollary 3.1

$$\frac{1}{T}\int_d^{d+T} \left[f(t)\right]^2 \mathrm{d}t = \sum_{n=1}^{\infty} |c_n|^2 = \tfrac{1}{4}a_0^2 + \tfrac{1}{4}\sum_{n=1}^{\infty}\left(a_n^2 + b_n^2\right), \tag{92}$$

where we call the left hand side the ***root mean square (RMS)*** *value of the periodic function.*

3.4 The Fourier Transform

3.4.1 *The Fourier transform pair*

The **Fourier transform**, of a function $f(x)$, defined for $-\infty < x < \infty$, is given by

$$\mathcal{F}\{f(x)\} = \bar{f}(k) = \int_{-\infty}^{\infty} f(x)\mathrm{e}^{-\mathrm{i}kx}\,\mathrm{d}x, \tag{93}$$

where k is usually taken to be real, provided the integral converges. Given that $\mathcal{F}\{f(x)\} = \bar{f}(k)$ we define the **inverse Fourier transform**, of $\bar{f}$ by

$$\mathcal{F}^{-1}\{\bar{f}(k)\} = f(x) = \frac{1}{2\pi}\int_{-\infty}^{\infty} \bar{f}(k)e^{ikx}\,dk. \tag{94}$$

Although this is a fairly standard definition, some authors divide and multiply Eqs. (93) and (94) by a factor $\sqrt{2\pi}$ respectively, while others take $e^{\pm ikx} \mapsto e^{\mp ikx}$.

3.4.2 *Fourier's theorem*

As regards convergence of Eq. (93) and (94), a function $f(x)$ is said to satisfy **Dirichlet's conditions** if f is absolutely integrable, namely

$$\int_{-\infty}^{\infty} |f(x)|\,dx < \infty, \tag{95}$$

and, in any finite interval, it has a finite number of maxima and minima and points of discontinuity.

Theorem 3.6 *If $f(x)$ is a function satisfying Dirichlet's conditions, then*

$$\frac{1}{2\pi}\int_{-\infty}^{\infty}\left[\int_{-\infty}^{\infty} f(r)e^{-ikr}\,dr\right]e^{ikx}dk = \begin{cases} f(x), \\ \frac{1}{2}[f(x+) + f(x-)], \end{cases} \tag{96}$$

depending on whether $f(x)$ is continuous or not.

3.4.3 *General comments*

Clearly many simple functions, such as $f(x) = 1$, $f(x) = x^n$ $(n \geq 0)$, $f(x) = e^{\pm\alpha x}$, $f(x) = \sin(\omega x)$ do not satisfy the requirement of absolute integrability Eq. (95) and so do not have a Fourier transform.

For both $f(x)$ and k real, the Fourier transform is, in general, a complex-valued function, $\bar{f}(k) = |\bar{f}(k)|e^{i\arg\bar{f}(k)}$, where the **Riemann–Lebesgue lemma** gives $|\bar{f}(k)| \to 0$ as $|k| \to \pm\infty$. For *causal* functions, where $f(x) = 0$ for $x < 0$, satisfying Eq. (95), we have

$$\mathcal{F}\{H(x)f(x)\} = \int_0^{\infty} f(x)e^{-ikx}dx = \mathcal{L}\{f(x)\}_{s=ik} = \bar{f}(ik). \tag{97}$$

Thus the Fourier transform can be obtained from the Laplace transform, if known.

Example 3.7 For $n \geq 0$ find the Fourier transform of the causal function

$$f(x) = H(x)x^n e^{-ax}, \qquad (a > 0). \tag{98}$$

Solution Clearly f satisfies Eq. (95). From Eq. (97) and Example 3.1 we have

$$\mathcal{F}\left\{H(x)x^n\mathrm{e}^{-ax}\right\} = \int_0^\infty x^n\mathrm{e}^{-(a+\mathrm{i}k)x}\,\mathrm{d}x = \mathcal{L}\left\{x^n\right\}_{s=\mathrm{i}k} = \frac{n!}{(a+\mathrm{i}k)^{n+1}}. \tag{99}$$

Example 3.8 Find the Fourier transform of the non-causal function

$$f(x) = \mathrm{e}^{-\alpha x^2}, \qquad \alpha > 0. \tag{100}$$

Solution Again f satisfies Eq. (95). We can interpret

$$\mathcal{F}\left\{\mathrm{e}^{-\alpha x^2}\right\} = \int_{-\infty}^\infty \mathrm{e}^{-(\alpha x^2+\mathrm{i}kx)}\,\mathrm{d}x = \mathrm{e}^{-k^2/4\alpha}\int_{-\infty}^\infty \mathrm{e}^{-\alpha(x+\mathrm{i}k/2\alpha)^2}\mathrm{d}x, \tag{101}$$

as a contour integral along the real-axis in the complex $z = x + \mathrm{i}y$-plane. Thus

$$\begin{aligned}\mathcal{F}\left\{\mathrm{e}^{-\alpha x^2}\right\} &= \mathrm{e}^{-k^2/4\alpha}\int_{-\infty}^\infty \mathrm{e}^{-\alpha(z+\mathrm{i}k/2\alpha)^2}\mathrm{d}z \\ &= \mathrm{e}^{-k^2/4\alpha}\int_{-\infty-\mathrm{i}k/2\alpha}^{\infty-\mathrm{i}k/2\alpha} \mathrm{e}^{-\alpha(z+\mathrm{i}k/2\alpha)^2}\mathrm{d}z.\end{aligned} \tag{102}$$

Here we have used Cauchy's theorem to lower the contour into $y < 0$ because the integrand is analytic (everywhere) and decays exponentially fast for $|z| \gg 1$. Thus

$$\begin{aligned}\mathcal{F}\left\{\mathrm{e}^{-\alpha x^2}\right\} &= \mathrm{e}^{-k^2/4\alpha}\int_{-\infty}^\infty \mathrm{e}^{-\alpha u^2}\,\mathrm{d}u \\ &= 2\mathrm{e}^{-k^2/4\alpha}\int_0^\infty \mathrm{e}^{-\alpha u^2}\,\mathrm{d}u \\ &= \frac{1}{\sqrt{\alpha}}\,\mathrm{e}^{-k^2/4\alpha}\int_0^\infty v^{-1/2}\mathrm{e}^{-v}\,\mathrm{d}v,\end{aligned} \tag{103}$$

where $u = z + \mathrm{i}k/2\alpha$, $v = \sqrt{\alpha}u$. Given the well-known result

$$\sqrt{\pi} = \Gamma(\tfrac{1}{2}) = \int_0^\infty v^{-1/2}\mathrm{e}^{-v}\,\mathrm{d}v = \int_{-\infty}^\infty \mathrm{e}^{-u^2}\,\mathrm{d}u, \tag{104}$$

where $\Gamma(z)$ is the gamma function, we have the transform

$$\mathcal{F}\left\{\mathrm{e}^{-\alpha x^2}\right\} = \sqrt{\frac{\pi}{\alpha}}\mathrm{e}^{-k^2/4\alpha}. \tag{105}$$

3.4.4 *The duality property*

For the **Fourier transform pair** Eq. (93), (94) symmetry yields the property

$$\mathcal{F}\{f(x)\} = \bar{f}(k), \qquad \text{so} \qquad \mathcal{F}\{\bar{f}(x)\} = 2\pi f(-k). \tag{106}$$

Example 3.9 Find the Fourier transform of

$$\text{(a)} \quad f(x) = C\,[\mathrm{H}(x+a) - \mathrm{H}(x-a)]\,, \qquad \text{(b)} \quad f(x) = \frac{2C}{x}\sin(ax).$$

Solution

(a) We have

$$\mathcal{F}\{f(x)\} = \int_{-a}^{a} C\mathrm{e}^{-\mathrm{i}kx}\,\mathrm{d}x = \left[-\frac{C}{\mathrm{i}k}\,\mathrm{e}^{-\mathrm{i}kx}\right]_{-a}^{a} = \frac{2C}{k}\sin(ka). \tag{107}$$

(b) From the symmetry property Eq. (106) we must then have

$$\begin{aligned}\mathcal{F}\left\{\frac{2C}{x}\sin(ax)\right\} &= 2\pi C\,[\mathrm{H}(-k+a) - \mathrm{H}(-k-a)] \\ &= 2\pi C\,[\mathrm{H}(k+a) - \mathrm{H}(k-a)]\,. \end{aligned} \tag{108}$$

3.4.5 *Further properties*

3.4.5.1 *The 1st shift property*

We see that

$$\mathcal{F}\{f(x)\mathrm{e}^{\mathrm{i}\alpha x}\} = \int_{-\infty}^{\infty} f(x)\mathrm{e}^{-\mathrm{i}(k-\alpha)x}\mathrm{d}x = \bar{f}(k-\alpha). \tag{109}$$

3.4.5.2 *The 2nd shift property*

We see that

$$\begin{aligned}\mathcal{F}\{f(ax-b)\} &= \int_{-\infty}^{\infty} f(ax-b)\mathrm{e}^{-\mathrm{i}kx}\mathrm{d}x \\ &= \frac{\mathrm{e}^{-\mathrm{i}kb/a}}{|a|}\int_{-\infty}^{\infty} f(r)\mathrm{e}^{-\mathrm{i}kr/a}\mathrm{d}r = \frac{\mathrm{e}^{-\mathrm{i}kb/a}}{|a|}\bar{f}(k/a). \end{aligned} \tag{110}$$

More simply $\mathcal{F}\{f(-x)\} = \bar{f}(-k)$ and $\mathcal{F}\{f(x-\alpha)\} = \mathrm{e}^{-\mathrm{i}k\alpha}\bar{f}(k)$.

3.4.5.3 *Multiplication by x^n*

We see that

$$\mathcal{F}\{x^n f(x)\} = \int_{-\infty}^{\infty} x^n f(x) \mathrm{e}^{-\mathrm{i}kx} \mathrm{d}x$$
$$= \mathrm{i}^n \frac{\mathrm{d}^n}{\mathrm{d}k^n} \left(\int_{-\infty}^{\infty} f(x) \mathrm{e}^{-\mathrm{i}kx} \mathrm{d}x \right) = \mathrm{i}^n \frac{\mathrm{d}^n \bar{f}}{\mathrm{d}k^n}. \tag{111}$$

Example 3.10 Obtain the Fourier transform of

$$f(x) = C\left[\mathrm{H}(x) - \mathrm{H}(x-2a)\right]. \tag{112}$$

Solution Using the 1st-shift property Eq. (109), we have

$$\mathcal{F}\{f(x)\} = \mathrm{e}^{-\mathrm{i}ka} \mathcal{F}\{C\left[\mathrm{H}(x+a) - \mathrm{H}(x-a)\right]\} = \frac{2C}{k} \mathrm{e}^{-\mathrm{i}ka} \sin(ka). \tag{113}$$

3.4.6 *Differential properties*

$$\mathcal{F}\{f'(x)\} = \int_{-\infty}^{\infty} \frac{\mathrm{d}f}{\mathrm{d}x} \mathrm{e}^{-\mathrm{i}kx} \mathrm{d}x$$
$$= \left[f(x)\mathrm{e}^{-\mathrm{i}kx}\right]_{-\infty}^{\infty} + \mathrm{i}k \int_{-\infty}^{\infty} f(x) \mathrm{e}^{-\mathrm{i}kx} \mathrm{d}x = (\mathrm{i}k)\bar{f}(k). \tag{114}$$

Repeating this argument n times, gives

$$\mathcal{F}\left\{\frac{\mathrm{d}^n f}{\mathrm{d}x^n}\right\} = (\mathrm{i}k)^n \bar{f}(k). \tag{115}$$

3.4.7 *Parseval's theorem*

Theorem 3.7 *Given that $\mathcal{F}\{f(x)\} = \bar{f}(k)$ and $\mathcal{F}\{g(x)\} = \bar{g}(k)$ we can show that*

(a)
$$\int_{-\infty}^{\infty} f(x)g(x)\,\mathrm{d}x = \frac{1}{2\pi} \int_{-\infty}^{\infty} \bar{f}(k)\bar{g}(-k)\,\mathrm{d}k, \tag{116}$$

(b)
$$\int_{-\infty}^{\infty} f(x)g(-x)\,\mathrm{d}x = \frac{1}{2\pi} \int_{-\infty}^{\infty} \bar{f}(k)\bar{g}(k)\,\mathrm{d}k. \tag{117}$$

Corollary 3.2 *By **Parseval's theorem** the constant*

$$E = \int_{-\infty}^{\infty} \left[f(x)\right]^2 \mathrm{d}x = \frac{1}{2\pi} \int_{-\infty}^{\infty} \bar{f}(k)\bar{f}(-k)\,\mathrm{d}k = \frac{1}{2\pi} \int_{-\infty}^{\infty} |\bar{f}(k)|^2 \,\mathrm{d}k, \tag{118}$$

since for f real, $\bar{f}(-k) = \bar{f}^*(k)$.

3.4.8 *The convolution theorem*

The **convolution** of two piecewise-continuous functions $f(x)$ and $g(x)$, defined for all x, is once again defined by Eq. (24), that is

$$(f*g)(x) = \int_{-\infty}^{\infty} f(r)g(x-r)\,\mathrm{d}r = \int_{-\infty}^{\infty} f(x-r)g(r)\,\mathrm{d}r = (g*f)(x). \quad (119)$$

Theorem 3.8 *The **convolution theorem** states that*

$$\begin{aligned}\mathcal{F}\{f(x)*g(x)\} &= \mathcal{F}\left\{\int_{-\infty}^{\infty} f(r)g(x-r)\,\mathrm{d}r\right\} \\ &= \mathcal{F}\{f(x)\}\,\mathcal{F}\{g(x)\} = \bar{f}(k)\bar{g}(k) \end{aligned} \quad (120)$$

and correspondingly

$$\mathcal{F}\{f(x)g(x)\} = \frac{1}{2\pi}\int_{-\infty}^{\infty} \bar{f}(r)\bar{g}(x-r)\,\mathrm{d}r = \frac{1}{2\pi}\bar{f}(k)*\bar{g}(k). \quad (121)$$

3.4.9 *The delta function*

The Dirac **delta function** is a *generalised function* having the properties

$$\delta(x-c) = \begin{cases} `\infty\text{'} & \text{if } x = c, \\ 0 & \text{if } x \neq c, \end{cases} \qquad \int_a^b \delta(x-c)f(x)\,\mathrm{d}x = \begin{cases} f(c) & \text{if } a < c < b, \\ 0 & \text{otherwise}, \end{cases} \quad (122)$$

for all functions f continuous at $x = c$. One way of representing this infinitely thin and tall spike, or impulse, is as the limit of standard functions. For example

$$\delta(x-c) = \lim_{\epsilon\to 0} \frac{1}{2\sqrt{\pi\epsilon}} \mathrm{e}^{-(x-c)^2/4\epsilon}, \quad (123)$$

which, as $\epsilon \to 0$, is exponentially small, for $x \neq c$ but algebraically large for $x = c$. The integral property follows since for any suitable function $f(x)$ we have

$$\frac{1}{2\sqrt{\pi\epsilon}}\int_{-\infty}^{\infty} f(x)\mathrm{e}^{-(x-c)^2/4\epsilon}\,\mathrm{d}x = \frac{1}{\sqrt{\pi}}\int_{-\infty}^{\infty} f\left[c+2\sqrt{\epsilon}u\right]\mathrm{e}^{-u^2}\,\mathrm{d}u \quad (124)$$

$$\to \frac{f(c)}{\sqrt{\pi}}\int_{-\infty}^{\infty} \mathrm{e}^{-u^2}\,\mathrm{d}u = f(c) \qquad \text{as } \epsilon \to 0+, \quad (125)$$

following Eq. (104). In addition we have

$$\mathcal{F}\{\delta(x-a)\} = \int_{-\infty}^{\infty} f(x)\mathrm{e}^{-\mathrm{i}kx}\,\mathrm{d}x = \mathrm{e}^{-\mathrm{i}ka} \quad \text{so} \quad \mathcal{F}\{1\} = 2\pi\delta(k-a), \quad (126)$$

by the duality property Eq. (106). Note that $f(x) \equiv 1$ is *not* absolutely integrable and does not have a Fourier Transform in the usual sense. We also see that

$$f(x) * \delta(x) = \int_{-\infty}^{\infty} f(r)\delta(x-r)\,\mathrm{d}r = f(x). \tag{127}$$

Example 3.11 Solve the differential equation

$$\frac{\mathrm{d}^2u}{\mathrm{d}x^2} + 3\frac{\mathrm{d}u}{\mathrm{d}x} + 7u = f(x) \qquad \text{in } -\infty < x < \infty, \tag{128}$$

for a general forcing function f, such that $u \to 0$, as $x \to \pm\infty$.

Solution Let us, instead, solve the equation

$$\frac{\mathrm{d}^2u}{\mathrm{d}x^2} + 3\frac{\mathrm{d}u}{\mathrm{d}x} + 7u = \delta(x). \tag{129}$$

Using Eq. (115), we can take the Fourier transform of the equation to obtain

$$\left[(\mathrm{i}k)^2 + 3(\mathrm{i}k) + 7\right]\bar{u}(k) = 1, \tag{130}$$

where $\mathcal{F}\{u(x)\} = \bar{u}(k)$, since $\mathcal{F}\{\delta(x)\} = 1$. This we see that

$$\bar{u} = \overline{G}(k) = \frac{1}{(\mathrm{i}k)^2 + 3(\mathrm{i}k) + 7} = \frac{1}{(7-k^2) + \mathrm{i}3k}. \tag{131}$$

If we can solve for $G(x)$, then the transform solution to our original problem is, $\bar{u} = \overline{G}(k)\bar{f}(k)$, where $\mathcal{F}\{f(x)\} = \bar{f}(k)$. Hence, by the convolution theorem (120), the solution, for any function f, is given by

$$u(x) = G(x) * f(x) = \int_{-\infty}^{\infty} G(x-r)f(r)\,\mathrm{d}r. \tag{132}$$

Example 3.12 Solve the wave equation

$$\frac{\partial^2 u}{\partial x^2} = \frac{\partial^2 u}{\partial t^2} \qquad \text{in } -\infty < x < \infty, \quad t > 0, \tag{133}$$

subject to the initial conditions

$$u = f(x), \qquad \text{and} \qquad \partial u/\partial t = 0, \qquad \text{at } t = 0, \qquad \text{for all } x. \tag{134}$$

Solution Taking a Fourier transform (in x) of the equation yields the problem

$$\frac{\partial^2 \bar{u}}{\partial t^2} + k^2\bar{u} = 0, \tag{135}$$

for $\bar{u}(k,t)$, subject to the initial conditions

$$\bar{u} = \mathcal{F}\{f(x)\} = \bar{f}(k), \qquad \frac{\partial \bar{u}}{\partial t} = \mathcal{F}\{0\} = 0 \qquad \text{at } t = 0. \tag{136}$$

The general solution of the transformed equation is given by

$$\bar{u}(k,t) = A(k)\cos(kt) + B(k)\sin(kt), \tag{137}$$

where A and B are arbitrary. To satisfy the initial conditions we require $A(k) = \bar{f}(k)$ and $B(k) \equiv 0$. Thus we have the transform solution

$$\bar{u}(k,t) = \bar{f}(k)\cos(kt) = \tfrac{1}{2}\bar{f}(k)\left[\mathrm{e}^{ikt} + \mathrm{e}^{-ikt}\right]. \tag{138}$$

Thus by the inversion theorem (94), we have the solution

$$\begin{aligned} u(x,t) &= \frac{1}{2\pi}\int_{-\infty}^{\infty} \bar{f}(k)\,\tfrac{1}{2}\left[\mathrm{e}^{ikt} + \mathrm{e}^{-ikt}\right]\mathrm{e}^{ikx}\,\mathrm{d}k \\ &= \frac{1}{2}\left\{\frac{1}{2\pi}\int_{-\infty}^{\infty} \bar{f}(k)\mathrm{e}^{\mathrm{i}k(x+t)}\,\mathrm{d}k + \frac{1}{2\pi}\int_{-\infty}^{\infty} \bar{f}(k)\mathrm{e}^{\mathrm{i}k(x-t)}\,\mathrm{d}k\right\}, \end{aligned} \tag{139}$$

which is d'Alembert's solution to the wave equation, namely

$$u(x,t) = \tfrac{1}{2}\left[f(x+t) + f(x-t)\right]. \tag{140}$$

3.4.10 *Further extensions*

So far we have taken the transform variable k to be real. A complex $k = \alpha + \mathrm{i}\tau$, where $\alpha = \mathrm{Re}(k)$, $\tau = \mathrm{Im}(k)$, allows us to consider functions which are not absolutely integrable in the usual sense.

Theorem 3.9 *Suppose that f is real and, for real constants $A_\pm$ and $\tau_\pm$, satisfies*

$$|f(x)| < A_\pm \exp(-\tau_\pm x) \qquad \text{as } x \to \pm\infty. \tag{141}$$

If, for some τ_0, $\tau_- < \tau_0 < \tau_+$, and $f(x)\mathrm{e}^{\tau_0 x}$ is absolutely integrable, then

$$\bar{f}(k) = \int_{-\infty}^{\infty} f(x)\mathrm{e}^{-ikx}\,\mathrm{d}x \tag{142}$$

is an analytic function of $k = \alpha + \mathrm{i}\tau$ in $\tau_- < \tau < \tau_+$, and for any fixed τ

$$f(x) = \frac{1}{2\pi}\int_{\mathrm{i}\tau-\infty}^{\mathrm{i}\tau+\infty} \bar{f}(k)\mathrm{e}^{\mathrm{i}kx}\,\mathrm{d}k \qquad \text{for } \tau_- < \tau < \tau_+. \tag{143}$$

Proof. This follows from the magnitude inequality, since for $\tau_- < \tau < \tau_+$

$$|f(x)\mathrm{e}^{-ikx}| = |f(x)|\mathrm{e}^{\tau x} < A_\pm \mathrm{e}^{(\tau-\tau_\pm)x} \to 0 \qquad \text{as } x \to \pm\infty. \tag{144}$$

□

Example 3.13 Show that the Fourier transform of

$$f(x) = \begin{cases} \mathrm{e}^x & \text{if } x < 0, \\ 1 & \text{if } x > 0, \end{cases} \tag{145}$$

which is not absolutely integrable, is analytic in the strip $-1 < \mathrm{Im}(k) < 0$.

Solution Note that $f(x)\mathrm{e}^{\tau x}$ is absolutely integrable for $-1 < \tau < 0$. Also

$$\begin{aligned} \mathcal{F}\{f(x)\} &= \int_{-\infty}^{0} \mathrm{e}^{(1-\mathrm{i}k)x}\,\mathrm{d}x + \int_{0}^{\infty} \mathrm{e}^{-\mathrm{i}kx}\,\mathrm{d}x \\ &= \left[\frac{1}{1-\mathrm{i}k}\mathrm{e}^{(1-\mathrm{i}k)x}\right]_{-\infty}^{0} + \left[-\frac{1}{\mathrm{i}k}\mathrm{e}^{-\mathrm{i}kx}\right]_{0}^{\infty} \\ &= \frac{1}{1-\mathrm{i}k} + \frac{1}{\mathrm{i}k} \end{aligned} \tag{146}$$

is analytic except for the simple poles at $k = 0$ and $k = -\mathrm{i}$.

3.5 The Heat Conduction Equation

Consider the problem

$$\frac{\partial^2 u}{\partial x^2} = \frac{\partial u}{\partial t} \qquad \text{in } -\infty < x < \infty, \quad t > 0, \tag{147}$$

subject to the conditions

$$u \to 0 \qquad \text{as } x \to \pm\infty, \qquad \text{and} \qquad u = f(x) \qquad \text{at } t = 0. \tag{148}$$

3.5.1 *Fourier transform*

Taking a Fourier transform (in x) of the differential equation yields

$$\frac{\partial \bar{u}}{\partial t} + k^2 \bar{u} = 0, \tag{149}$$

subject to the initial condition

$$\bar{u} = \mathcal{F}\{f(x)\} = \bar{f}(k), \qquad \text{at } t = 0. \tag{150}$$

From Example 3.9 the transform solution satisfying the initial condition is given by

$$u(k,t) = \bar{f}(k)\mathrm{e}^{-k^2 t} = \mathcal{F}\{f(x)\}\,\mathcal{F}\left\{\frac{1}{2\sqrt{\pi t}}\,\mathrm{e}^{-x^2/4t}\right\}. \tag{151}$$

Thus by the convolution theorem (120), we have

$$u(x,t) = \frac{1}{2\sqrt{\pi t}} \int_{-\infty}^{\infty} f(r) \mathrm{e}^{-(x-r)^2/4t} \mathrm{d}r. \tag{152}$$

3.5.2 *Green's function*

If $f(x) = \delta(x-a)$, then the solution is given by $u = G(x-a,t)$, where

$$G(x-a,t) = \mathcal{F}^{-1}\left\{ \mathrm{e}^{-\mathrm{i}ka - k^2 t} \right\} = \frac{1}{2\sqrt{\pi t}}\, \mathrm{e}^{-(x-a)^2/4t}, \tag{153}$$

where G is called the **Green's function**. The initial condition, namely $G(x-a,0) = \delta(x-a)$ confirms Eq. (123). Further the general solution Eq. (152) is given by

$$u(x,t) = \int_{-\infty}^{\infty} f(r) G(x-r)\, \mathrm{d}r. \tag{154}$$

Indeed it is often the case that the general solution can be immediately obtained once the Green's function is determined.

3.5.3 *Laplace transform*

Taking a Laplace transform (in t) of the differential equation yields

$$\frac{\partial^2 \hat{u}}{\partial x^2} - s\hat{u} = -f(x) = -\delta(x-a) = \begin{cases} 0 & \text{for } x \neq a, \\ `\infty' & \text{for } x = a, \end{cases} \tag{155}$$

subject to $\hat{u} \to 0$ as $x \to \pm\infty$. Indeed any function G satisfying $LG(x-a) = -\delta(x-a)$ where L is some linear differential operator, is referred to as a **Green's function**. Solving in $x-a>0$ and $x-a<0$ separately, yields

$$\hat{u}(x,s) = A_{\pm}(s) \mathrm{e}^{-s^{1/2}|x-a|}. \tag{156}$$

where we take $s^{1/2}$ to have a branch cut along the negative real axis, so $s = r\,\mathrm{e}^{\mathrm{i}\theta}$, $-\pi < \theta < \pi$. The coefficients $A_{\pm}(s)$ are found by requiring $\hat{u}$ to be continuous at $x = a$, and by taking the limit $\epsilon \to 0+$

$$\left[\frac{\partial \hat{u}}{\partial x}\right]_{a-0-}^{a+0+} = \lim_{\epsilon\to 0+} \int_{a-\epsilon}^{a+\epsilon} \left(\frac{\partial^2 \hat{u}}{\partial x^2} - s\hat{u} \right) \mathrm{d}x = -\lim_{\epsilon\to 0+} \int_{a-\epsilon}^{a+\epsilon} \delta(x-a)\, \mathrm{d}x = -1. \tag{157}$$

Thus we have $A_+(s) = A_-(s) = \frac{1}{2}s^{-1/2}$ so that the inversion integral yields

$$G(x-a,t) = \frac{1}{2\pi\mathrm{i}} \int_{\gamma-\mathrm{i}\infty}^{\gamma+\mathrm{i}\infty} \hat{u} \mathrm{e}^{st}\, \mathrm{d}s = \frac{1}{2\pi\mathrm{i}} \int_{\gamma-\mathrm{i}\infty}^{\gamma+\mathrm{i}\infty} \frac{1}{2s^{\frac{1}{2}}} \mathrm{e}^{-s^{1/2}|x-a|} \mathrm{e}^{st}\, \mathrm{d}s. \tag{158}$$

For $t > 0$ we close to the left as shown in Fig. 3.2 to obtain the integral

$$\frac{1}{2\pi \mathrm{i}} \oint_{C_R} \hat{u} e^{st} \mathrm{d}s = \frac{1}{2\pi \mathrm{i}} \left(\int_{\gamma - \mathrm{i}R}^{\gamma + \mathrm{i}R} + \int_{S_R^U} + \int_{s = r e^{\mathrm{i}\pi}} + \int_{C_\epsilon} + \int_{s = r e^{-\mathrm{i}\pi}} + \int_{S_R^L} \right) \hat{u} e^{st} \mathrm{d}s, \quad (159)$$

where on S_R, $s = Re^{\mathrm{i}\theta}$, $\frac{\pi}{2} < \theta < \pi$, $-\pi < \theta < -\frac{\pi}{2}$ and on C_ϵ: $s = \epsilon e^{\mathrm{i}\theta}$, $-\pi < \theta < \pi$.

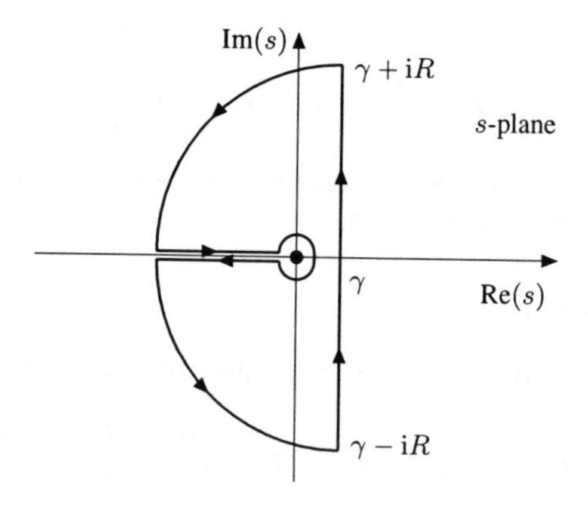

Fig. 3.2 The contour of integration.

Because there are no singularities within C_R Cauchy's theorem again yields

$$\frac{1}{2\pi \mathrm{i}} \oint_{C_R} \hat{u} e^{st} \, \mathrm{d}s = 0. \quad (160)$$

If the contributions $\int_{S_R^U}$, $\int_{S_R^L}$ as $R \to \infty$ and $\int_{C_\epsilon}$ as $\epsilon \to 0$, are negligible, then

$$G(x - a, t) = -\frac{1}{2\pi \mathrm{i}} \lim_{R \to \infty} \left(\int_{s = r e^{\mathrm{i}\pi}} + \int_{s = r e^{-\mathrm{i}\pi}} \right) \frac{1}{2 s^{1/2}} e^{-s^{1/2} |x - a|} e^{st} \, \mathrm{d}s. \quad (161)$$

Taking $s = r\mathrm{e}^{\pm \mathrm{i}\pi} = -r$, so $\mathrm{d}s = -\mathrm{d}r$, then yields

$$G(x-a,t) =$$
$$-\frac{1}{2\pi\mathrm{i}}\left[\int_{\infty}^{0} \frac{\mathrm{e}^{-\sqrt{r}\,\mathrm{e}^{\mathrm{i}\pi/2}|x-a|}}{2\sqrt{r}\,\mathrm{e}^{\mathrm{i}\pi/2}}\,\mathrm{e}^{-rt}\,(-\mathrm{d}r) + \int_{0}^{\infty} \frac{\mathrm{e}^{-\sqrt{r}\,\mathrm{e}^{-\mathrm{i}\pi/2}|x-a|}}{2\sqrt{r}\,\mathrm{e}^{-\mathrm{i}\pi/2}}\,\mathrm{e}^{-rt}\,(-\mathrm{d}r)\right]$$
$$= -\frac{1}{2\pi\mathrm{i}}\int_{0}^{\infty} \frac{1}{2\mathrm{i}\sqrt{r}}\left[\mathrm{e}^{\mathrm{i}\sqrt{r}|x-a|} + \mathrm{e}^{-\mathrm{i}\sqrt{r}|x-a|}\right]\mathrm{e}^{-rt}\,\mathrm{d}r. \quad (162)$$

Further taking $r = k^2$, so $\mathrm{d}r = 2k\,\mathrm{d}k$, then gives

$$\begin{aligned} G(x-a,t) &= \frac{1}{2\pi}\int_{0}^{\infty}\left[\mathrm{e}^{\mathrm{i}k|x-a|} + \mathrm{e}^{-\mathrm{i}k|x-a|}\right]\mathrm{e}^{-k^2 t}\,\mathrm{d}k \\ &= \frac{1}{2\pi}\int_{-\infty}^{\infty} \mathrm{e}^{\mathrm{i}k|x-a|}\mathrm{e}^{-k^2 t}\,\mathrm{d}k \\ &= \frac{1}{2\pi}\mathrm{e}^{-(x-a)^2/4t}\int_{-\infty}^{\infty} \mathrm{e}^{-[k-\mathrm{i}|x-a|/2t]^2 t}\,\mathrm{d}k \\ &= \frac{1}{2\pi}\mathrm{e}^{-(x-a)^2/4t}\int_{-\infty}^{\infty} \mathrm{e}^{-u^2 t}\,\mathrm{d}u. \end{aligned} \quad (163)$$

Finally taking $v = \sqrt{t}\,u$, so $\mathrm{d}v = \sqrt{t}\,\mathrm{d}u$, gives the result

$$G(x-a,t) = \frac{1}{2\pi\sqrt{t}}\,\mathrm{e}^{-(x-a)^2/4t}\int_{-\infty}^{\infty} \mathrm{e}^{-v^2}\,\mathrm{d}v = \frac{1}{2\sqrt{\pi t}}\,\mathrm{e}^{-(x-a)^2/4t}. \quad (164)$$

3.6 The Wave Equation

For an unbounded fluid at rest, subject to a pressure source $F(\mathbf{r},t)$, the pressure fluctuation, p, satisfies the 3-D inhomogeneous wave equation

$$\frac{1}{c_0^2}\frac{\partial^2 p}{\partial t^2} - \nabla^2 p = F(\mathbf{r},t). \quad (165)$$

Here t is the time and $\mathbf{r} = (x,y,z)$, or $\mathbf{r} = (x_1,x_2,x_3)$, is the position vector $\mathbf{r} = \overrightarrow{OP}$ of a general point P in the fluid.

3.6.1 *Time-harmonic motion*

If the source is time-harmonic of radian frequency ω, then the substitution

$$p(\mathbf{r},t) = \hat{p}(\mathbf{r},\omega)\mathrm{e}^{-\mathrm{i}\omega t}, \qquad F(\mathbf{r},t) = \hat{F}(\mathbf{r},\omega)\mathrm{e}^{-\mathrm{i}\omega t} \quad (166)$$

turns Eq. (165) into the inhomogeneous **Helmholtz equation**

$$-\left(\nabla^2+\kappa^2\right)\hat{p}(\mathbf{r},\omega)=\hat{F}(\mathbf{r},\omega), \tag{167}$$

where $\kappa = \omega/c_0$ is the **acoustic wavenumber**. The frequency domain Green's function $G_R(\mathbf{r},\mathbf{y},\omega)=G_R(\mathbf{r}-\mathbf{y},\omega)$ here satisfies

$$-\left(\nabla^2+\kappa^2\right)G_R(\mathbf{r},\mathbf{y},\omega)=\delta(\mathbf{r}-\mathbf{y}), \tag{168}$$

which, in addition to outgoing wave behaviour, must have the singular form

$$G_R(\mathbf{r},\mathbf{y},\omega)\sim\frac{1}{4\pi}\frac{1}{|\mathbf{r}-\mathbf{y}|}\qquad \text{as } \mathbf{r}\to\mathbf{y}, \tag{169}$$

as Helmholtz's equation effectively reduces to Laplace's equation near $\mathbf{r}=\mathbf{y}$.

3.6.2 *Fourier transform*

To solve Eq. (168) we define the 3-D Fourier transform of a function $f(\mathbf{r})$ by

$$\bar{f}(\mathbf{k})=\mathcal{F}_3\{f(\mathbf{r})\}=\iiint_{-\infty}^{\infty} f(\mathbf{r})\mathrm{e}^{-\mathrm{i}\mathbf{k}\cdot\mathbf{r}}\,\mathrm{d}x_1\,\mathrm{d}x_2\,\mathrm{d}x_3, \tag{170}$$

where $\mathbf{k}=(k_1,k_2,k_3)$ is the 3-D wavenumber vector. We have a corresponding inverse transform

$$f(\mathbf{r})=\mathcal{F}_3^{-1}\{\bar{f}(\mathbf{k})\}=\frac{1}{(2\pi)^3}\iiint_{-\infty}^{\infty}\bar{f}(\mathbf{k})\mathrm{e}^{\mathrm{i}\mathbf{k}\cdot\mathbf{r}}\,\mathrm{d}k_1\,\mathrm{d}k_2\,\mathrm{d}k_3. \tag{171}$$

The transformed counterpart to Eq. (168) is then given by

$$\left(k^2-\kappa^2\right)\overline{G}_R(\mathbf{k},\omega)=\mathrm{e}^{-\mathrm{i}\mathbf{k}\cdot\mathbf{y}}, \tag{172}$$

where $k^2=|\mathbf{k}|=k_1^2+k_2^2+k_3^2$. Hence by the inversion theorem

$$G_R(\mathbf{r},\mathbf{y},\omega)=\frac{1}{(2\pi)^3}\iiint_{-\infty}^{\infty}\frac{\mathrm{e}^{\mathrm{i}\mathbf{k}\cdot(\mathbf{r}-\mathbf{y})}}{k^2-\kappa^2}\,\mathrm{d}k_1\,\mathrm{d}\,k_2\,\mathrm{d}k_3. \tag{173}$$

3.6.2.1 *The Lighthill–Landau procedure*

For an *outgoing wave* we temporarily assume that $\omega\mapsto\omega+\mathrm{i}\epsilon$, where $0<\epsilon\ll 1$. This displaces the singularities away from the contour of integration and on restoring the exponential time-factor, we see that

$$\psi(\mathbf{r},\mathbf{y},t)=G_R(\mathbf{r},\mathbf{y},\omega)\mathrm{e}^{-\mathrm{i}(\omega+\mathrm{i}\epsilon)t}=\left[G_R(\mathbf{r},\mathbf{y},\omega)\mathrm{e}^{-\mathrm{i}\omega t}\right]\mathrm{e}^{\epsilon t} \tag{174}$$

is equivalent to radiation from a source that was *switched on* at $t=-\infty$.

Example 3.14 Obtain the 1-D frequency domain Green's function that satisfies

$$-\left(\frac{\partial^2}{\partial x^2}+\kappa^2\right)G_R(x,y,\omega)=\delta(x-y). \tag{175}$$

Solution From Eq. (173) we have

$$G_R(x,y,\omega)=\frac{1}{2\pi}\int_{-\infty}^{\infty}\overline{G}_R(k,y,\omega)\mathrm{e}^{\mathrm{i}kx}\,\mathrm{d}k=\frac{1}{2\pi}\int_{-\infty}^{\infty}\frac{\mathrm{e}^{\mathrm{i}k(x-y)}}{k^2-\kappa^2}\,\mathrm{d}k. \tag{176}$$

The integral is undefined for $\kappa > 0$ because of the poles on the real axis at

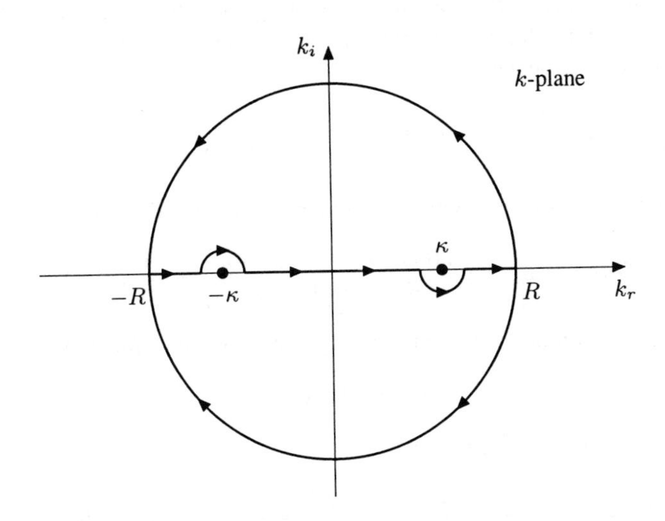

Fig. 3.3 The contour of integration for Example 3.14.

$k = \pm\kappa$. These poles can be avoided by indenting the path of integration into the complex k-plane to pass either above or below the singularity. Four different solutions can be obtained, but only one exhibits *outgoing wave behaviour*. If $k = k_r + \mathrm{i}k_i$ then

$$\mathrm{e}^{\mathrm{i}k(x-y)}\to 0 \qquad \text{as } |k|\to\infty, \qquad \text{if} \quad \begin{cases} k_i \ge 0 & \text{for } x>y,\\ k_i \le 0 & \text{for } x<y. \end{cases} \tag{177}$$

Further the residue at the two simple poles is given by

$$\mathrm{Res}(\pm\kappa)=\lim_{k\to\pm\kappa}\left[\frac{(k\mp\kappa)}{k^2-\kappa^2}\,\mathrm{e}^{\mathrm{i}k(x-y)}\right]=\pm\frac{1}{2\kappa}\,\mathrm{e}^{\pm\mathrm{i}\kappa(x-y)} \tag{178}$$

Applying the Lighthill–Landau procedure displaces the pole at $k = \kappa + \mathrm{i}\epsilon$ into the upper half plane and the pole at $k = -(\kappa + \mathrm{i}\epsilon)$ into the lower half-plane. Thus we must indent the contour above the pole at $k = -\kappa$ and below the pole at $k = +\kappa$ as shown in Fig. 3.3.

For $x > y$ we can close the contour by a large semicircle in the upper half-plane, where the integrand decays exponentially. Only the pole at $k = \kappa$ is enclosed by the contour, and hence

$$\int_{C_R} \frac{\mathrm{e}^{\mathrm{i}k(x-y)}}{k^2 - \kappa^2} \mathrm{d}k = 2\pi\mathrm{i}\,\mathrm{Res}(\kappa) = 2\pi\mathrm{i} \times \frac{1}{2\kappa}\, \mathrm{e}^{\mathrm{i}\kappa(x-y)} \tag{179}$$

For $x < y$ we can close the contour by a large semicircle in the lower half-plane, where the integrand decays exponentially. Only the pole at $k = -\kappa$ is enclosed by the contour, and hence

$$\int_{C_R} \frac{\mathrm{e}^{\mathrm{i}k(x-y)}}{k^2 - \kappa^2} \mathrm{d}k = -2\pi\mathrm{i}\,\mathrm{Res}(-\kappa) = -2\pi\mathrm{i} \times \frac{1}{2\kappa}\, \mathrm{e}^{-\mathrm{i}\kappa(x-y)} \tag{180}$$

where the negative sign is due to the contour orientation. Therefore

$$G_R(x, y, \omega) = \begin{Bmatrix} \mathrm{i}\,\mathrm{Res}(\kappa) & \text{for } x > y \\ -\mathrm{i}\,\mathrm{Res}(-\kappa) & \text{for } x < y \end{Bmatrix} = \frac{\mathrm{i}}{2\kappa}\, \mathrm{e}^{\mathrm{i}\kappa|x-y|} \tag{181}$$

Example 3.15 Obtain the 3-D frequency domain Green's function that satisfies

$$-\left(\nabla^2 + \kappa^2\right) G_R(\mathbf{r}, \mathbf{y}) = \delta(\mathbf{r} - \mathbf{y}). \tag{182}$$

Solution From Eq. (173) we have

$$G_R(\mathbf{r}, \mathbf{y}, \omega) = \frac{1}{(2\pi)^3} \int_{-\infty}^{\infty} \int_{-\infty}^{\infty} \int_{-\infty}^{\infty} \frac{\mathrm{e}^{\mathrm{i}\mathbf{k}\cdot(\mathbf{r}-\mathbf{y})}}{k^2 - \kappa^2}\, \mathrm{d}k_1\, \mathrm{d}k_2\, \mathrm{d}k_3. \tag{183}$$

Taking spherical coordinates (k, θ, ϕ), defined by

$$k_1 = k \sin\theta \cos\phi, \qquad k_2 = k \sin\theta \sin\phi, \qquad k_3 = k \cos\theta, \tag{184}$$

with θ measured from the source-observer direction so that

$$\mathbf{k} \cdot (\mathbf{r} - \mathbf{y}) = k|\mathbf{r} - \mathbf{y}| \cos\theta, \quad \text{and} \quad \mathrm{d}k_1\, \mathrm{d}k_2\, \mathrm{d}k_3 = k^2 \sin\theta\, \mathrm{d}k\, \mathrm{d}\theta\, \mathrm{d}\phi. \tag{185}$$

Hence we have

$$G_R(\mathbf{r}, \mathbf{y}, \omega) =$$
$$\frac{1}{(2\pi)^3} \left(\int_{\phi=0}^{\phi=2\pi} \mathrm{d}\phi \right) \int_{k=0}^{k=\infty} \frac{k^2}{(k^2 - \kappa^2)} \left\{ \int_{\theta=0}^{\theta=\pi} \sin\theta \mathrm{e}^{\mathrm{i}k|\mathbf{r}-\mathbf{y}|\cos\theta}\, \mathrm{d}\theta \right\} \mathrm{d}k. \tag{186}$$

Integrating with respect to θ and ϕ yields

$$G_R(\mathbf{r},\mathbf{y},\omega) = \frac{1}{(2\pi)^3}\left[\phi\right]_0^{2\pi}\int_{k=0}^{k=\infty}\frac{k^2}{(k^2-\kappa^2)}\left[\frac{\mathrm{i}e^{\mathrm{i}k|\mathbf{r}-\mathbf{y}|\cos\theta}}{k|\mathbf{r}-\mathbf{y}|}\right]_{\theta=0}^{\theta=\pi}\mathrm{d}k,$$
$$= \frac{-\mathrm{i}}{(2\pi)^2|\mathbf{r}-\mathbf{y}|}\int_0^{\infty}\frac{k}{(k^2-\kappa^2)}\left(e^{\mathrm{i}k|\mathbf{r}-\mathbf{y}|}-e^{-\mathrm{i}k|\mathbf{r}-\mathbf{y}|}\right)\mathrm{d}k,$$
$$= \frac{-\mathrm{i}}{(2\pi)^2|\mathbf{r}-\mathbf{y}|}\int_{-\infty}^{\infty}\frac{k}{(k^2-\kappa^2)}e^{\mathrm{i}k|\mathbf{r}-\mathbf{y}|}\,\mathrm{d}k. \tag{187}$$

Following an identical approach to that adopted for the previous example, we avoid the poles at $k=\pm\kappa$ by indenting the path of integration to pass below the pole at $k=\kappa$ and above the pole at $k=-\kappa$. Here, though,

$$\mathrm{Res}(\pm\kappa) = \lim_{k\to\pm\kappa}\left[\frac{k(k\mp\kappa)}{k^2-\kappa^2}e^{\mathrm{i}k|\mathbf{r}-\mathbf{y}|}\right] = \tfrac{1}{2}e^{\pm\mathrm{i}\kappa|\mathbf{r}-\mathbf{y}|}. \tag{188}$$

Since $|\mathbf{r}-\mathbf{y}|>0$ we can close the contour by a large semicircle in the upper half-plane, where the integrand decays exponentially. Only the pole at $k=\kappa$ is enclosed by the contour, and hence

$$\int_{C_R}\frac{k}{(k^2-\kappa^2)}e^{\mathrm{i}k|\mathbf{r}-\mathbf{y}|}\mathrm{d}k = 2\pi\mathrm{i}\,\mathrm{Res}(\kappa) = 2\pi\mathrm{i}\times\tfrac{1}{2}e^{\pm\mathrm{i}\kappa|\mathbf{r}-\mathbf{y}|}. \tag{189}$$

Therefore

$$G_R(\mathbf{r},\mathbf{y},\omega) = \frac{-\mathrm{i}}{(2\pi)^2|\mathbf{r}-\mathbf{y}|}\left(\pi\mathrm{i}e^{\mathrm{i}\kappa|\mathbf{r}-\mathbf{y}|}\right) = \frac{1}{4\pi|\mathbf{r}-\mathbf{y}|}e^{\mathrm{i}\kappa|\mathbf{r}-\mathbf{y}|}. \tag{190}$$

Example 3.16 Obtain the 3-D Green's function that satisfies

$$-\left(\nabla^2+\kappa^2\right)G_R(\mathbf{r},\mathbf{y}) = \delta(\mathbf{r}-\mathbf{y}) = \delta(x-x_s)\delta(y-y_s)\delta(z-z_s), \quad \text{in } z>0, \tag{191}$$

where here $\mathbf{y}=(x_s,y_s,z_s)$ and $z_s>0$. In addition we must satisfy the Sommerfeld radiation condition of no incoming waves, and at the solid interface

$$\frac{\partial G_R}{\partial z} = 0 \qquad \text{at } z=0. \tag{192}$$

Solution (i) The simplest approach is to recall that

$$G_R(\mathbf{r},\mathbf{y}) = \frac{1}{4\pi}\frac{e^{\mathrm{i}\kappa|\mathbf{r}-\mathbf{y}|}}{|\mathbf{r}-\mathbf{y}|} + H(\mathbf{r},\mathbf{y}), \tag{193}$$

where

$$\left(\nabla^2+\kappa^2\right)H(\mathbf{r},\mathbf{y}) = 0 \qquad \text{in} \qquad z>0, \tag{194}$$

due to the presence of a monopole source located at $\mathbf{y}$. Suppose we consider the presence of an **image source** located at $\mathbf{y}^\star = (x_s, y_s, -z_s)$, then we note that

$$G_R(\mathbf{r},\mathbf{y}) = \frac{1}{4\pi}\left\{\frac{e^{i\kappa|\mathbf{r}-\mathbf{y}|}}{|\mathbf{r}-\mathbf{y}|} + \frac{e^{i\kappa|\mathbf{r}-\mathbf{y}^\star|}}{|\mathbf{r}-\mathbf{y}^\star|}\right\} = \frac{1}{4\pi}\left\{\frac{e^{i\kappa R}}{R} + \frac{e^{i\kappa R^\star}}{R^\star}\right\}, \tag{195}$$

where $R = |\mathbf{r}-\mathbf{y}|$ and $R^\star = |\mathbf{r}-\mathbf{y}^\star|$. This solution does indeed satisfy the equation and radiating condition. Moreover we can show that

$$\frac{\partial G_R}{\partial z} = -\frac{1}{4\pi}\left\{(z-z_s)\left(\frac{1}{R^3} - \frac{i\kappa}{R^2}\right)e^{i\kappa R} + (z+z_s)\left(\frac{1}{R^{\star 3}} - \frac{i\kappa}{R^{\star 2}}\right)e^{i\kappa R^\star}\right\}. \tag{196}$$

Further on $z = 0$ we have that

$$R = R^\star = \sqrt{(x-x_s)^2 + (y-y_s)^2 + z_s^2} \qquad \text{so} \qquad \frac{\partial G_R}{\partial z} = 0. \tag{197}$$

(ii) Alternatively we can use Fourier transforms but, due to the presence of the interface at $z = 0$, we shall only use the 2-D Fourier transform, namely

$$\overline{G}(\mathbf{k}, z; \mathbf{y}) = \frac{1}{(2\pi)^2}\iint_{-\infty}^{\infty} G_R(\mathbf{r},\mathbf{y})e^{-i\mathbf{k}\cdot\mathbf{r}}\,dx\,dy, \tag{198}$$

where $\mathbf{k} = (k_1, k_2, 0)$ so that $k = |\mathbf{k}| = k_1^2 + k_2^2$. Thus the transformed equation is

$$\frac{d^2\overline{G}}{dz^2} + \left(\kappa^2 - k^2\right)\overline{G} = -\delta(z-z_s)e^{-i\mathbf{k}\cdot\mathbf{y}} \qquad \text{in} \qquad z > 0, \tag{199}$$

subject to $\partial\overline{G}/\partial z = 0$ on $z = 0$, and outgoing wave behaviour. Solving for $\overline{G}$ in $z > z_s$ and $z < z_s$ separately then requires imposing the conditions

$$\left[\overline{G}\right]_{z_s-}^{z_s+} = 0, \qquad \left[\frac{\partial \overline{G}}{\partial z}\right]_{z_s-}^{z_s+} = -e^{-i\mathbf{k}\cdot\mathbf{y}}. \tag{200}$$

The first condition ensures continuity and the second comes from an integration of the equation for $\overline{G}$ from $z_s - \epsilon$ to $z_s + \epsilon$ in the limit as $\epsilon \to 0+$. A solution satisfying the condition of outgoing waves is given by

$$\overline{G} = \begin{cases} A_u e^{i[\gamma z - \mathbf{k}\cdot\mathbf{y}]} & \text{for } z > z_s, \\ A_d\left\{e^{-i\gamma z} + U e^{i\gamma z}\right\}e^{-i\mathbf{k}\cdot\mathbf{y}} & \text{for } 0 < z < z_s, \end{cases} \tag{201}$$

where A_u, A_d and U are constants and

$$\gamma = \sqrt{\kappa^2 - k^2} = \sqrt{\kappa^2 - (k_1^2 + k_2^2)}. \tag{202}$$

The conditions at $z = z_s$ yield

$$A_u = A_d\left[U + \mathrm{e}^{-2\mathrm{i}\gamma z_s}\right] = A_d\left[U - \mathrm{e}^{-2\mathrm{i}\gamma z_s}\right] + \frac{\mathrm{i}}{2\gamma}\,\mathrm{e}^{-\mathrm{i}\gamma z_s}. \tag{203}$$

This leads to the result

$$\overline{G} = \begin{cases} \dfrac{\mathrm{i}}{2\gamma}\left[U\mathrm{e}^{\mathrm{i}\gamma z_s} + \mathrm{e}^{-\mathrm{i}\gamma z_s}\right]\mathrm{e}^{\mathrm{i}[\gamma z - \mathbf{k}\cdot\mathbf{y}]} & \text{for } z > z_s, \\ \dfrac{\mathrm{i}}{2\gamma}\,\mathrm{e}^{\mathrm{i}\gamma z_s}\left\{\mathrm{e}^{-\mathrm{i}\gamma z} + U\mathrm{e}^{\mathrm{i}\gamma z}\right\}\mathrm{e}^{-\mathrm{i}\mathbf{k}\cdot\mathbf{y}} & \text{for } 0 < z < z_s, \end{cases} \tag{204}$$

which we can combine into a single formula

$$\overline{G} = \frac{\mathrm{i}}{2\gamma}\left\{\mathrm{e}^{\mathrm{i}\gamma|z-z_s|} + U\mathrm{e}^{\mathrm{i}\gamma(z+z_s)}\right\}\mathrm{e}^{-\mathrm{i}\mathbf{k}\cdot\mathbf{y}}, \tag{205}$$

where the condition at $z = 0$ requires $U = -1$. Thus we have

$$\begin{aligned} G_R &= \frac{1}{4\pi^2}\iint_{-\infty}^{\infty} \overline{G}\mathrm{e}^{\mathrm{i}\mathbf{k}\cdot\mathbf{r}}\,\mathrm{d}k_1\,\mathrm{d}k_2 \\ &= \frac{1}{4\pi^2}\iint_{-\infty}^{\infty} \frac{\mathrm{i}}{2\gamma}\left\{\mathrm{e}^{\mathrm{i}\gamma|z-z_s|} - \mathrm{e}^{\mathrm{i}\gamma(z+z_s)}\right\}\mathrm{e}^{\mathrm{i}\mathbf{k}\cdot(\mathbf{r}-\mathbf{y})}\,\mathrm{d}k_1\,\mathrm{d}k_2. \end{aligned} \tag{206}$$

This is a difference of two Sommerfeld integrals, whose evaluation is

$$\frac{1}{4\pi^2}\iint_{-\infty}^{\infty} \frac{\mathrm{i}}{2\gamma}\mathrm{e}^{\mathrm{i}[k_1(x-x_s)+k_2(y-y_s)+\gamma|z-z_s|]}\,\mathrm{d}k_1\,\mathrm{d}k_2 = \frac{1}{4\pi}\frac{\mathrm{e}^{\mathrm{i}\kappa|\mathbf{r}-\mathbf{y}|}}{|\mathbf{r}-\mathbf{y}|}, \tag{207}$$

$$\frac{1}{4\pi^2}\iint_{-\infty}^{\infty} \frac{\mathrm{i}}{2\gamma}\mathrm{e}^{\mathrm{i}[k_1(x-x_s)+k_2(y-y_s)+\gamma(z+z_s)]}\,\mathrm{d}k_1\,\mathrm{d}k_2 = \frac{1}{4\pi}\frac{\mathrm{e}^{\mathrm{i}\kappa|\mathbf{r}-\mathbf{y}^\star|}}{|\mathbf{r}-\mathbf{y}^\star|}, \tag{208}$$

which recovers the solution we obtained earlier using image sources.

3.6.3 *Time-domain Green's function*

The time-domain *Green's function* $G(\mathbf{r}, \mathbf{y}, t, \tau) = G(\mathbf{r} - \mathbf{y}, t - \tau)$ for the wave equation satisfies

$$\left(\frac{1}{c_0^2}\frac{\partial^2}{\partial t^2} - \nabla^2\right)G(\mathbf{r}, \mathbf{y}, t, \tau) = \delta(\mathbf{r} - \mathbf{y})\delta(t - \tau), \tag{209}$$

where the right-hand side represents an impulsive point force at $\mathbf{r} = \mathbf{y}$ that vanishes except at time $t = \tau$. We can define a time–Fourier (Laplace) transform by

$$\hat{f}(\mathbf{r},\omega) = \int_{-\infty}^{\infty} f(\mathbf{r},t)\mathrm{e}^{\mathrm{i}\omega t}\mathrm{d}t \quad \text{where} \quad f(\mathbf{r},t) = \frac{1}{2\pi}\int_{-\infty+\mathrm{i}\epsilon}^{\infty+\mathrm{i}\epsilon} \hat{f}(\mathbf{r},\omega)\mathrm{e}^{-\mathrm{i}\omega t}\mathrm{d}\omega, \tag{210}$$

then Eq. (209) is transformed into

$$-\left(\nabla^2+\lambda^2\right)\hat{G}(\mathbf{r},\mathbf{y},\omega,\tau) = \delta(\mathbf{r}-\mathbf{y})\mathrm{e}^{\mathrm{i}\omega\tau}. \tag{211}$$

Thus we see that $\overline{G}(\mathbf{r},\mathbf{y},\omega,\tau) = G_R(\mathbf{r},\mathbf{y},\omega)\mathrm{e}^{\mathrm{i}\omega\tau}$ and correspondingly

$$G(\mathbf{r},\mathbf{y},t,\tau) = \frac{1}{(2\pi)^4}\int_{-\infty+\mathrm{i}\epsilon}^{\infty+\mathrm{i}\epsilon}\left(\iiint_{-\infty}^{\infty}\frac{\mathrm{e}^{\mathrm{i}\mathbf{k}\cdot(\mathbf{r}-\mathbf{y})-\mathrm{i}\omega(t-\tau)}}{k^2-\lambda^2}\,\mathrm{d}k_1\,\mathrm{d}k_2\,\mathrm{d}k_3\right)\mathrm{d}\omega. \tag{212}$$

Problems

Exercise 3.1 Show the inverse Laplace transform of the function

$$\hat{f}(s) = s^{-a}, \qquad \text{where } 0 < a < 1,$$

where we take a branch cut along the negative real s-axis to account for the branch point at $s = 0$, is given by

$$\mathrm{H}(t)f(t) = \frac{\sin(a\pi)}{\pi}t^{a-1}\Gamma(1-a),$$

where $\Gamma(z)$ is the **Gamma function** defined by

$$\Gamma(z) = \int_0^\infty r^{z-1}\mathrm{e}^{-r}\,\mathrm{d}r \quad \text{where} \quad \Gamma(0) = \int_0^\infty \mathrm{e}^{-r}\,\mathrm{d}r = \left[-\mathrm{e}^{-r}\right]_0^\infty = 1.$$

Moreover for integer $n > 0$ $\Gamma(n+1) = n!$, the factorial function. Indeed upon integration by parts we see for $\mathrm{Re}(z) > 0$

$$\Gamma(z+1) = \int_0^\infty r^z\mathrm{e}^{-r}\,\mathrm{d}r = \left[-r^z\mathrm{e}^{-r}\right]_0^\infty + z\int_0^\infty r^{z-1}\mathrm{e}^{-r}\,\mathrm{d}r = z\Gamma(z)$$

Exercise 3.2 Solve the following model for sound propagation near to outdoor ground surfaces, by describing the field due to a point source, located at the point $\mathbf{y} = (0,0,z_s)$, in a semi-infinite fluid in $z > 0$, with a semi-infinite porous medium in $z < 0$.

In the upper medium the equation for the acoustic pressure, p

$$\left(\nabla^2+\kappa^2\right)G(\mathbf{r}) = -\delta(x)\delta(y)\delta(z-z_s) \qquad \text{in } z > 0,$$

where $\kappa = \omega/c_0$ is the acoustic wavenumber, and here the time-dependent factor $e^{-i\omega t}$ is understood but suppressed throughout. In the lower medium, the governing equation for the acoustic pressure is

$$\left(\nabla^2 + \kappa_1^2\right) G_1 = 0 \qquad \text{in } z < 0,$$

The boundary conditions at the plane interface are

$$G = G_1 \qquad \text{and} \qquad \frac{1}{\rho}\frac{\partial G}{\partial z} = \frac{1}{\rho_1}\frac{\partial G_1}{\partial z} \qquad \text{at } z = 0,$$

and the sound field should contain no incoming waves when $z \to \pm\infty$. To solve this problem use the 2-D Fourier transform pair

$$\overline{G}(k_1, k_2, z) = \iint_{-\infty}^{\infty} G(x, y, z) e^{-i(k_1 x + k_2 y)}\, dx\, dy.$$

Chapter 4

Asymptotic Expansion of Integrals

R. H. Self
Institute of Sound and Vibration Research
University of Southampton

4.1 Introduction

Many problems in acoustics are posed in the form of differential equations and, if the equation is a simple one then a closed form solution may be possible. However, this is not generally the case and we must be content with an integral representation of the solution. Often such integral representations arise through the use of integral transforms and we are naturally led to consider integrals of the following forms:

$$I(x) = \int_0^\infty \mathrm{e}^{-xf(t)} g(t)\,\mathrm{d}t, \qquad (1)$$

and

$$I(x) = \int_{-\infty}^\infty \mathrm{e}^{\mathrm{i}xf(t)} g(t)\,\mathrm{d}t. \qquad (2)$$

The first of these arises when Laplace transforms have been used and might naturally be called a Laplace integral. Similarly, the second integral is a type of generalised Fourier integral.

If it is possible to find an exact analytic form for the integral then the problem is (rather obviously) solved, but if such a form cannot be found then a different way of proceeding must be sought. While a numerical solution might always be available for specific cases, we often wish to consider what happens for a whole range of parameter values and this leads us to consider approximate analytic solutions. The above integrals have been written in a way which explicitly indicates their dependence on a parameter x, and the methods described below will consider approximate solutions where this parameter becomes very large (i.e. $x \to \infty$). Many problems which arise in acoustics can be posed in such a way.

Asymptotic methods seek to find the dominant behaviour of the solution to problems for regions where a variable or parameter tends to some limit (finite or infinite). We begin by considering order symbols and what is meant by an asymptotic expansion.

Given two real valued functions $f(x)$ and $g(x)$ defined in some neighbourhood of $x = x_0$ we say that:

$$f(x) = o(g(x)) \qquad \text{as } x \to x_0, \qquad \text{if} \qquad \lim_{x \to x_0} \frac{f(x)}{g(x)} = 0. \tag{3}$$

This is pronounced as 'f is little o of g as x tends to x_0'. We also define a relation 'f is big O of g as x tends to x_0', written in symbols as

$$f(x) = O(g(x)) \qquad \text{as } x \to x_0, \tag{4}$$

and which means that there exists a constant, M, such that $|f(x)| < M|g(x)|$ for x 'close' to x_0 (i.e. $|x - x_0|$ near 0).

We also define the important symbol $\sim$. For two functions $f(x)$ and $g(x)$ we write $f(x) \sim g(x)$ as $x \to x_0$ if $\lim_{x \to x_0} |f(x)/g(x)| = 1$.

Note that x_0 may be zero, finite or infinite.

- $f(x) = o(1)$ as $x \to \infty$ means that $f(x) \to 0$ as $x \to \infty$.
- $f(x) = O(1)$ as $x \to \infty$ means that $f(x)$ is bounded.
- If $f(x) = 5x^2 - 7x + 9$ then as $x \to \infty$ we have:

$$f(x) = O(x^2), \qquad f(x) \sim 5x^2, \qquad f(x) = o(x^3). \tag{5}$$

The symbols O, o and $\sim$ allow us to compare functions in terms of ratios rather than in terms of differences. If we use $5x^2$ to approximate the function $5x^2 - 7x + 9$ (as $x \to \infty$), then we have captured the 'dominant' behaviour. Notice that the absolute error involved tends to infinity, while the percentage error tends to zero. This is typical of asymptotic approximations.

A sequence $\{g_n(x)\}$ is an **asymptotic sequence** as $x \to x_0$ if for all n:

$$g_{n+1}(x) = o(g_n(x)) \qquad \text{as } x \to x_0. \tag{6}$$

Likewise, a series $\sum_{n=0}^{N} a_n g_n(x)$ is an **asymptotic expansion** of the function $f(x)$ as $x \to x_0$ if

$$\left| f(x) - \sum_{n=0}^{N} a_n g_n(x) \right| = o(g_N(x)) \qquad \text{as } x \to x_0. \tag{7}$$

If $\sum_{n=1}^{N} a_n g_n(x)$ is an asymptotic expansion of $f(x)$ then

$$f(x) \sim \sum_{n=1}^{N} a_n g_n(x). \tag{8}$$

In the remainder of this chapter we will consider methods for obtaining asymptotic expansions to integrals. Generally we will be content with finding only the first, or leading order, term in such series.

4.2 Two Elementary Methods

Two simple techniques that can often be used involve expanding the integrand (or part of the integrand) in a series, or the repeated use of integration by parts. We illustrate these methods by example.

4.2.1 *Term by term integration*

Consider the integral

$$I(x) = \int_0^\infty \frac{\mathrm{e}^{-xt}}{1+t}\,\mathrm{d}t, \qquad \text{as } x \to +\infty. \tag{9}$$

The integrand is small everywhere apart from t close to 0. Since

$$\frac{1}{1+t} = \sum_{n=0}^{\infty}(-1)^n t^n, \qquad \text{for } t < 1, \tag{10}$$

we might try and *formally* expand the integrand:

$$\begin{aligned} I(x) &\sim \int_0^\infty \sum_{n=0}^{\infty}(-1)^n t^n \mathrm{e}^{-xt}\,\mathrm{d}t \\ &\sim \sum_{n=0}^{\infty}\int_0^\infty (-1)^n t^n \mathrm{e}^{-xt}\,\mathrm{d}t \\ &\sim \sum_{n=0}^{\infty}(-1)^n \frac{n!}{x^{n+1}}. \end{aligned} \tag{11}$$

Notice that this is an example of an asymptotic series that diverges. The result makes sense, however, because for $x \to \infty$ the dominant contribution clearly comes from the region close to $t = 0$. The result can also be obtained by repetitive use of integration by parts.

4.2.2 *Integration by parts*

Consider the integral

$$I(x) = \int_0^x e^{-t^2}\, dt, \qquad \text{as } x \to \infty. \tag{12}$$

Clearly $I(x) \to \sqrt{\pi}/2$ as $x \to \infty$ so write:

$$I(x) = \frac{\sqrt{\pi}}{2} - \int_x^\infty e^{-t^2}\, dt \tag{13}$$

and use integration by parts:

$$\begin{aligned}\int_x^\infty e^{-t^2}\, dt &= \int_x^\infty -2t\, e^{-t^2} \frac{dt}{-2t} \\ &= \left[\frac{e^{-t^2}}{-2t}\right]_x^\infty - \int_x^\infty \frac{e^{-t^2}}{2t^2}\, dt \\ &= \frac{e^{-x^2}}{2x} - \int_x^\infty \frac{e^{-t^2}}{2t^2}\, dt.\end{aligned} \tag{14}$$

Since

$$\left|\int_x^\infty \frac{e^{-t^2}}{2t^2}\, dt\right| \leq \frac{1}{4x^3}\left|\int_x^\infty -2te^{-t^2}\, dt\right| = \frac{e^{-x^2}}{4x^3}, \tag{15}$$

we have

$$I(x) = \frac{\sqrt{\pi}}{2} - \frac{e^{-x^2}}{2x} + O\left(\frac{e^{-x^2}}{4x^3}\right), \tag{16}$$

or

$$I(x) = \frac{\sqrt{\pi}}{2} + e^{-x^2}\left(-\frac{1}{x^2} + O\left(\frac{1}{4x^3}\right)\right). \tag{17}$$

4.2.3 *Watson's lemma*

Watson's lemma is an important tool in the analysis of Laplace type integrals:

$$I(x) = \int_0^\infty e^{-xt} f(t)\, dt, \tag{18}$$

where a solution is sought in the limit $x \to \infty$.

Consider a graph of $\exp(-xt)$ as shown in Fig. 4.1. As x increases the integrand will be close to zero everywhere except near $t = 0$, and thus the main contribution to the integral comes from the region close to $t = 0$.

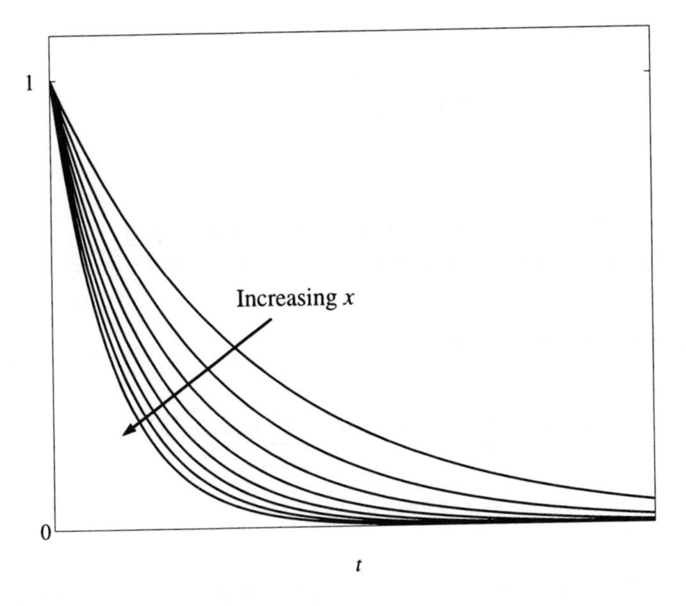

Fig. 4.1 Graphs of e^{-xt} against t for varying x.

Example 4.1 Consider the integral

$$I(x) = \int_0^\infty e^{-xt} \sin t \, dt. \tag{19}$$

Solution The exact solution is

$$I(x) = \frac{1}{1+x^2} \tag{20}$$

and this can be expanded to give

$$I(x) = \frac{1}{x^2} + \frac{1}{x^4} + \frac{1}{x^6} + \frac{1}{x^8} + \cdots . \tag{21}$$

We can also obtain an approximate solution to the integral directly by noticing that, near $t = 0$, $\sin t \sim t$. If we approximate the integral by

$$I(x) \sim \int_0^\infty t e^{-xt} \, dt, \tag{22}$$

then integration by parts recovers the correct leading order behaviour $I(x) \sim 1/x^2$.

This type of argument can be made more rigorous through Watson's lemma.

Lemma 4.1 *Let*

$$I(x) = \int_0^\infty \mathrm{e}^{-xt} f(t)\,\mathrm{d}t, \tag{23}$$

then if

(i) the integral $\int_0^\infty \mathrm{e}^{-\alpha t}|f(t)|\,\mathrm{d}t$ *exists for some* $\alpha > 0$ *and if*
(ii) $f(t)$ *has the asymptotic expansion* $f(t) \sim \sum_{n=1}^\infty a_n t^{\lambda_n - 1}$ *as* $t \to 0$ *and* $0, \lambda_1 < \lambda_2 < \ldots$

then ***Watson's lemma*** *states that:*

$$I(x) = \int_0^\infty \mathrm{e}^{-xt} f(t)\,\mathrm{d}t \sim \sum_{n=0}^N a_n \Gamma(\lambda_n) x^{-\lambda_n}, \tag{24}$$

where the Gamma function is defined by $\Gamma(\lambda) = \int_0^\infty \mathrm{e}^{-t} t^{\lambda-1}\,\mathrm{d}t$ *and* $\Gamma(n+1) = n!$ *for integer* n.

A special case of this result is when $f(t)$ can be expanded as a power series about $t = 0$ in which case:

$$I(x) = \int_0^\infty \mathrm{e}^{-xt} f(t)\,\mathrm{d}t \sim \sum_{k=0}^\infty \frac{f^{(k)}(0)}{x^{k+1}}. \tag{25}$$

Proof. For $x \to \infty$ the dominant contribution to the integral is from the region close to $t = 0$. This motivates us to split the integral in two by writing:

$$I(x) = \int_0^\eta \mathrm{e}^{-xt} f(t)\,\mathrm{d}t + \int_\eta^\infty \mathrm{e}^{-xt} f(t)\,\mathrm{d}t = I_1(x) + I_2(x) \tag{26}$$

where η is a small positive number. Now:

$$\begin{aligned} |I_2(x)| &= |\int_\eta^\infty \mathrm{e}^{-xt} f(t)\,\mathrm{d}t| \leq \int_\eta^\infty \mathrm{e}^{-xt}|f(t)|\,\mathrm{d}t \\ &= \int_\eta^\infty \mathrm{e}^{-(x-\alpha)t}\mathrm{e}^{-\alpha t}|f(t)|\,\mathrm{d}t \end{aligned} \tag{27}$$

and this gives

$$|I_2(x)| \leq \mathrm{e}^{-(x-\alpha)\eta} \int_\eta^\infty \mathrm{e}^{-\alpha t}|f(t)|\,\mathrm{d}t \leq \mathrm{e}^{-(x-\alpha)t} M \tag{28}$$

for some number $M > 0$. Hence we have proved that $|I_2(x)|$ is exponentially small and negligible compared to any power of x.

Now consider the first integral and use the asymptotic approximation for $f(t)$:

$$f(t) = \sum_{n=1}^{N} a_n t^{\lambda_n - 1} + \varepsilon_N, \qquad (\text{so} \quad \varepsilon_N = o(t^{\lambda_N - 1})). \tag{29}$$

Then

$$I_1(x) = \int_0^{\eta} \mathrm{e}^{-xt} \left\{ \sum_{n=1}^{N} a_n t^{\lambda_n - 1} \right\} \mathrm{d}t + E_N, \qquad E_N = \int_0^{\eta} \mathrm{e}^{-xt} \varepsilon_N \, \mathrm{d}t. \tag{30}$$

By making the substitution $u = xt$ we obtain:

$$\begin{aligned} I_1(x) &= \int_0^{x\eta} \mathrm{e}^{-u} \sum_{n=1}^{N} a_n \left(\frac{u}{x}\right)^{\lambda_n - 1} \frac{\mathrm{d}u}{x} + E_N \\ &= \sum_{n=1}^{N} a_n x^{\lambda_n} \int_0^{x\eta} \mathrm{e}^{-u} u^{\lambda_n - 1} \, \mathrm{d}u + E_N. \end{aligned} \tag{31}$$

and

$$E_N = \int_0^{\eta} \mathrm{e}^{-xt} \varepsilon_N \, \mathrm{d}t \leq \int_0^{\eta} \mathrm{e}^{-xt} \beta t^{\lambda_N - 1} \, \mathrm{d}t$$

for some β since $\varepsilon_N = o(t^{\lambda_N - 1})$

$$\begin{aligned} &= \int_0^{x\eta} \mathrm{e}^{-u} \beta \left(\frac{u}{x}\right)^{\lambda_N - 1} \frac{\mathrm{d}u}{x} \\ &= \frac{\beta}{x^{\lambda_N}} \int_0^{x\eta} \mathrm{e}^{-u} u^{\lambda_N - 1} \, \mathrm{d}u. \end{aligned} \tag{32}$$

By writing

$$\int_0^{x\eta} \equiv \int_0^{\infty} - \int_{x\eta}^{\infty} \tag{33}$$

we can now express $I_1(x)$ as

$$I_1(x) = \sum_{n=1}^{N} a_n x^{-\lambda_n} \left\{ \int_0^{\infty} \mathrm{e}^{-u} u^{\lambda_n - 1} \, \mathrm{d}u - \int_{x\eta}^{\infty} \mathrm{e}^{-u} u^{\lambda_n - 1} \, \mathrm{d}u \right\} + E_N \tag{34}$$

and where we note that the second integral is asymptotically small because

$$\left| \int_{x\eta}^{\infty} \mathrm{e}^{-u} u^{\lambda_n - 1} \, \mathrm{d}u \right| \leq \left| \mathrm{e}^{-x\eta/2} \int_{x\eta}^{\infty} \mathrm{e}^{-u/2} u^{\lambda_n - 1} \, \mathrm{d}u \right|. \tag{35}$$

Similarly we find that E_N is a sum of an integral of order $x^{-\lambda_N}$ and one which is asymptotically small. Hence we have shown that:

$$I(x) = I_1(x) + I_2(x) = \sum_{n=1}^{N} a_n x^{-\lambda_n} \int_0^\infty \mathrm{e}^{-u} u^{\lambda_n - 1}\,\mathrm{d}u + \text{e.s.t.} + O(x^{-\lambda_N}), \tag{36}$$

i.e.

$$I(x) \sim \sum_{n=1}^{N} a_n x^{-\lambda_n} \Gamma(\lambda_n). \tag{37}$$

□

Example 4.2 Consider again the example

$$I(x) = \int_0^\infty \mathrm{e}^{-xt} \sin t\,\mathrm{d}t \tag{38}$$

as $x \to \infty$.

Solution Since $\sin t = t - t^3/3! + t^5/5! + \ldots$ we have $a_n = (-1)^{n+1}/(2n-1)!$, $\lambda_n = 2n$ and Watson's lemma gives:

$$\begin{aligned} I(x) &\sim a_1\Gamma(\lambda_1)x^{-\lambda_1} + a_2\Gamma(\lambda_2)x^{-\lambda_2} + \ldots \\ &= \Gamma(2)x^{-2} - \frac{1}{3!}\Gamma(4)x^{-4} + \ldots \\ &= \frac{1}{x^2} - \frac{1}{x^4} + \cdots, \end{aligned} \tag{39}$$

which agrees with the approximation found above.

Example 4.3 Consider the behaviour of

$$I(x) = \int_0^\infty \mathrm{e}^{-xt^3}(1+t)\,\mathrm{d}t \tag{40}$$

as $x \to \infty$.

Solution By introducing the new variable $u = t^3$ we obtain

$$I(x) = \int_0^\infty \mathrm{e}^{-xu} \frac{(1+u^{1/3})}{3u^{2/3}}\,\mathrm{d}u \tag{41}$$

and Watson's lemma then gives $I(x) \sim \frac{1}{3}\Gamma(\frac{1}{3})x^{-1/3} + \frac{1}{3}\Gamma(\frac{2}{3})x^{-2/3}$.

4.3 Method of Laplace

In the last example a simple substitution was used to reduce the integral to one where Watson's lemma was applicable. Sometimes substitutions are hard to find or do not work and in these cases Laplace's method is useful. Thus we look at integrals of the form

$$I(x) = \int_a^b f(t)\mathrm{e}^{xh(t)}\,\mathrm{d}t, \tag{42}$$

and seek the asymptotic behaviour as $x \to \infty$.

If $h(t)$ has a maximum then we may expect the major contribution to the integral to come from a subregion surrounding this maximum. The essence of Laplace's method is to replace $h(t)$ by the first few terms of its Taylor series about this maximum. Suppose then that $h(t)$ obtains its maximum value at $t = c$. There are three cases to consider. If $c = a$ or $c = b$ then we can have $h'(c) \neq 0$ and $h(t)$ is approximated by

$$h(c) + (t - c)h'(c). \tag{43}$$

If $a \leq c \leq b$ and $h'(c) = 0$, $h''(c) \neq 0$ then approximate $h(t)$ by

$$h(c) + \frac{(t-c)^2}{2}h''(c), \tag{44}$$

or more generally by

$$h(c) + \frac{(t-c)^n}{n!}h^{(n)}(c), \tag{45}$$

when $h^{(n)}(c)$ is the first non-vanishing derivative. In each case we obtain an integral to which Watson's lemma can be applied after a suitable substitution.

Consider, for example, the second of these cases. We make the approximation

$$\begin{aligned} I(x) &\sim \int_a^b f(c)\mathrm{e}^{x(h(c)+\frac{1}{2}(t-c)^2h''(c))}\,\mathrm{d}t, \\ &\sim \mathrm{e}^{xh(c)}\int_a^b f(c)\mathrm{e}^{x(t-c)^2h''(c)/2}\,\mathrm{d}t. \end{aligned} \tag{46}$$

Now extend the range of integration to the whole of the real line, (since we have $h''(c) < 0$ this change leads only to exponentially small terms), split the range of integration at $t = c$, and make the substitution

$$u = -\tfrac{1}{2}h''(c)(t-c)^2 \tag{47}$$

to give

$$I(x) \sim \mathrm{e}^{xh(c)} f(c)\sqrt{-2/h''(c)} \int_0^\infty \frac{\mathrm{e}^{-xu}}{u^{1/2}}\,\mathrm{d}u \tag{48}$$

and apply Watson's lemma to obtain

$$I(x) \sim \mathrm{e}^{xh(c)} f(c)\sqrt{-2\pi/h''(c)}. \tag{49}$$

4.4 Method of Stationary Phase

The method of stationary phase is of use when considering generalised Fourier integrals with complex integrands of the form

$$I(x) = \int_a^b f(t)\mathrm{e}^{\mathrm{i}x\phi(t)}\,\mathrm{d}t, \qquad \text{as } x \to \infty. \tag{50}$$

Integrals such as these are governed by a very general result known as the Riemann–Lebesgue lemma. This states that $I(x) \to 0$ as $x \to \infty$ so long as $|f(t)|$ is integrable, $\phi(t)$ is continuously differentiable on $[a, b]$, and $\phi(t)$ is not constant on any part of $[a, b]$. Assuming these conditions are satisfied, the leading order behaviour of $I(x)$ can be found by integrating by parts:

$$\begin{aligned} I(x) &= \int_a^b \frac{f(t)}{\mathrm{i}x\phi'(t)} \frac{\mathrm{d}}{\mathrm{d}t}(\mathrm{e}^{\mathrm{i}x\phi(t)})\,\mathrm{d}t \\ &= \left[\frac{f(t)}{\mathrm{i}x\phi'(t)}\mathrm{e}^{\mathrm{i}x\phi(t)}\right]_a^b - \int_a^b \mathrm{e}^{\mathrm{i}x\phi(t)} \frac{\mathrm{d}}{\mathrm{d}t}\left(\frac{f(t)}{\mathrm{i}x\phi'(t)}\right)\mathrm{d}t. \end{aligned} \tag{51}$$

Both terms are $O(1/x)$ as $x \to \infty$, but a second integration by parts quickly establishes that the second term is actually $O(1/x^2)$. Assuming the boundary terms do not vanish, they give the leading order asymptotic behaviour of the integral.

We will not prove the Riemann–Lebesgue lemma but the result can be understood as arising from a destructive interference. The integrand is an oscillatory function, and the greater x the more rapidly it oscillates, with contributions from neighbouring regions nearly cancelling. The least effective cancellation occurs at the end points of the interval.

If the conditions on $\phi(t)$ are relaxed then integration by parts may not work because other subregions may exist where the cancellation is inefficient and these will also contribute (or dominate) the leading order asymptotic behaviour of the integral. Consider the integral

$$I(x) = \int_a^b (1+t^2)\mathrm{e}^{\mathrm{i}kt^2}\,\mathrm{d}t. \tag{52}$$

The graph of the real part of the integrand about the origin is shown in Fig. 4.2 for the case $k = 50$.

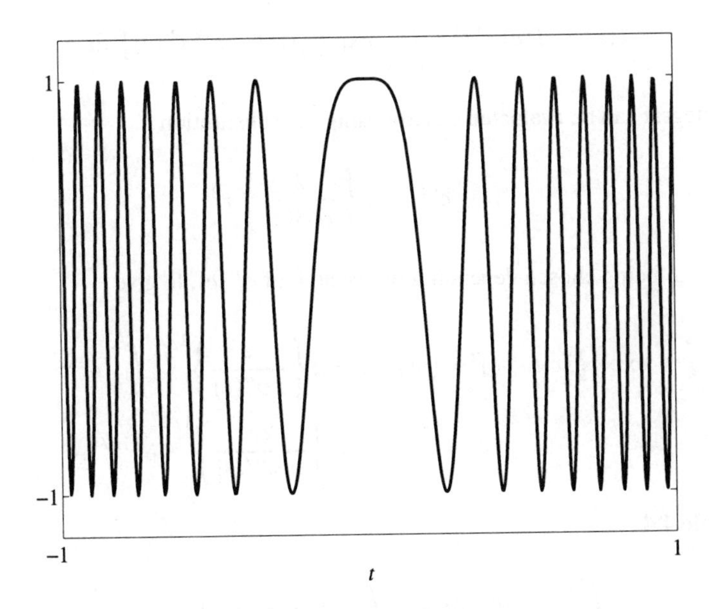

Fig. 4.2 Graph of $\mathrm{Re}\{\mathrm{e}^{\mathrm{i}kt^2}\}$ against t for $k = 50$.

Because the phase of the integrand kt^2 has a minimum at $t = 0$, the real part varies slowly at the origin, cancellation is inefficient, and this region contributes significantly to the asymptotic behaviour of the integral.

If there is only one point c for which $a \leq c \leq b$ and $\phi'(c) = 0$ then so long as $\phi''(c) \neq 0$ the method of stationary phase states that

$$I(x) \sim f(c)\left(\frac{2\pi}{x|\phi''(c)|}\right)^{1/2} \mathrm{e}^{(\mathrm{i}x\phi(c) \pm \mathrm{i}\pi/4)} \qquad \text{as } x \to \infty, \tag{53}$$

with the + or − sign according as $\phi''(c) > 0$ or $\phi''(c) < 0$.

4.4.1 *Outline proof*

We expect that the major contribution comes from the region near c so extend the region of integration to the whole of the real line and make the approximations:

$$f(t) \sim f(c), \qquad \phi(t) \sim \phi(c) + \tfrac{1}{2}(t-c)^2\phi''(c) \tag{54}$$

giving

$$I(x) \sim f(c)\mathrm{e}^{\mathrm{i}x\phi(c)} \int_{-\infty}^{\infty} \exp\left\{\tfrac{1}{2}\mathrm{i}x(t-c)^2\phi''(c)\right\} \,\mathrm{d}t. \tag{55}$$

This integral can be evaluated exactly using the substitution

$$(t-c)^2 = \pm\left\{\frac{2}{x\phi''(c)}\right\} x^2 \tag{56}$$

with the $\pm$ being chosen depending on the sign of $\phi''(c)$, to give

$$\begin{aligned}\int_{-\infty}^{\infty} \exp\left\{\tfrac{1}{2}\mathrm{i}x(t-c)^2\phi''(c)\right\} \,\mathrm{d}t &= 2\left\{\frac{2}{x\phi''(c)}\right\}^{1/2} \int_0^{\infty} \mathrm{e}^{\pm\mathrm{i}x^2} \,\mathrm{d}x \\ &= \left\{\frac{2\pi}{x|\phi''(c)|}\right\}^{1/2} \mathrm{e}^{\pm\mathrm{i}\pi/4}.\end{aligned} \tag{57}$$

Example 4.4

$$\begin{aligned}\int_{-\pi/2}^{\pi/2} \mathrm{e}^{\mathrm{i}x\cos t} \,\mathrm{d}t &\sim 2\int_0^{\infty} \mathrm{e}^{\mathrm{i}x(1-t^2/2)} \,\mathrm{d}t \\ &= 2\mathrm{e}^{\mathrm{i}x} \int_0^{\infty} \mathrm{e}^{-\mathrm{i}xt^2/2} \,\mathrm{d}t \\ &= 2\mathrm{e}^{\mathrm{i}x}\mathrm{e}^{-\mathrm{i}\pi/4} \int_0^{\infty} \mathrm{e}^{-xu^2/2} \,\mathrm{d}u \\ &= 2\mathrm{e}^{\mathrm{i}x}\mathrm{e}^{-\mathrm{i}\pi/4} \int_0^{\infty} \mathrm{e}^{-s^2} \sqrt{\frac{2}{x}} \,\mathrm{d}s \\ &= \mathrm{e}^{\mathrm{i}x}\mathrm{e}^{-\mathrm{i}\pi/4}\sqrt{\frac{2\pi}{x}}\end{aligned} \tag{58}$$

The result proved here is easily generalised to cases where $\phi^{(p)}(c)$ is the first non-vanishing derivative.

4.5 Method of Steepest Descents

This is a far more general method and encompasses both types of integral we have considered above. It is a technique for finding the asymptotic behaviour of integrals of the form

$$I(x) = \int_C f(z)\mathrm{e}^{x\phi(z)}\,\mathrm{d}z \tag{59}$$

where $f(z)$ and $\phi(z)$ are complex analytic functions, C is a contour in the complex plane, and we are again interested in the limit as (real) $x \to \infty$.

If we decompose $\phi(z)$ as $\phi(z) = u(z) + \mathrm{i}v(z)$ where u and v are the (real valued) real and imaginary parts we may suspect that the major contribution to the integral comes from points close to the maximum value of u along the contour and seek an approximation by applying Laplace's method. However, this might well lead to an overestimate since the imaginary part, $v(z)$, may vary rapidly along the contour, giving rise to rapid oscillations of the integrand and significant cancellation.

The method of steepest descents gets around this difficulty by exploiting the analyticity of the integrand. In such cases the value of the integral depends only upon the end points and not on the contour joining them. If the original contour C is deformed to a contour C' where the phase $v(z)$ varies as little as possible, then an application of Laplace's method will give a far better asymptotic approximation.

Because $\phi(z)$ is analytic, it satisfies the Cauchy–Riemann equations:

$$\frac{\partial u}{\partial x} = \frac{\partial v}{\partial y} \qquad \frac{\partial u}{\partial y} = -\frac{\partial v}{\partial x} \tag{60}$$

and these imply that $\boldsymbol{\nabla} u \cdot \boldsymbol{\nabla} v = 0$. Hence the directional derivative of $v(z)$ along a contour with tangent vector parallel to $\boldsymbol{\nabla} u$ must be zero, such a contour is a contour of constant phase. It is also the contour along which $u(z)$ varies most rapidly, and since $\boldsymbol{\nabla}|\mathrm{e}^{x\phi(z)}| = \boldsymbol{\nabla}\mathrm{e}^{xu(z)}$, it is the contour along which the magnitude of the integrand varies the most rapidly. In other words, it is a contour of steepest descent.

Once the contour C has been deformed to the contour C' which has constant phase, the asymptotic behaviour of the integral is determined by the value of the integrand close to the maximum of $v(z)$ along C'. This maximum may occur at one of the end points but it may occur at a local interior point. Because $\phi(z)$ is assumed analytic, maxima of $v(z)$ must correspond to saddle points of $\phi(z)$. At saddle points, curves of steepest descent can intersect and care must be taken to ensure that a curve of steepest *ascent* is not chosen.

Example 4.5 The Bessel function $\mathrm{J}_0(x)$ can be represented as

$$\mathrm{J}_0(x) = \mathrm{Re}\left\{\frac{1}{\mathrm{i}\pi}\int_{-i\pi/2}^{+\mathrm{i}\pi/2} \exp(\mathrm{i}x\cosh z)\,\mathrm{d}z\right\}, \tag{61}$$

and we seek the asymptotic behaviour as $x \to \infty$.

Solution Here $\phi(z) = \mathrm{i}\cosh(z)$ and $\phi'(z) = \mathrm{i}\sinh(z) = 0$ at $z = 0$. Since

$$\cosh(x + \mathrm{i}y) = -\sinh(x)\sin(y) + \mathrm{i}\cosh(x)\cos(y) \tag{62}$$

the imaginary part is equal to 1 at the origin, and along the contour of constant phase we must have

$$\cosh(x)\cos(y) = 1. \tag{63}$$

We can now apply steepest descents extending the limits of integration to infinity and writing

$$\mathrm{J}_0(x) = \mathrm{Re}\left\{\frac{1}{\mathrm{i}\pi}\int_C \exp(\mathrm{i}x\cosh z)\mathrm{d}z\right\}, \tag{64}$$

where C is any contour joining $-\infty - \mathrm{i}\pi/2$ and $+\infty + \mathrm{i}\pi/2$ and passing through the saddle point at the origin. The gradient, $\mathrm{d}y/\mathrm{d}x$, of the contour of steepest descent is equal to 1 at the origin and we can therefore approximate $z = (1 + \mathrm{i})t$ where t is real. Likewise, $\cosh z \sim 1 + \mathrm{i}t^2$ and

$$\mathrm{J}_0(x) \sim \mathrm{Re}\left\{\frac{1+\mathrm{i}}{\mathrm{i}\pi}\int_{-\infty}^{\infty} \exp(\mathrm{i}x - xt^2)\,\mathrm{d}t\right\}. \tag{65}$$

This integral is easily evaluated to give

$$\mathrm{J}_0(x) \sim \sqrt{2/\pi x}\cos(x - \pi/4), \qquad \text{as } x \to \infty. \tag{66}$$

Further Reading

Leppington, F. G. (1992) Asymptotic Evaluation of Integrals, chapter 4 of *Modern Methods in Analytical Acoustics* by Crighton, D. G. *et al.* Springer Verlag.

Carrier, G. F., Krook, M. and Pearson, C. E. (1966) *Functions of a Complex Variable.* McGraw–Hill.

Problems

Exercise 4.1 Show that as $x \to 0^+$

$$\int_x^\infty e^{-t^2}\,dt \sim \frac{\sqrt{\pi}}{2} + \sum_{n=0}^{\infty} \frac{(-1)^{n+1} x^{2n+1}}{(2n+1)n!}$$

Exercise 4.2 By using term by term integration, show that the elliptic integral

$$K(m) = \int_0^{\pi/2} (1 - m\cos^2\theta)^{-1/2}\,d\theta$$

can be approximated for $m \ll 1$ by

$$K(m) = \frac{\pi}{2}\left(1 + \frac{m}{4} + \frac{9}{64}m^2\right) + O(m^3)$$

Exercise 4.3 Under the change of variable $u = \sin t$ the integral

$$I(x) = \int_0^{\pi/2} \sqrt{\sin t}\, e^{-x(\sin t)^4}\,dt$$

becomes

$$I(x) = \int_0^1 \sqrt{\frac{u}{1-u^2}}\, e^{-xu^4}\,du.$$

Use a further change of variable, $s = u^4$ and apply Watson's lemma to show that

$$I(x) \sim \frac{\Gamma(3/8)}{4x^{3/8}} \qquad \text{as } x \to \infty.$$

Exercise 4.4 Use the method of steepest descent show that

$$H_\nu^{(1)}(\rho) \sim \sqrt{\frac{2}{\pi\rho}} e^{i\rho - i\pi(\nu+1/2)/2}$$

PART II
Wave Motion

Chapter 5

The Wiener–Hopf Technique

M. C. M. Wright

Institute of Sound and Vibration Research
University of Southampton

5.1 Introduction

The Wiener–Hopf technique is, for acousticians, a way to determine wavefields in situations with separable geometry but mixed boundary conditions. Here 'mixed boundary conditions' means different conditions on different regions of the boundary (as opposed to what are otherwise known as Robin boundary conditions, a 'mixture' of Neumann and Dirichlet conditions.) The technique works by applying our knowledge of the behaviour of complex functions to complex Fourier transforms of the wavefield. We shall illustrate the technique by applying it to the problem of diffraction of plane waves by a semi-infinite screen. Similar treatments of the same problem are given by Noble (1988) and Leppington (1992).

5.2 Rigid Screen Diffraction

Consider, then, a rigid screen located at $x < 0$, $y = 0$ extending to infinity in the positive x-direction, all quantities being independent of z. Plane waves with wavenumber k are incident at an angle θ to the screen, as shown in Fig. 5.1. The total velocity potential field is

$$\phi(x, y, t) = \mathrm{Re}[\Phi \mathrm{e}^{-\mathrm{i}\omega t}]. \tag{1}$$

We can write the total field as the sum of the incident and scattered fields

$$\Phi(x, y) = \mathrm{e}^{-\mathrm{i}k(x\cos\theta + y\sin\theta)} + f(x, y), \tag{2}$$

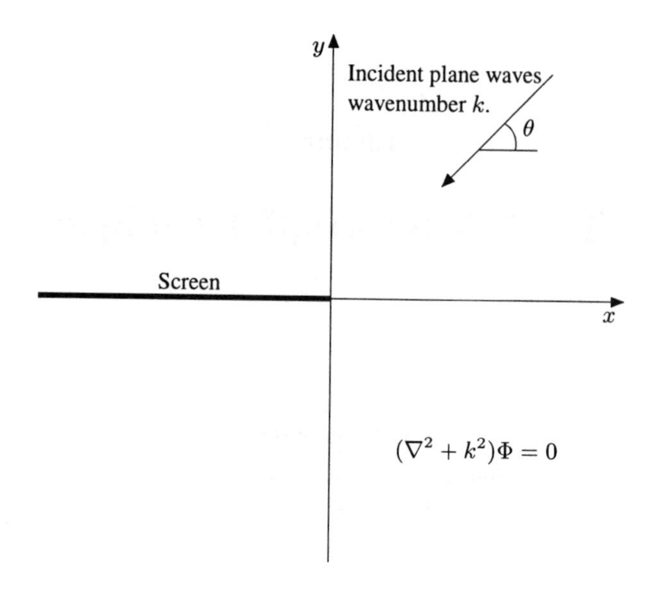

Fig. 5.1 Geometry of the rigid screen diffraction problem.

where f is the scattered field that we wish to find and $k = \omega/c_0$.

Since the total field satisfies the two-dimensional Helmholtz equation

$$\left(\frac{\partial^2}{\partial x^2} + \frac{\partial^2}{\partial y^2}\right)\Phi + k^2\Phi = 0, \tag{3}$$

then so must the scattered field f, for which we must now determine the boundary conditions. On the screen we have the usual vanishing normal velocity so the scattered y-velocity field is equal and opposite to the incident y-velocity field

$$\frac{\partial f}{\partial y} = \mathrm{i}k \sin\theta\, \mathrm{e}^{-\mathrm{i}kx\cos\theta}, \qquad \text{for } x < 0 \text{ and } y = 0. \tag{4}$$

This is not enough, we need a boundary condition for the rest of $y = 0$, which we can get by recalling that the problem of finding the field scattered by a rigid body is equivalent to that of finding the field radiated by "a vibrating body of the same size and shape whose normal velocity is the negative of what is associated with the incident wave" (Pierce, 1981). Since the screen has no thickness, but its normals have opposite signs on opposite sides, the scattered field must have also have opposite signs on opposite sides, in other words $f(x, y)$ is an odd function in y and continuous for $x > 0$, hence

$$f(x, 0) = 0, \qquad \text{for } x > 0. \tag{5}$$

In order to ensure that our waves are physically realistic we let $k \mapsto k + \mathrm{i}\epsilon$, where $\epsilon > 0$ for the time being but we can let it vanish once we are satisfied with our solution. We expect the scattered field to account for the ordinary reflection from the screen away from its edge and the cancellation of the incident field on the other side, and it can be seen that equal and opposite plane wave behaviour on each side is consistent with this. It is perhaps also worth remarking that the problem doesn't have an intrinsic length scale other than the wavelength of the incident wave so there is no need to rescale variables.

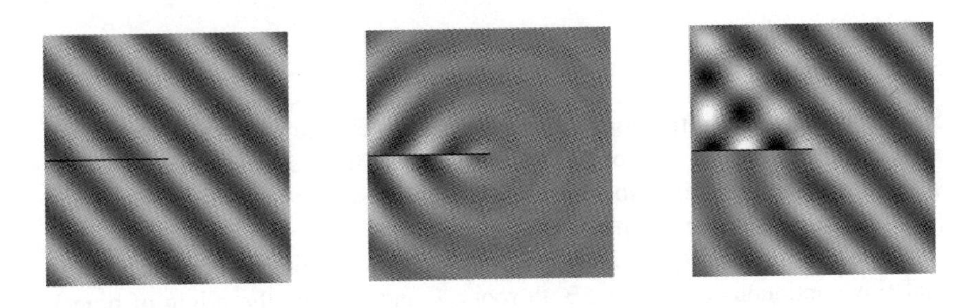

Fig. 5.2 The incident, scattered and total potential fields.

Figure 5.2 shows snapshots of the incident, scattered and total wavefields that we are trying to find, calculated using the results of this chapter. Ordinarily, of course, we won't know the wavefield in advance, but it may be useful to refer to these pictures in what follows.

5.2.1 *Generalised Fourier transforms*

We can see that the geometry is separable, so we can hope to treat the x and y dependencies separately and this suggests that Fourier transforming a spatial variable will be beneficial. Since we are allowing complex wavenumbers we must also allow the transform variable to be complex, giving

$$F(s, y) = \int_{-\infty}^{\infty} f(x, y)\mathrm{e}^{\mathrm{i}sx}\,\mathrm{d}x, \tag{6}$$

where $s = s_1 + \mathrm{i}s_2$. Assuming for the time being that this transform is valid the transformed wave equation takes the form

$$\left(\frac{\partial^2}{\partial y^2} - s^2 + k^2\right) F(s, y) = 0, \tag{7}$$

or

$$\left(\frac{\partial^2}{\partial y^2} - \gamma^2\right) F(s, y) = 0, \tag{8}$$

having defined $\gamma = (s^2 - k^2)^{1/2}$. Whereas we used Φ to denote the spatial component of a time-harmonic instance of ϕ (in effect a Fourier transform with respect to time) we use F to denote the Fourier transform with respect to space (in the x-direction) of f.

5.2.2 *Branch cuts*

Since γ is defined as a fractional power we have to choose which root to take for each complex value of s whenever we move away from $s = \pm k$ where $\gamma = 0$ without ambiguity. Remember that k is complex so $\pm k$ lie in the $(+,+)$ and $(-,-)$ quadrants of the complex plane.

Recall from §2.2.3 (p. 40) that it is impossible to choose roots in such a way that γ is continuous over all s. Both roots, k and $-k$, are the origin of branch cuts. A single, finite branch cut joining the two roots would be possible. But for reasons which will become apparent we will prefer to make a separate branch cut from each zero of γ to infinity in its own half-plane. In this way the strip $-\epsilon < s_2 < \epsilon$ will not contain any branch cuts, and since it doesn't contain any other singularities either we can ascertain that γ so defined is analytic (i.e. single-valued and differentiable) in this strip. It is perhaps worth remarking that the statement that a complex function is differentiable is considerably more restrictive than it is when applied to a scalar function.

5.2.3 *One-sided transforms*

With γ thus defined we can solve the ODE obtained from Eq. (8) by separating variables as

$$F(s, y) = \begin{cases} A(s)\mathrm{e}^{-\gamma y}, & \text{for } y > 0, \\ -A(s)\mathrm{e}^{\gamma y}, & \text{for } y < 0, \end{cases} \tag{9}$$

because $\mathrm{Re}(\gamma) > 0$ between the two branch cuts. Note that $\lim_{y \to 0+} F(s, y) = F(s, 0+) \neq F(s, 0-)$, which should not be surprising since f is discontinuous across the screen. The form of $A(s)$ remains to be determined from the boundary conditions.

It is at this point that the difficulty raised by the mixed boundary conditions becomes evident, and we have to (a) find a way to transform the two regions separately and (b) solve the resulting equations. Step (a) is a matter of defining so-called generalised Fourier transforms, step (b) is the Wiener–Hopf technique (or more strictly Jones's method for solving Wiener–Hopf problems.)

It is simple enough to define two half-transforms thus

$$\begin{aligned} F(s,y) &= \int_{-\infty}^{0} f\mathrm{e}^{\mathrm{i}sx}\,\mathrm{d}x + \int_{0}^{-\infty} f\mathrm{e}^{\mathrm{i}sx}\,\mathrm{d}x, \\ &= F_{-}(s,y) + F_{+}(s,y), \end{aligned} \tag{10}$$

but the time has now come to consider the implications of letting s be complex. Each integrand contains a factor $\exp(-s_2 x)$ so for each integral there will be values of s_2 for which it will grow exponentially. This can only be accommodated as long as f shrinks at a greater rate as $x \to \pm\infty$, implying a *maximum* value of s_2 for which F_- can exist and a *minimum* s_2 for F_+. Specifically, as $x \to \infty$ the behaviour of f will be dominated by cylindrical waves (due to the line sources on the screen, which satisfy the equivalent radiation problem) with an exponential decay factor of ϵ, implying that $s_2 > -\epsilon$ for F_+. As $x \to -\infty$ there will similarly be cylindrical radiation but there will also be cancelling/reflecting incident plane waves, whose exponential decay in the $-x$-direction will be $\epsilon\cos\theta$ requiring $s_2 < \epsilon\cos\theta$. As long as the relevant condition is satisfied each transform exists and is furthermore analytic since f is continuous in the x-direction. Crucially there is an overlapping strip in which both F_- and F_+ are analytic. Any equation containing both quantities must have s_2 in this strip to be valid.

Since we're doing one-sided transforms with a complex transform variable (and even calling it s) it might be asked why we don't just use Laplace transforms as indeed many authors do. There is no strong reason not to, other than that the references quoted here use Fourier transforms and, perhaps, going from left and right hand spatial domains to upper and lower transform domains might help to keep the two domains distinct in our minds.

5.2.4 *Transformed boundary conditions*

We can now write our boundary conditions in terms of transformed quantities so that the condition off the screen gives

$$F_{+}(s,0) = 0, \tag{11}$$

(continuous across $y = 0$) whereas the condition on the screen becomes a statement about the transform's y-derivative (denoted by a dash):

$$F'_{-}(s,0) = \frac{k\sin\theta}{s - k\cos\theta}, \tag{12}$$

(also continuous across $y = 0$ because $\partial f/\partial y$, the scattered vertical particle velocity in the field, reaches zero on both sides of the screen). Since we know that f, and hence F, is odd with respect to y we can work with the boundary conditions on $y = 0+$ and then recover the other half field. We shall therefore write F_+ for $F_+(s, 0+)$ etc. hereafter, noting that the '+' in F_+ refers to the half of the x-domain from which it was transformed, not the side of the screen on which we're letting y vanish. If this symmetry argument were not available we would have to work with new functions defined to be the sum and difference of F_+ on either side of the screen. As it is, Eq. (9) gives us

$$F_- + F_+ = A, \tag{13}$$
$$F'_- + F'_+ = -\gamma A, \tag{14}$$

with which we can eliminate A and substitute in our transformed boundary conditions to get

$$\gamma F'_+ + F_- = \frac{-k \sin\theta}{s - k\cos\theta}. \tag{15}$$

Both sides will be analytic as long as s lies in the strip between $-\epsilon$ and $\epsilon\cos\theta$. Figure 5.3 summarises the situation in the complex plane.

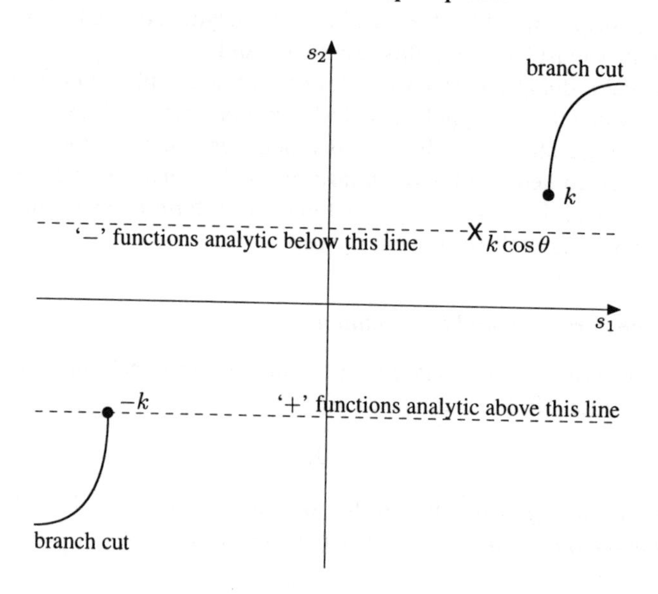

Fig. 5.3 Argand diagram showing domains of analyticity for Eq. (15).

5.3 The General Wiener–Hopf Equation

Equation (15) is in the form of the general Wiener–Hopf equation

$$K(s)U_+(s) + V_-(s) = P(s), \qquad \alpha < s_2 < \beta, \tag{16}$$

with K (often called the 'kernel' of the problem) and P known and U_+ and V_- unknown. The plus and minus subscripts arose from the fact that these quantities are generalised Fourier transforms of functions over the positive and negative values of their dependent variable (x in our example.) We shall recognise other functions as being 'plus' or 'minus' (or neither) according to whether they share their domains of analyticity with U_+ so that $s_2 > \alpha$, or with V_- so that $s_2 < \beta$. Whereas the domains from which U_+ and V_- were transformed do not overlap, the domains of analyticity of the transforms do (or at least should for the Wiener–Hopf technique to be applicable.) The procedure is then to rearrange this equation so that one side is all 'plus' and the other all 'minus'. This can be done as follows (assuming, of course, that it can be done at all): factorise K as K_+K_- and divide through by K_-

$$K_+U_+ + \frac{V_-}{K_-} = \frac{P}{K_-} \equiv R. \tag{17}$$

Obviously K_- must be non-zero for this to work. Then decompose R as a sum, $R = R_- + R_+$ and rearrange to get

$$K_+U_+ - R_+ = R_- - \frac{V_-}{K_-}, \tag{18}$$

where the left hand side is analytic for $s_2 > \alpha$ and the right hand side is analytic for $s_2 < \beta$.

5.3.1 *Analytic continuation*

Recall now the earlier comment that analyticity is a more severe restriction for a complex function than it is for a scalar one. This is borne out in a surprising result, utterly different to that of real functions, that if a complex function can be differentiated once it can be differentiated any number of times. One of the consequences of this is a theorem that can be stated colloquially as saying that an analytic function can't have a flat spot without being flat everywhere. That is to say, if a function is known to be analytic over a domain and is also known to be zero everywhere in a subdomain then it must be zero everywhere in the whole domain of analyticity (and hence likewise for any other constant value). This is because if the function is zero over a subdomain then the coefficients of its Taylor series expansion in that subdomain must all be zero. And if this is true in the subdomain it must also be true in the domain containing it, throughout which the function is analytic, since each analytic function only has one unique Taylor series expansion about a particular point.

A consequence of this was given in §2.6 (p. 52), namely that if two functions are analytic over domains that overlap, and are equal in that overlap region, then there exists a unique function that is analytic over both domains, called the 'analytic continuation' of one function into the other's domain. This follows immediately from applying the 'no flat spots' rule to the difference between the two functions in the overlap region. This is exactly the case for the two sides of Eq. (18); each side forms an analytic continuation of the other, so there must be a unique function $E(s)$ analytic over the entire complex plane, defined by

$$K_+(s)U_+(s) - R_+(s) = R_-(s) - \frac{V_-(s)}{K_-(s)} \equiv E(s), \qquad \text{for all } s. \tag{19}$$

We shall now do this for our diffraction problem.

5.3.2 *Application to the diffraction problem*

We have the following correspondence between the general Wiener–Hopf equation and the quantities in the half-plane diffraction problem:

$$\begin{aligned} K(s) &\longleftrightarrow \gamma(s) = (s^2 - k^2)^{1/2}, \\ U_+(s) &\longleftrightarrow F'_+(s), \\ V_-(s) &\longleftrightarrow F_-(s), \\ P(s) &\longleftrightarrow -k\sin\theta/(s - k\cos\theta), \\ \alpha &\longleftrightarrow -\epsilon, \\ \beta &\longleftrightarrow \epsilon\cos\theta. \end{aligned}$$

We can factorise the kernel γ by inspection to give

$$(s^2 - k^2)^{1/2} = \underbrace{(s-k)^{1/2}}_{-}\underbrace{(s+k)^{1/2}}_{+}, \tag{20}$$

where each factor has one of the two branch cuts in γ so that our version of Eq. (17) is

$$(s-k)^{1/2}F_- + \frac{F'_+}{(s+k)^{1/2}} = \frac{-k\sin\theta}{(s - k\cos\theta)(s+k)^{1/2}}. \tag{21}$$

The right hand side of this equation is our $R(s)$, which we need to decompose into the sum of a '+' and a '−' function. It has a pole at $s = k\cos\theta$ and a branch cut from $-k$ to ∞ in the lower half plane. By inspection the pole due to the factor

$1/(s - k\cos\theta)$ can be removed by multiplying by a function with a zero at that point as follows:

$$\frac{-k\sin\theta}{(s-k\cos\theta)(s+k)^{1/2}} = \underbrace{\underbrace{\frac{-k\sin\theta}{s-k\cos\theta}}_{\text{pole at } k\cos\theta}\left[\overbrace{\frac{1}{\underbrace{(s+k)^{1/2}}_{\text{branch from } -k}} - \frac{1}{(k+k\cos\theta)^{1/2}}}^{\text{zero at } k\cos\theta}\right]}_{\text{analytic above } k\cos\theta} + \overbrace{\frac{-k\sin\theta}{(k+k\cos\theta)^{1/2}\underbrace{(s-k\cos\theta)}_{\text{pole at } k\cos\theta}}}^{\text{analytic below } k}, \tag{22}$$

so that the two terms on the right hand side are $R_+(s)$ and $R_-(s)$ respectively. Using these we can finally write our Wiener–Hopf equation in the form of Eq. (18):

$$(s-k)^{1/2}F_- + \frac{k\sin\theta}{(s-k\cos\theta)(k+k\cos\theta)^{1/2}} = \frac{-k\sin\theta}{s-k\cos\theta}\left[\frac{1}{(s+k)^{1/2}} - \frac{1}{(k+k\cos\theta)^{1/2}}\right] - \frac{F'_+}{(s+k)^{1/2}} \equiv E(s), \tag{23}$$

so that since the two sides are analytic and equal in the overlapping strip they are equal everywhere that either side is analytic, in other words over the entire complex plane.

Product factorisations are more straightforward, partly because the roots of the product of two functions is just the union of the roots of each factor. The form of $K(s)$ influences the difficulty of both decompositions because $R(s) = P(s)/K_-(s)$.

5.3.3 *Behaviour at infinity in the s-plane*

The function $E(s)$ is still defined in terms of the unknown functions F'_- and F_+ but it turns out that we can deduce all we need to know about its form from its asymptotic behaviour as $s \to \infty$ in the half plane for which the right hand side of each equation is analytic. Starting with the first equation, and hence considering the upper half plane where $s_2 > -\epsilon\cos\theta$, the second term obviously tends to zero as $s \to \infty$, but we also need to know the asymptotic behaviour of F_+. Even though we are using asymptotic behaviour we will still get a final result that is exact.

The so-called Abelian theorem relates the large s asymptotic behaviour of the '+' transform of a function to that function's small x asymptote. For a function $g(x)$ that satisfies $|g(x)| < A\mathrm{e}^{\alpha x}$ so that

$$G_+(s) = \int_0^\infty f(x)\mathrm{e}^{\mathrm{i}sx}\,\mathrm{d}x, \tag{24}$$

is analytic for $s_2 > \alpha$ then as $|s| \to \infty$ in the upper half plane (written $|s| \to \infty\uparrow$) then the integral will be dominated by the values of f when x is small so that

$$\lim_{|s|\to\infty\uparrow} G_+(s) = \int_0^\infty \left(\lim_{x\to 0+} g(x)\right)\mathrm{e}^{\mathrm{i}sx}\,\mathrm{d}x, \tag{25}$$

and if $g(x) \to x^\lambda$ as $x \to 0+$ for some $\lambda > -1$ then this gives

$$\begin{aligned}\lim_{|s|\to\infty\uparrow} G_+(s) &= \int_0^\infty x^\lambda \mathrm{e}^{\mathrm{i}sx}\,\mathrm{d}x \\ &= \frac{\lambda!}{(-\mathrm{i}s)^{\lambda+1}}.\end{aligned} \tag{26}$$

Note that $-1 < \lambda < 0$ corresponds to some sort of 'blowing-up' at $x = 0+$. If it is known that this doesn't happen i.e. x is bounded at $0+$ then λ must be positive and G_+ must diminish at least as fast as $O(s^{-1})$ as $|s|$ goes to infinity in the upper half plane.

Similarly if G_- is analytic in a lower half plane and $g(x)$ tends to $(-x)^\mu$ with $\mu > -1$ as $x \to 0-$ then

$$\lim_{|s|\to\infty\downarrow} G_-(s) = \frac{\mu!}{(\mathrm{i}s)^{\mu+1}}. \tag{27}$$

These results also hold when $|s| \to \infty$ in the relevant half plane of analyticity with s_2 remaining finite, as long as $g(x)$ is infinitely differentiable.

5.3.4 *Behaviour at the edge*

The Abelian theorem tells us that we can deduce the behaviour of $F'_+(s)$ at infinity in the upper half plane if we know the behaviour of $f(x,0)\partial f/\partial y$ as $x \to 0+$, and of $F'_-(s)$ at infinity in the lower half plane if we know the behaviour of $f(x,0)$ as $x \to 0-$ with $y = 0$. We can infer this from the requirement that the kinetic energy of the field should be finite since we don't have any energy sources. If we write $q = |\nabla f|$, which varies with distance from the edge r according to $q \sim r^\alpha$

as $r \to 0+$ then the kinetic energy in a region Σ surrounding $x = y = 0$ will be proportional to

$$\iint_\Sigma q^2 \, \mathrm{d}x \, \mathrm{d}y \sim \iint_\Sigma r^{2\alpha} \, r \, \mathrm{d}r \, \mathrm{d}\theta = \iint_\Sigma r^{2\alpha+1} \, \mathrm{d}r \, \mathrm{d}\theta, \tag{28}$$

and for this quantity to remain finite we need $2\alpha + 1 > -1$ so $\alpha > -1$. We therefore know that $\partial f/\partial y = O(x^\alpha)$ as $x \to 0+$ and this implies, by the Abelian theorem, that $F'_+ = O(s^{-1-\alpha})$ as $|s| \to \infty \uparrow$. Similar arguments require that f be bounded as $x \to 0-$ so that $F_- = O(s^{-1})$ as $|s| \to \infty \downarrow$. We can therefore conclude that $E(s) \to 0$ as $|s| \to \infty$ in any direction.

Once again we can exploit the analyticity of $E(s)$, this time by observing that it must have a Taylor series

$$E(s) = e_0 + e_1 s + e_2 s^2 + \cdots, \tag{29}$$

within a circle of radius r about the origin. Since $E(s)$ is analytic it must be bounded everywhere, and it can be shown that if $|E(s)| < A$ then these coefficients are bounded by

$$|e_n| \le A r^{-n}. \tag{30}$$

And this must still be true when we let $r \to \infty$, which leads us to conclude that all the e_n apart from e_0 must vanish. This is Liouville's theorem, which tells us that an analytic function that is bounded at infinity in all directions can only be a constant. A generalisation of Liouville's theorem that is often useful says that if $E(s) < A|s|^N$ everywhere then $E(s)$ must be a polynomial of degree N or less. In our case since we know that $E(s) \to 0$ as $|s| \to \infty$ we must conclude that $e_0 = 0$ as well, leaving $E(s) = 0$ everywhere.

We therefore have from Eq. (23)

$$F_- = \frac{k \sin\theta}{(k + k\cos\theta)^{1/2}(s - k\cos\theta)(s - k)^{1/2}}, \tag{31}$$

and a corresponding equation for F'_+, but if we just solve for F_- we can determine f completely because $A(s) = F_-(s)$. This expression must then be inverse Fourier transformed with $\epsilon \to 0$. The result can be written

$$f(x, y) = \frac{1 - \mathrm{i}}{\sqrt{2\pi}} \left\{ \mathrm{e}^{-\mathrm{i}kr\cos(\Theta+\theta)} \tilde{\mathcal{F}}\left[\sqrt{2kr}\cos\left(\frac{\Theta+\theta}{2}\right)\right] \right.$$
$$\left. - \mathrm{e}^{-\mathrm{i}kr\cos(\Theta-\theta)} \tilde{\mathcal{F}}\left[\sqrt{2kr}\cos\left(\frac{\Theta-\theta}{2}\right)\right] \right\}, \tag{32}$$

where $x = r\cos\Theta$ and $y = r\sin\Theta$ (Θ defined to have a discontinuity where the screen lies), and where $\tilde{\mathcal{F}}$ is a Fresnel function defined by

$$\tilde{\mathcal{F}}(x) = \int_x^\infty e^{it^2}\,dt. \tag{33}$$

The more usual definition of a Fresnel function is (Abramowitz & Stegun, 1972)

$$\mathcal{F}(x) = \int_0^x e^{it^2}\,dt. \tag{34}$$

The two are related by

$$\mathcal{F}(x) + \tilde{\mathcal{F}}(x) = \frac{\sqrt{2\pi}}{2}(1+i). \tag{35}$$

5.4 A Systematic Decomposition Procedure

The factorisation and sum decomposition in the half-plane diffraction problem were obtained in a distinctly *ad hoc* fashion, a more systematic approach is desirable. In fact if we can find such an approach to the sum decomposition we will also have found one for the factorisation since using it to find

$$\ln K = \ln K_- + \ln K_+, \tag{36}$$

where K is the kernel gives K_+ and K_- that satisfy $K = K_+K_-$, at least as long as $\ln K$ is analytic in the strip in which we require both '+' and '−' functions to be analytic.

So we wish to find R_- and R_+ such that $R = R_- + R_+$ for a given R, analytic in the strip $\alpha < s_2 < \beta$. Recall that Cauchy's integral formula gives us

$$R(s) = \frac{1}{2\pi i}\oint \frac{R(\zeta)\,d\zeta}{\zeta - s}, \tag{37}$$

where the integral is around a closed path in the plane of $\zeta = \zeta_1 + i\zeta_2$, that encloses $\zeta = s$ for whichever value of s we are evaluating it. The integral around a closed path is the sum of integrals over any set of path segments that make up the closed path so if we can find a way to cut the path into two segments so that the integral over one is a '+' function and the other is a '−' function we will have achieved our goal. Since $R(\zeta)$ is analytic in $\alpha < \zeta_2 < \beta$ the integrand won't have any singularities other than $\zeta = s$ in this strip so we are free to deform the path as we please as long as it stays within the strip. But since we want the result to work for ζ equal to any and all of the values of s within the strip we would like to make it as large as possible within the strip.

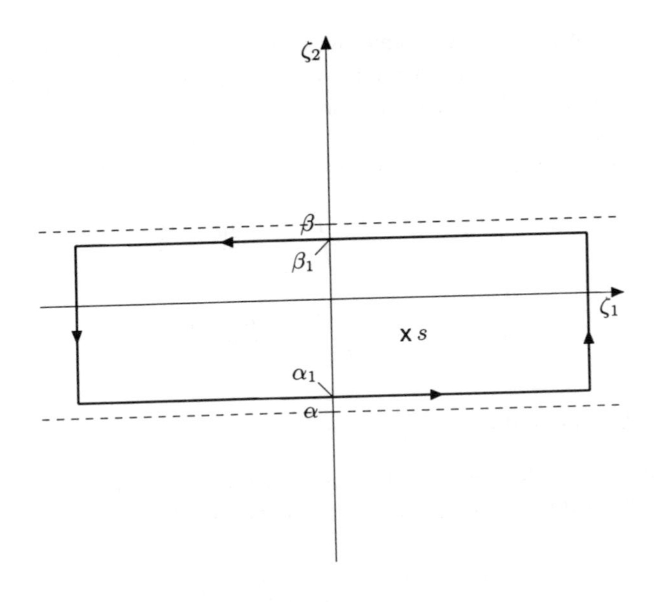

Fig. 5.4 Integration path in the ζ-plane for sum decomposition.

Start by fixing s at an arbitrary position in the strip in the ζ-plane and draw a rectangular path enclosing it, which crosses the ζ_2 axis at α_1 and β_1. At this stage we have got four integrals that add up to $R(s)$ rather than the two we want, but in order to include all s in the strip we are going to want to expand the rectangle infinitely at the sides. If R behaves suitably there then the contributions from integrating along the sides will vanish and we will be left with two terms as required. For this to be the case requires $R(\zeta) = O(\zeta^{-\delta})$ as $|\zeta| \to \infty$ in the strip for some positive δ, since then

$$\left|\frac{R(\zeta)}{\zeta - s}\right| = \mathcal{O}(\zeta^{-1-\delta}), \tag{38}$$

and this is a condition for the sum decomposition method to work. If it is satisfied then we have

$$R(s) = \frac{1}{2\pi \mathrm{i}} \int_{\mathrm{i}\alpha_1 - \infty}^{\mathrm{i}\alpha_1 + \infty} \frac{R(\zeta)\,\mathrm{d}\zeta}{\zeta - s} + \frac{1}{2\pi \mathrm{i}} \int_{\mathrm{i}\beta_1 + \infty}^{\mathrm{i}\beta_1 - \infty} \frac{R(\zeta)\,\mathrm{d}\zeta}{\zeta - s}, \tag{39}$$

which is *one* sum decomposition but for it to be the one we want we must examine the analyticity of each integral. As long as s lies within our (infinitely wide) rectangle then the integrand will be analytic along each of the paths so the integrals will both be analytic. If s lies outside the rectangle then the integrals will still be analytic functions, but not the same ones, since they will now add up to zero instead of $R(s)$ by the residue theorem. We can therefore think of the first integral as defining a function that is analytic for $s_2 > \alpha_1$ and the second one defining a function that is analytic for $s_2 < \beta_1$. We can then let α_1 and β_1 get arbitrarily close to α and β respectively so that we have

$$R_-(s) = -\frac{1}{2\pi \mathrm{i}} \int_{\mathrm{i}\beta_1-\infty}^{\mathrm{i}\beta_1+\infty} \frac{R(\zeta)\,\mathrm{d}\zeta}{\zeta - s} \tag{40}$$

$$R_+(s) = \frac{1}{2\pi \mathrm{i}} \int_{\mathrm{i}\alpha_1-\infty}^{\mathrm{i}\alpha_1+\infty} \frac{R(\zeta)\,\mathrm{d}\zeta}{\zeta - s}. \tag{41}$$

5.4.1 *Application to the screen problem*

We should demonstrate this procedure on the sum decomposition that occurs in the half-plane diffraction problem, which is proportional to

$$R(s) = \frac{1}{(s - k\cos\theta)(s+k)^{1/2}}, \tag{42}$$

which has a simple pole at $s = k\cos\theta$ and a branch cut at $s = -k$ extending to infinity in the lower half plane. Then by our formula we have

$$R_+(s) = -\frac{1}{2\pi \mathrm{i}} \int_{\mathrm{i}\alpha_1-\infty}^{\mathrm{i}\alpha_1+\infty} \frac{\mathrm{d}\zeta}{(\zeta - k\cos\theta)(\zeta+k)^{1/2}(\zeta - s)}. \tag{43}$$

Ironically, having gone to the trouble of breaking a closed path integral into a sum of line integrals to achieve the sum decomposition formula, we now find that to evaluate the resulting line integral we have to turn it back into a (different) closed path integral. We do this by joining the end points of the integration path with a semicircle at infinity in the upper half plane that encloses the poles at $\zeta = k\cos\theta$ and at $\zeta = s$ but excludes the branch cut. It can be shown that the integral along the semicircle vanishes but because the path is now closed we can apply the residue theorem to evaluate the integral. Recall that the integral around a closed path is $2\pi\mathrm{i}$ times the sum of the residues at the singular points enclosed, and that the residue at ζ_0 of $g(\zeta) = p(\zeta)/q(\zeta)$ (where p and q are analytic at ζ_0 and $q(\zeta_0) = 0$) is $p(\zeta_0)/q'(\zeta_0)$ as long as the order of the zero is one. In our case

$$q'(\zeta) = [(\zeta - s) + (\zeta - k\cos\theta)](\zeta + k)^{1/2} + \frac{(\zeta - s)(\zeta - k\cos\theta)}{2(\zeta+k)^{1/2}}, \tag{44}$$

so the residues are

$$\mathrm{Res}(\zeta = s) = \frac{1}{q'(s)} = \frac{1}{(s - k\cos\theta)(s+k)^{1/2}}, \tag{45}$$

and

$$\mathrm{Res}(\zeta = k\cos\theta) = \frac{-1}{q'(-k\cos\theta)} = \frac{1}{(s - k\cos\theta)(k + k\cos\theta)^{1/2}}, \tag{46}$$

which add up to give

$$R_+(s) = \frac{1}{s - k\cos\theta}\left[\frac{1}{(s+k)^{1/2}} - \frac{1}{(k + k\cos\theta)^{1/2}}\right], \tag{47}$$

which corresponds to the answer we obtained previously. The same procedure, and indeed the same integrand, yields R_- wherein the only singularity is at $\zeta = -k\cos\theta$ so that we already have an expression for $R_-(s)$ in Eq. (46). Straightforward addition confirms that $R(s) = R_-(s) + R_+(s)$.

5.5 Summary

The Wiener–Hopf procedure can be summarised as follows:

(i) Determine the boundary conditions, including far-field conditions and behaviour at the boundary between specified regions.
(ii) Apply appropriate transforms so as to obtain an algebraic instead of differential equation.
(iii) Apply transformed boundary conditions to plus and minus functions whose domains of analyticity overlap.
(iv) Manipulate these relations into the form of the Wiener–Hopf equation.
(v) Factorise and decompose functions to get an equality of plus functions and minus functions.
(vi) Use analytic continuation to infer an entire function.
(vii) Use the Abelian theorem to deduce its behaviour at infinity from the expected physical behaviour at the edge where the boundary condition changes.
(viii) Use Liouville's theorem to deduce the form of the entire function from its behaviour at infinity.
(ix) Solve for the transformed field variables.
(x) Invert the transform to obtain the required field.

Each of these steps has to work for the whole procedure to work. Its use has been widespread and persists, it is widely used in solving diffraction problems such as the one we have examined in detail, and more generally for finding solutions to the wave equation when mixed boundary conditions are specified. It can also be used to solve the Laplace equation $\nabla^2\phi = 0$ with mixed boundary conditions by first solving the Helmholtz equation $(\nabla^2 + k^2)\phi = 0$ and then letting $k \to 0$. This can allow the calculation of potential flow in situations where no suitable conformal mapping can be found. Vorticity can also be calculated in some situations. And although the previous derivation was for a steady state problem it is always possible to include the time factor throughout and solve transient problems in the same way.

Other configurations can still be treated as long as the geometry of the problem leads to a separable solution of the wave equation, for example the radiation from a circular pipe, or diffraction by parallel plates can be obtained by the Wiener–Hopf procedure. It is also necessary that the transform chosen allows the formulation of the Wiener–Hopf equation. The complex Fourier transform is not the only choice, others that may be necessary are the Mellin transform or occasionally the Kontorovich–Lebedev transform, details of which are given in Noble (1988).

Some problems, such as diffraction by a plate with finite thickness or finite extent, or radiation from a pipe with a finite flange, cannot be solved exactly by this procedure but are amenable to approximate methods based on it. In some cases, however exact results can be obtained by other methods, such as matched asymptotic expansion. Difficulties also arise in problems involving matrix-valued functions since the fact that matrix multiplication is not, in general, commutative can prevent the factorisation of the kernel. Recent work by Abrahams (2000) has used Padé approximants, which effectively replace branch cuts with arrays of poles, to good effect.

The Wiener–Hopf technique can be applied to Fourier transforms of time as well as space. An obvious example of a semi-infinite interval is time divided into the past and the future. In this way it is fundamental to much of the signal processing literature concerning optimal prediction.

References

Abrahams, I. D. (2000) The application of Padé approximants to Wiener–Hopf factorization, *IMA Journal of Applied Mathematics* **65**, pp. 257–281.

Abramowitz, M. and Stegun, I. (1972) *Handbook of Mathematical Functions.* Dover.

Leppington, F. G. (1992) Wiener–Hopf Technique, chapter 5 of *Modern Methods in Analytical Acoustics* by Crighton, D. G. *et al.* Springer Verlag.

Noble, B. (1988) *Methods Based on the Wiener–Hopf Technique for the Solution of Partial Differential Equations.* American Mathematical Society.

Pierce, A. D. (1989) *Acoustics: An Introduction to its Physical Principles and Applications.* Acoustical Society of America, Woodbury, NY, p. 426.

Chapter 6

Waveguides

M. McIver & C. M. Linton
Department of Mathematical Sciences
Loughborough University

6.1 Basic Theory

6.1.1 *Introduction*

Consider a small disturbance of a uniform, stationary, acoustic medium. The total pressure p_{T} is

$$p_{\mathrm{T}} = p_0 + p, \qquad |p| \ll p_0, \tag{1}$$

where p_0 is the pressure of the undisturbed medium. It can be shown that, in the linear approximation,

$$\nabla^2 p = \frac{1}{c_0^2}\frac{\partial^2 p}{\partial t^2}, \tag{2}$$

where c_0 is the speed of sound. The velocity field associated with the disturbance is $\mathbf{u}$ and we can write

$$\mathbf{u} = \boldsymbol{\nabla}\phi, \qquad p = -\rho_0\frac{\partial \phi}{\partial t}, \tag{3}$$

where ρ_0 is the density of the undisturbed medium. Here ϕ is called the acoustic velocity potential and it too satisfies the wave equation:

$$\nabla^2 \phi = \frac{1}{c_0^2}\frac{\partial^2 \phi}{\partial t^2}. \tag{4}$$

6.1.2 *Boundary conditions*

At a rigid boundary there is no flow through the boundary and so

$$\mathbf{u} \cdot \mathbf{n} \equiv \frac{\partial \phi}{\partial n} = 0. \tag{5}$$

A more general boundary condition is the so-called impedance condition

$$Z \frac{\partial \phi}{\partial n} = p \tag{6}$$

in which the acoustic impedance Z is the ratio of the pressure to the normal velocity at the surface. A rigid boundary corresponds to Z tending to infinity. The case of zero impedance leads to what is known as a pressure-release boundary, on which

$$p = 0. \tag{7}$$

For a case of a rigid boundary moving with velocity $\mathbf{v}$ we would have

$$\mathbf{u} \cdot \mathbf{n} \equiv \frac{\partial \phi}{\partial n} = \mathbf{v} \cdot \mathbf{n}. \tag{8}$$

6.1.3 *Time harmonic motion*

If we want to consider disturbances with circular frequency ω, it is often convenient to remove the time-dependence at the expense of introducing complex-valued quantities into the theory by writing

$$\phi(\mathbf{x}, t) = \mathrm{Re}[\Phi(\mathbf{x}) \mathrm{e}^{-\mathrm{i}\omega t}]. \tag{9}$$

The frequency in Hertz is $f = \omega/2\pi$. Since ϕ satisfies the wave equation we have

$$\nabla^2 \phi = \frac{1}{c_0^2} \frac{\partial^2 \phi}{\partial t^2}, \tag{10}$$

so that

$$\mathrm{Re}[\nabla^2 \Phi \, \mathrm{e}^{-\mathrm{i}\omega t}] = -\frac{\omega^2}{c_0^2} \mathrm{Re}[\Phi \, \mathrm{e}^{-\mathrm{i}\omega t}], \tag{11}$$

and therefore

$$\mathrm{Re}[(\nabla^2 \Phi + \frac{\omega^2}{c_0^2} \Phi) \, \mathrm{e}^{-\mathrm{i}\omega t}] = 0, \tag{12}$$

giving

$$\nabla^2 \Phi + k^2 \Phi = 0. \tag{13}$$

Thus the time-dependence has been removed and the wave equation has become the Helmholtz equation (or reduced wave equation).

From Φ we can get p. Let

$$p = \mathrm{Re}[P\,\mathrm{e}^{-\mathrm{i}\omega t}]. \tag{14}$$

Then

$$\mathrm{Re}[P\,\mathrm{e}^{-\mathrm{i}\omega t}] = -\rho_0 \frac{\partial \phi}{\partial t} = \mathrm{Re}[\mathrm{i}\omega\rho_0\Phi\,\mathrm{e}^{-\mathrm{i}\omega t}]. \tag{15}$$

Hence

$$P = \mathrm{i}\omega\rho_0\Phi, \tag{16}$$

and

$$\nabla^2 P + k^2 P = 0. \tag{17}$$

The general impedance boundary condition becomes

$$Z\frac{\partial \Phi}{\partial n} = \mathrm{i}\omega\rho_0\Phi. \tag{18}$$

6.1.4 *Plane waves*

Suppose that

$$\Phi = A\mathrm{e}^{\mathrm{i}\mathbf{k}\cdot\mathbf{x}} \tag{19}$$

for some real constant A and some constant vector $\mathbf{k} = k\hat{\mathbf{k}}$. Then

$$\phi = \mathrm{Re}[A\mathrm{e}^{\mathrm{i}(\mathbf{k}\cdot\mathbf{x}-\omega t)}] = A\cos(k(\hat{\mathbf{k}}\cdot\mathbf{x} - c_0 t)), \qquad c_0 = \omega/k. \tag{20}$$

This represents a wave of amplitude A travelling in the $\hat{\mathbf{k}}$ direction with speed $c_0 = \omega/k$.

6.1.5 *Acoustic waveguides*

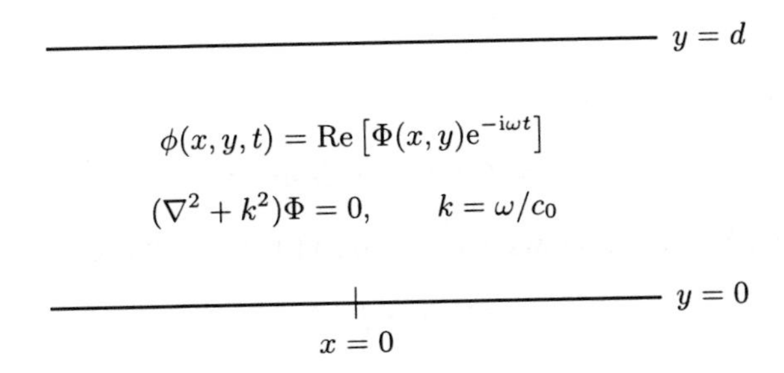

Fig. 6.1 A two-dimensional acoustic waveguide.

A standard method for determining the acoustic velocity potential in a wave-guide like that sketched in Fig. 6.1 is the technique of separation of variables. Let

$$\Phi(x, y) = X(x)Y(y). \tag{21}$$

Then $\partial^2\Phi/\partial x^2 + \partial^2\Phi/\partial y^2 + k^2\Phi = 0$ implies

$$X''Y + XY'' + k^2XY = 0, \tag{22}$$

from which

$$\frac{X''}{X} + k^2 = -\frac{Y''}{Y}. \tag{23}$$

Now the left-hand side is a function of x alone and the right-hand side is a function of y alone. The only way that this can hold for all x and y is if both sides have a constant value. Hence

$$\frac{X''}{X} + k^2 = -\frac{Y''}{Y} = \lambda, \qquad \lambda \text{ a constant.} \tag{24}$$

For the case of rigid walls we have $\partial\Phi/\partial n = 0$ and this implies that $Y'(0) = Y'(d) = 0$.

If we assume that $\lambda < 0$ and write $\lambda = -\gamma^2$ we get

$$Y'' - \gamma^2 Y = 0, \tag{25}$$

which has general solution

$$Y = Ae^{\gamma y} + Be^{-\gamma y}. \tag{26}$$

The boundary conditions then show that $A = B = 0$, so this is no good.

Hence assume that $\lambda = \gamma^2 > 0$. We get

$$Y'' + \gamma^2 Y = 0, \tag{27}$$

which has general solution

$$Y = C\cos\gamma y + D\sin\gamma y. \tag{28}$$

The condition $Y'(0) = 0$ shows that $D = 0$ and $Y'(d) = 0$ leads to

$$C\gamma\sin\gamma d = 0. \tag{29}$$

We can't have $C = 0$, so $\sin\gamma d = 0$. Hence we must have

$$\gamma d = n\pi, \qquad n = 1, 2, \ldots . \tag{30}$$

Finally we need to consider the case $\gamma = 0$. Then $Y'' = 0$ so that $Y = E + Fy$ and the boundary conditions show that $F = 0$. Hence Y is a constant, which is just what Eq. (28) gives when $\gamma = 0$.

The solution in the cross-guide variable must therefore be of the form

$$Y(y) = C_n \cos \gamma_n y, \qquad \gamma_n = n\pi/d, \qquad n = 0, 1, 2, \ldots. \tag{31}$$

The equation for X is now

$$X'' + \alpha_n^2 X = 0, \qquad \alpha_n = \sqrt{k^2 - \gamma_n^2}. \tag{32}$$

Assuming that $\alpha_n \neq 0$, it is helpful to write the solutions in the form of complex exponentials:

$$X = \mathrm{e}^{\pm \mathrm{i}\alpha_n x}. \tag{33}$$

We have shown that

$$C_n \mathrm{e}^{\pm \mathrm{i}\alpha_n x} \cos \gamma_n y$$

is a solution to our problem for any $n = 0, 1, 2, \ldots$. Since the equations we are solving are linear and homogeneous, any linear combination of these solution is also a solution. Hence we write the *general* solution to our problem as

$$\Phi = \sum_{n=0}^{\infty} \left(C_n \mathrm{e}^{\mathrm{i}\alpha_n x} + D_n \mathrm{e}^{-\mathrm{i}\alpha_n x} \right) \cos \gamma_n y. \tag{34}$$

6.1.6 *Cut-off frequencies*

The nature of the solutions we have just derived depends crucially on whether $\alpha_n = \sqrt{k^2 - \gamma_n^2}$ is real or imaginary. Consider the solution

$$\left(C_n \mathrm{e}^{\mathrm{i}\alpha_n x} + D_n \mathrm{e}^{-\mathrm{i}\alpha_n x} \right) \cos \gamma_n y.$$

Suppose first that $k < \gamma_n$. Then $\alpha_n = \mathrm{i}\sqrt{\gamma_n^2 - k^2}$ and the C_n term decays exponentially as $x \to \infty$ but grows exponentially as $x \to -\infty$. The D_n term has the reverse behaviour. In any physical application we must set the constants that multiply solutions that grow at infinity to zero.

Suppose next that $k = \gamma_n$. Then $\alpha_n = 0$ and the solutions look like

$$(C_n + D_n x) \cos \gamma_n y.$$

The resulting solutions with $D_n = 0$ are called standing waves.

Suppose finally that $k > \gamma_n$. Then α_n is real (and positive). The C_n term represents a wave travelling in the direction of increasing x whereas the D_n term represents a wave travelling in the direction of decreasing x.

Now $k = \omega/c_0$ so $\omega = kc_0$. We define the quantities $\omega_n = \gamma_n c_0$; these are the cut-off frequencies for the waveguide. Note that they depend on both the guide geometry and the boundary conditions on the guide walls. For $\gamma_{n-1} < k < \gamma_n$, $n = 1, 2, \ldots$, there are n possible wave modes in the guide.

6.1.7 *Cylindrical guides*

In cylindrical polar coordinates (r, θ, x), the Helmholtz equation is

$$\frac{\partial^2 \Phi}{\partial r^2} + \frac{1}{r}\frac{\partial \Phi}{\partial r} + \frac{1}{r^2}\frac{\partial^2 \Phi}{\partial \theta^2} + \frac{\partial^2 \Phi}{\partial x^2} + k^2 \Phi = 0. \tag{35}$$

Consider a circular cylindrical guide of radius a. If we have pressure-release walls then

$$\Phi = 0, \qquad \text{on } r = a. \tag{36}$$

Separate variables. Let

$$\Phi(r, \theta, x) = R(r)\Theta(\theta)X(x), \tag{37}$$

then

$$R''\Theta X + \frac{1}{r}R'\Theta X + \frac{1}{r^2}R\Theta'' X + R\Theta X'' + k^2 R\Theta X = 0, \tag{38}$$

from which

$$\frac{R''}{R} + \frac{R'}{rR} + \frac{\Theta''}{r^2\Theta} = -\frac{X''}{X} - k^2 = -\gamma^2, \qquad \text{say.} \tag{39}$$

Next

$$r^2\frac{R''}{R} + r\frac{R'}{R} + \gamma^2 r^2 = -\frac{\Theta''}{\Theta} = \lambda^2, \qquad \text{say.} \tag{40}$$

The equation for Θ is

$$\Theta'' + \lambda^2\Theta = 0 \tag{41}$$

and in order that the solution is single valued we require Θ to be periodic with period 2π. Hence we must have $\lambda = m$, where m is an integer and then

$$\Theta = A\cos m\theta + B\sin m\theta. \tag{42}$$

The equation for R is then

$$r^2 R'' + rR' + (\gamma^2 r^2 - m^2)R = 0. \tag{43}$$

This is a linear second-order ODE and hence it has two linearly-independent solutions. They are called Bessel functions and are represented by the symbols

$J_m(\gamma r)$ and $Y_m(\gamma r)$. Properties of these functions can be looked up in standard reference books and the functions J_0 and Y_0 are plotted in Fig. 6.2. One of the most important facts about them is that $Y_m(\gamma r)$ is singular at $\gamma r = 0$ whereas J_m is not. Hence if $r = 0$ is within the physical domain (as it is in our case), only the J_m functions can be present in the solution.

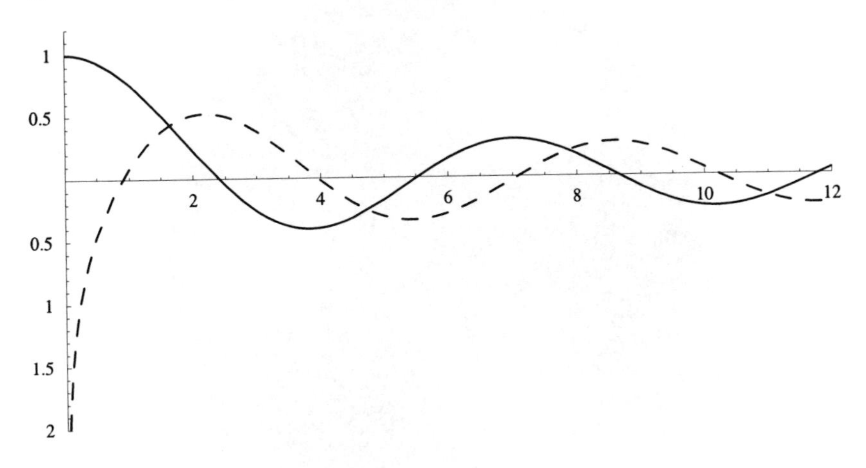

Fig. 6.2 Bessel functions J_0 (solid line) and Y_0 (dashed line).

If the guide wall has zero impedance we would have $R(a) = 0$ and hence

$$J_m(\gamma a) = 0. \tag{44}$$

The Bessel function J_m is an oscillatory function with a set of zeros (much like the sine or cosine function). These zeros are labelled $j_{m,n}$, $n = 1, 2, \ldots$, and can be easily computed. We must then have

$$\gamma = \gamma_{m,n} = j_{m,n}/a. \tag{45}$$

Finally, the x-variation is of the form

$$e^{\pm \alpha_{m,n} x}, \tag{46}$$

where

$$\alpha_{m,n} = \sqrt{k^2 - \gamma_{m,n}^2}. \tag{47}$$

The cut-off frequencies are now

$$\omega_{m,n} = \gamma_{m,n} c_0 = j_{m,n}\, c_0/a. \tag{48}$$

One way to visualise the modes is to draw contour plots of their cross section, i.e. curves of constant $\Phi(r,\theta,0)$. As an example, the mode with r,θ-dependence $\mathrm{J}_3(\mathrm{j}_{3,1}\,r)\cos 3\theta$, where $\mathrm{j}_{3,1}$ is the first zero of J_3, is illustrated in Fig. 6.3.

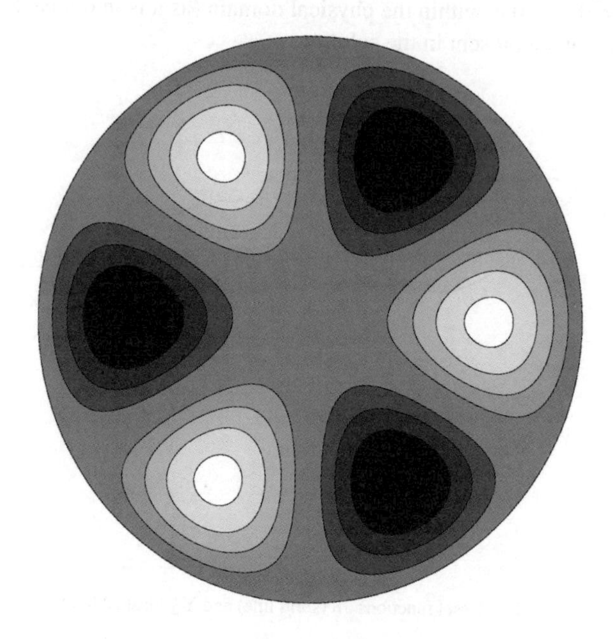

Fig. 6.3 Curves of constant $\mathrm{J}_3(\mathrm{j}_{3,1}\,r)\cos 3\theta$. The light areas correspond to positive values and the dark areas to negative values.

6.2 Eigenfunction Expansions

6.2.1 *A rolling piston*

We can represent a rigid piston, making small oscillations about its midpoint, at one end of a waveguide, by the boundary condition

$$\mathbf{u}|_{x=0} = \left((y - \tfrac{1}{2}d)\cos\omega t, 0\right) \tag{49}$$

and then we have the boundary-value problem shown in Fig. 6.4.

As we have seen, the general solution, satisfying the boundary conditions on the guide walls, can be written

$$\Phi = \sum_{n=0}^{\infty} \left(C_n \mathrm{e}^{\mathrm{i}\alpha_n x} + D_n \mathrm{e}^{-\mathrm{i}\alpha_n x}\right)\cos\gamma_n y, \tag{50}$$

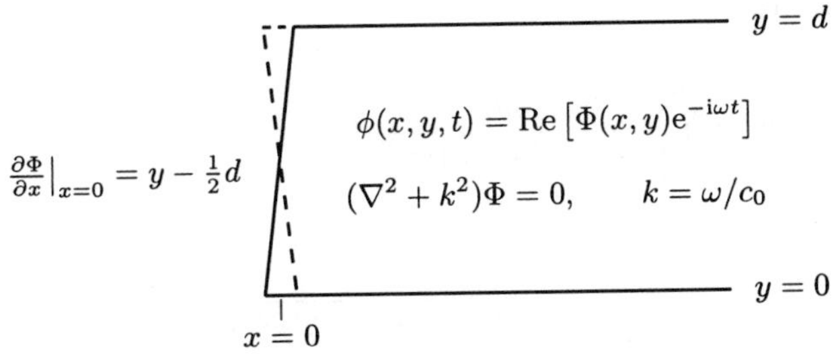

$y = d$

$\phi(x, y, t) = \mathrm{Re}\left[\Phi(x, y)\mathrm{e}^{-\mathrm{i}\omega t}\right]$

$\left.\frac{\partial \Phi}{\partial x}\right|_{x=0} = y - \frac{1}{2}d$

$(\nabla^2 + k^2)\Phi = 0, \qquad k = \omega/c_0$

$y = 0$

$x = 0$

Fig. 6.4 A boundary-value problem for a rolling piston in a waveguide.

where

$$\alpha_n = \sqrt{k^2 - \gamma_n^2} = \mathrm{i}\sqrt{\gamma_n^2 - k^2}, \qquad \gamma_n = n\pi/d. \tag{51}$$

For $k < \gamma_n$ the C_n terms decay as $x \to \infty$, but the D_n terms grow. For $k > \gamma_n$ the C_n terms represent waves travelling to the right and the D_n terms are waves travelling to the left. Hence, since we can only have outgoing waves as $x \to \infty$, $D_n = 0$ for all n. Thus

$$\Phi = \sum_{n=0}^{\infty} C_n \mathrm{e}^{\mathrm{i}\alpha_n x} \cos \gamma_n y. \tag{52}$$

The coefficients C_n must be determined from the boundary condition on the piston, which is

$$\sum_{n=0}^{\infty} \mathrm{i}\alpha_n C_n \cos \gamma_n y = y - \tfrac{1}{2}d. \tag{53}$$

We can think of this as a Fourier cosine series for the $2d$-periodic extension of the function $y - d/2$. Multiply by $\cos m\pi/d$, $m = 0, 1, 2, \ldots$ and use the fact that

$$\int_0^d \cos \frac{m\pi y}{d} \cos \frac{n\pi y}{d}\, \mathrm{d}y = \frac{d}{\epsilon_m}\delta_{mn}, \tag{54}$$

where $\epsilon_0 = 1$, $\epsilon_m = 2$, $m \geq 1$, and δ_{mn} is the Kronecker delta (see §1.8.1, p. 27.) This yields

$$\frac{d}{\epsilon_m}\mathrm{i}\alpha_m C_m = \int_0^d (y - \tfrac{1}{2}d) \cos \frac{m\pi y}{d}\, \mathrm{d}y = \frac{d^2}{m^2\pi^2}(-1 + \cos m\pi). \tag{55}$$

Hence if m is even, C_m is zero. On the other hand, if m is odd we have $C_m =$

$4\mathrm{i}d/\alpha_m m^2\pi^2$. Thus

$$\Phi = \sum_{q=1}^{\infty} \frac{4\mathrm{i}d}{\alpha_{2q-1}(2q-1)^2\pi^2} \mathrm{e}^{\mathrm{i}\alpha_{2q-1}x} \cos \gamma_{2q-1} y. \tag{56}$$

Note that

$$\alpha_{2q-1} = \sqrt{k^2 - (2q-1)^2\pi^2/d^2} \tag{57}$$

and so if $k < \pi/d$ there is no propagation down the guide. For

$$(2q-1)\frac{\pi}{d} < k < (2q+1)\frac{\pi}{d} \tag{58}$$

there are q possible propagating modes.

6.3 An Abrupt Change in Boundary Condition

Consider the problem illustrated in Fig. 6.5. In the diagram the heavier lines represent boundaries on which Neumann conditions ($\partial\Phi/\partial n = 0$) are applied and the thin line represents a Dirichlet boundary ($\Phi = 0$). To completely specify the problem we need to add conditions on Φ as $x \to \pm\infty$.

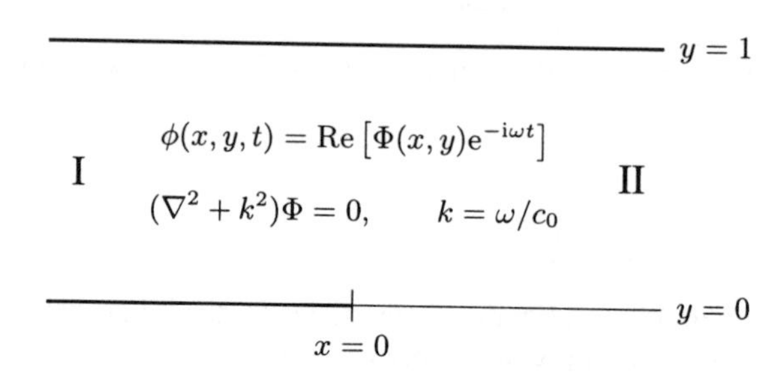

Fig. 6.5 An abrupt change in boundary condition on one side of a waveguide.

Separation of variables shows that solutions in $x < 0$ (region I) and $x > 0$ (region II) can be written as linear combinations of the terms $\exp(\pm k_n x)\psi_n(y)$, $n \geq 0$, and $\exp(\pm\kappa_n x)\Psi_n(y)$, $n \geq 1$, respectively, where

$$\psi_n(y) = \epsilon_n^{1/2} \cos \lambda_n y, \qquad k_n = (\lambda_n^2 - k^2)^{1/2}, \qquad \lambda_n = n\pi, \tag{59}$$

$$\Psi_n(y) = 2^{1/2} \sin \mu_n y, \qquad \kappa_n = (\mu_n^2 - k^2)^{1/2}, \qquad \mu_n = (n - \tfrac{1}{2})\pi, \tag{60}$$

and $\epsilon_0 = 1$, $\epsilon_n = 2$, $n \geq 1$. Both $\{\psi_n(y)\}_{n=0}^{\infty}$ and $\{\Psi_n(y)\}_{n=0}^{\infty}$ form complete orthonormal sets on $(0,1)$. In particular,

$$\int_0^1 \psi_n(y)\psi_m(y)\,\mathrm{d}y = \delta_{mn} = \int_0^1 \Psi_n(y)\Psi_m(y)\,\mathrm{d}y. \tag{61}$$

For simplicity we assume that the wavenumber k lies in the range $0 < k < \pi/2$ and then $k_0 = -\mathrm{i}k$ is purely imaginary whereas k_n, κ_n, $n \geq 1$ are real. Without loss of generality k, k_n, κ_n, $n \geq 1$, are all taken to be positive. The fact that k_0 is imaginary but that all other values of k_n and κ_n are real shows that one possible wave mode can exist in region I whereas no such modes are possible in region II. Thus we can define a scattering problem by imposing the radiation conditions

$$\Phi \sim (\mathrm{e}^{-k_0 x} + R\mathrm{e}^{k_0 x})\psi_0(y) = \mathrm{e}^{\mathrm{i}kx} + R\mathrm{e}^{-\mathrm{i}kx} \qquad \text{as } x \to -\infty, \tag{62}$$

$$\Phi \to 0 \qquad \text{as } x \to \infty. \tag{63}$$

Here R, the reflection coefficient, is an unknown complex constant but energy considerations show that $|R| = 1$ and so we can write $R = \exp(\mathrm{i}\delta)$. Then

$$\phi \sim \cos(kx - \omega t) + \cos(kx + \omega t - \delta) \qquad \text{as } x \to -\infty. \tag{64}$$

6.3.1 *Matched eigenfunction expansions*

In region I we expand Φ as

$$\Phi = (\mathrm{e}^{-k_0 x} + \mathrm{e}^{k_0 x})\psi_0(y) + \sum_{n=0}^{\infty} \frac{U_n^I}{k_n}\,\mathrm{e}^{k_n x}\psi_n(y) \tag{65}$$

where U_n^I, $n \geq 0$ are complex constants to be determined and $U_0^I = k_0(R-1)$. Similarly, in $x > 0$, we write

$$\Phi = \sum_{n=1}^{\infty} \frac{U_n^{II}}{-\kappa_n}\,\mathrm{e}^{-\kappa_n x}\Psi_n(y). \tag{66}$$

Physical considerations require that the pressure and velocity are continuous across $x = 0$. Continuity of Φ and $\partial\Phi/\partial x$ ensures this and yields

$$2\psi_0(y) + \sum_{n=0}^{\infty} \frac{U_n^I}{k_n}\,\psi_n(y) = -\sum_{n=1}^{\infty} \frac{U_n^{II}}{\kappa_n}\Psi_n(y), \qquad 0 < y < 1, \tag{67}$$

$$\sum_{n=0}^{\infty} U_n^I\psi_n(y) = \sum_{n=1}^{\infty} U_n^{II}\Psi_n(y), \qquad 0 < y < 1. \tag{68}$$

There are a number of ways of converting Eqs. (67) and (68) into an infinite system of linear algebraic equations. Each equation can be multiplied by each of the functions $\psi_m(y)$ and $\Psi_m(y)$ and integrated over $(0,1)$. Thus we can derive

$$\frac{U_m^I}{k_m}, = -2\delta_{m0} - \sum_{n=1}^{\infty} \frac{U_n^{II}}{\kappa_n} c_{mn}, \qquad m \geq 0, \tag{69}$$

$$U_m^I = \sum_{n=1}^{\infty} U_n^{II} c_{mn}, \qquad m \geq 0, \tag{70}$$

$$\frac{U_m^{II}}{\kappa_m} = -2c_{0m} - \sum_{n=0}^{\infty} \frac{U_n^I}{k_n} c_{nm}, \qquad m \geq 1, \tag{71}$$

$$U_m^{II} = \sum_{n=0}^{\infty} U_n^I c_{nm}, \qquad m \geq 1, \tag{72}$$

where

$$c_{mn} = \int_0^1 \psi_m(y)\Psi_n(y)\,\mathrm{d}y = \frac{(2\epsilon_m)^{1/2}\mu_n}{\kappa_n^2 - k_m^2}. \tag{73}$$

There are now a number of different strategies. We can:

(a) eliminate U_m^I using Eqs. (69) and (70);
(b) eliminate U_m^{II} using Eqs. (71) and (72);
(c) eliminate U_m^I using Eqs. (70) and (71);
(d) eliminate U_m^{II} using Eqs. (69) and (72).

In each case we use one equation from the continuity of Φ and one from the continuity of $\partial\Phi/\partial x$.

For example, if we take option (c) from this list we obtain the infinite system of equations

$$U_m^{II} + \sum_{n=1}^{\infty} U_n^{II} \left(\sum_{s=0}^{\infty} \frac{\kappa_m}{k_s} c_{sm} c_{sn} \right) = -2\kappa_m c_{0m}, \qquad m \geq 1 \tag{74}$$

and option (d) leads to a similar type of system for the unknowns U_n^I. This approach has the advantage that the unknown appears both inside and outside the summation, which tends to lead to better conditioned matrices when attempting a numerical solution.

If we define a matrix $\mathbf{M}$ with elements

$$M_{mn} = \delta_{mn} + \sum_{s=0}^{\infty} \frac{\kappa_m}{k_s} c_{sm} c_{sn}, \qquad m, n \geq 1, \tag{75}$$

then our system can be written

$$\mathbf{M}\mathbf{x} = \mathbf{b}, \tag{76}$$

where the unknown column vector $\mathbf{x}$ has nth element U_n^{II} and the nth element of the column vector $\mathbf{b}$ is $-2\kappa_n c_{0n}$. In practice this system of equations must be solved by first truncating it to an $N \times N$ system

$$\mathbf{M}^{(N)}\mathbf{x}^{(N)} = \mathbf{b}^{(N)}, \tag{77}$$

and then numerically solving the resulting approximate system.

One disadvantage of this approach is that the solution to the boundary-value problem has singular derivatives near to $x = y = 0$ due to the discontinuous boundary condition. The appropriate behaviour for the velocity near the origin is in fact given by

$$\partial\Phi/\partial r = O(r^{-1/2}) \qquad \text{as } r = (x^2 + y^2)^{1/2} \to 0. \tag{78}$$

This has not been built in to the solution procedure and as a consequence one could expect that a large value of N would have to be used to obtain accurate results.

6.4 Complex Variable Methods

6.4.1 *Cauchy's residue theorem*

Let D be a simply connected domain and C a simple closed contour lying entirely within D. If a function f is analytic within and on C, except at a finite number of isolated singular points z_n, $n = 1, 2, \ldots, N$, then

$$\oint_C f(z)\,\mathrm{d}z = 2\pi\mathrm{i} \sum_{n=1}^{N} \mathrm{Res}(f : z_n), \tag{79}$$

where $\mathrm{Res}(f : z_n)$ is the residue of $f(z)$ at $z = z_n$. If we apply this result to the integrand $f(z)/(z - w)$ we get

$$\frac{1}{2\pi\mathrm{i}} \oint_C \frac{f(z)}{z - w}\,\mathrm{d}z = f(w) + \sum_{n=1}^{N} \frac{\mathrm{Res}(f : z_n)}{z_n - w}. \tag{80}$$

We will use this in the next section.

6.4.2 *Abrupt change in boundary condition: solution using residue calculus theory*

In order to use the residue calculus theory we need to proceed using either option (a) (eliminate U_m^I using Eqs. (69) and (70)) or (b) (eliminate U_m^{II} using Eqs. (71) and (72)). Since the main objective is to calculate R, which is related to U_0^I, the simplest approach is (b). Thus, for $m \geq 1$,

$$\sum_{n=0}^{\infty} \frac{U_n^I}{\kappa_m} c_{nm} + \sum_{n=0}^{\infty} \frac{U_n^I}{k_n} c_{nm} = -2c_{0m}, \tag{81}$$

so from Eq. (73)

$$\sum_{n=0}^{\infty} U_n^I \frac{(2\epsilon_n)^{1/2}\mu_m}{\kappa_m^2 - k_n^2}\left(\frac{1}{\kappa_m} + \frac{1}{k_n}\right) = -2\frac{(2\epsilon_0)^{1/2}\mu_m}{\kappa_m^2 - k_0^2}, \tag{82}$$

or, from the definitions of μ_m and ϵ_0

$$\sum_{n=0}^{\infty} \frac{U_n^I \epsilon_n^{1/2}}{k_n(\kappa_m - k_n)} = -\frac{2\kappa_m}{\kappa_m^2 - k_0^2} = \frac{1}{k_0 - \kappa_m} - \frac{1}{k_0 + \kappa_m}, \tag{83}$$

which can be written as

$$\sum_{n=0}^{\infty} \frac{V_n}{k_n - \kappa_m} = \frac{1}{k_0 + \kappa_m}, \tag{84}$$

where

$$V_0 = 1 + \frac{U_0^I}{k_0} = R, \qquad V_n = \frac{2^{1/2} U_n^I}{k_n}, \qquad n \geq 1. \tag{85}$$

The behaviour of the velocity near the origin can be used to show that

$$U_n^I, U_n^{II} = O(n^{-1/2}) \qquad \text{as } n \to \infty, \tag{86}$$

and hence V_n must satisfy $V_n = O(n^{-3/2})$ as $n \to \infty$.

We can solve Eq. (84) as follows. Consider the numbers

$$I_m = \lim_{N\to\infty} \frac{1}{2\pi \mathrm{i}} \int_{C_N} \frac{f(z)}{z - \kappa_m}\,\mathrm{d}z, \qquad m \geq 1, \tag{87}$$

where C_N is a sequence of contours (to be determined) on which $z \to \infty$ as $N \to \infty$ and $f(z)$ is a meromorphic function that has the following properties:

(P1) $f(z)$ has simple poles at $z = k_n$, $n \geq 0$, and at $-k_0$,
(P2) $f(z)$ has simple zeros at $z = \kappa_n$, $n \geq 1$,

(P3) $f(z) = o(1)$ as $z \to \infty$ on C_N as $N \to \infty$.

The last of these conditions ensures that $I_m = 0$, $m \geq 1$, and then Cauchy's residue theorem gives

$$\sum_{n=0}^{\infty} \frac{\text{Res}(f : k_n)}{k_n - \kappa_m} + \frac{\text{Res}(f : -k_0)}{-k_0 - \kappa_m} = 0, \qquad m \geq 1. \tag{88}$$

If we compare this with Eq. (84) we see that the solution to Eq. (84) is given by

$$V_n = \text{Res}(f : k_n) \tag{89}$$

provided $\text{Res}(f : -k_0) = 1$.

Can a suitable f be found? Let

$$f(z) = \frac{A}{z^2 - k_0^2} \prod_{m=1}^{\infty} \frac{1 - z/\kappa_m}{1 - z/k_m}, \tag{90}$$

where A is a constant. The infinite product converges uniformly on any compact set excluding $z = k_m$, $m \geq 0$ and $z = -k_0$ since

$$\prod_{m=1}^{\infty} \frac{1 - z/\kappa_m}{1 - z/k_m} = \prod_{m=1}^{\infty} \left(1 + O(m^{-2})\right). \tag{91}$$

It can be shown that if we choose C_N to be a circle centred on the origin with radius $(n - \frac{1}{4})\pi$, then as $z \to \infty$ on the curves C_N as $N \to \infty$ we have

$$\prod_{m=1}^{\infty} \frac{1 - z/\kappa_m}{1 - z/k_m} = O(z^{1/2}) \tag{92}$$

and so, from Eq. (90), $f(z) = O(z^{-3/2})$, which certainly satisfies condition (P3).

The function $f(z)$ defined by Eq. (90) clearly satisfies conditions (P1) and (P2) and we can make sure that $\text{Res}(f : -k_0) = 1$ by choosing

$$A = -2k_0 \prod_{m=1}^{\infty} \frac{1 + k_0/k_m}{1 + k_0/\kappa_m}. \tag{93}$$

We then have

$$V_0 = R = \text{Res}(f : k_0) = -\prod_{m=1}^{\infty} \frac{(1 + k_0/k_m)(1 - k_0/\kappa_m)}{(1 - k_0/k_m)(1 + k_0/\kappa_m)} \tag{94}$$

and the condition $V_n = O(n^{-3/2})$ can be checked directly. The numerator within the product is the complex conjugate of the denominator since k_0 is purely imaginary and so it immediately follows that $|R| = 1$ as required. If we write $R = -\exp(2\mathrm{i}\chi)$ then

$$2\mathrm{i}\chi = \sum_{m=1}^{\infty}\left[\ln\left(\frac{1+\mathrm{i}k/\kappa_m}{1-\mathrm{i}k/\kappa_m}\right) - \ln\left(\frac{1+\mathrm{i}k/k_m}{1-\mathrm{i}k/k_m}\right)\right] \tag{95}$$

so

$$\chi = \sum_{m=1}^{\infty}\left(\tan^{-1}\frac{k}{\kappa_m} - \tan^{-1}\frac{k}{k_m}\right). \tag{96}$$

6.4.3 *The Wiener–Hopf technique*

In the problem with the abrupt change in boundary condition we can extract the incident wave field by setting

$$\Phi(x,y) = f(x,y) + \mathrm{e}^{\mathrm{i}kx}. \tag{97}$$

Taking Fourier transforms as in the previous chapter leads to the Wiener–Hopf equation

$$-\frac{\mathrm{i}}{s+k} + F_-(s,0) = -\frac{\cosh\gamma}{\gamma\sinh\gamma}F'_+(s,0), \tag{98}$$

where

$$\gamma = (s^2 - k^2)^{1/2} \tag{99}$$

and the prime indicates differentiation with respect to y. To ensure the appropriate convergence of the Fourier transforms we assume that k has a small positive imaginary part ϵ; the final result is obtained by letting $\epsilon \to 0$.

Equation (98) can be rearranged to give

$$-\mathrm{i}\frac{s-k}{s+k} + F_-(s,0)(s-k) = -K(s)\frac{F'_+(s,0)}{s+k}, \tag{100}$$

where

$$K(s) = \gamma\coth\gamma. \tag{101}$$

The next step is to write

$$K(s) = K_+(s)K_-(s), \tag{102}$$

where K_+ is a plus function and K_- is a minus function, and each is analytic and non-zero in its respective half-plane. This can be done by noting that $K(s)$ has an infinite number of simple poles and zeros (but no branch points) and expanding it as an infinite product. The poles and zeros in the respective half-planes can then be easily separated. (It turns out that $K_-(-s) = K_+(s)$.) Suppose that this has been done. Then we can rearrange Eq. (100) to give

$$-\frac{\mathrm{i}}{s+k}\left(\frac{s-k}{K_-(s)} + \frac{2k}{K_+(k)}\right) + \frac{F_-(s,0)}{K_-(s)}(s-k) = \\ -K_+(s)\frac{F'_+(s,0)}{s+k} - \frac{2\mathrm{i}k}{(s+k)K_+(k)}. \tag{103}$$

Using the appropriate conditions on the solution near the origin we can show that both sides of Eq. (103) must be identically zero and hence $F'_+(s,0)$ and $F_-(s,0)$ can be determined. Knowledge of either of these functions is sufficient to calculate $F(s,y)$ and then Fourier inversion yields

$$f(x,y) = \frac{\mathrm{i}k}{\pi K_+(k)} \int \frac{\cosh\gamma(y-1)}{K_+(s)\gamma\sinh\gamma} \mathrm{e}^{-\mathrm{i}sx}\,\mathrm{d}s. \tag{104}$$

The contour of integration is from $-\infty$ to ∞ in the strip of analyticity. On letting $\epsilon \to 0$ this reduces to the real axis, but we must indent the contour above any singularities on $(-\infty, 0)$ and below any singularities on $(0,\infty)$.

For $x < 0$ we can close the contour in the upper-half plane and we will pick up modes corresponding to the zeros of $\sinh\gamma$. On the other hand, for $x > 0$ we would have to close the contour in the lower-half plane. Since

$$\frac{1}{K_+(s)} = \frac{\tanh\gamma}{\gamma} K_-(s) \tag{105}$$

we could rewrite the solution as

$$f(x,y) = \frac{\mathrm{i}k}{\pi K_+(k)} \int \frac{\cosh\gamma(y-1)}{\gamma^2\cosh\gamma} K_-(s)\,\mathrm{e}^{-\mathrm{i}sx}\,\mathrm{d}s \tag{106}$$

and so we would pick up modes from the zeros of $\cosh\gamma$.

We will explicitly solve the $x < 0$ case, for which we will have to find $K_+(s)$. The only poles of the integrand in Eq. (104) that lead to waves in the far field are those with s real and positive (the contour is indented above poles on the imaginary axis) and if s has positive imaginary part then $\exp(-\mathrm{i}sx)$ decays exponentially as $x \to -\infty$. Thus the reflection coefficient is determined by the zeros of $\gamma\sinh\gamma$. Note that

$$\sinh z = z\prod_{n=1}^{\infty}(1 + z^2/\lambda_n^2), \qquad \lambda_n = n\pi, \tag{107}$$

$$\cosh z = \prod_{n=1}^{\infty}(1 + z^2/\mu_n^2), \qquad \mu_n = (n - \tfrac{1}{2})\pi, \tag{108}$$

so

$$\gamma \sinh \gamma = \gamma^2 \prod_{n=1}^{\infty} (1 + \gamma^2/\lambda_n^2) = (s^2 - k^2) \prod_{n=1}^{\infty} (1 + (s^2 - k^2)/\lambda_n^2). \quad (109)$$

The only real positive zero is at $s = k$, and

$$\frac{s - k}{\gamma \sinh \gamma} \to \frac{1}{2k}, \qquad \text{as } s \to k. \quad (110)$$

Hence, from Cauchy's residue theorem

$$f(x, y) \sim 2\pi \mathrm{i} \frac{\mathrm{i}k \mathrm{e}^{-\mathrm{i}kx}}{\pi [K_+(k)]^2 2k} = -\frac{\mathrm{e}^{-\mathrm{i}kx}}{[K_+(k)]^2}, \qquad \text{as } x \to -\infty, \quad (111)$$

so

$$R = -[K_+(k)]^{-2}. \quad (112)$$

Next,

$$\begin{aligned} K(s) = \frac{\gamma \cosh \gamma}{\sinh \gamma} &= \prod_{n=1}^{\infty} (1 + \gamma^2/\mu_n^2) \prod_{n=1}^{\infty} (1 + \gamma^2/\lambda_n^2) \\ &= \prod_{n=1}^{\infty} \left(\frac{s^2 + \kappa_n^2}{\mu_n^2} \right) \left(\frac{s^2 + k_n^2}{\lambda_n^2} \right)^{-1} = \prod_{n=1}^{\infty} \frac{\lambda_n^2 (s + \mathrm{i}\kappa_n)(s - \mathrm{i}\kappa_n)}{\mu_n^2 (s + \mathrm{i}k_n)(s - \mathrm{i}k_n)}. \end{aligned} \quad (113)$$

Hence $K_+(s)$, which is analytic and non-zero in the upper-half plane, is given by

$$K_+(s) = \prod_{n=1}^{\infty} \frac{\lambda_n (s + \mathrm{i}\kappa_n)}{\mu_n (s + \mathrm{i}k_n)}, \quad (114)$$

with $K_-(s)$ just given by changing the plus signs to minus signs. In this way we ensure that $K_+(s) = K_-(-s)$. The reflection coefficient is then

$$\begin{aligned} R = -[K_+(k)]^{-2} &= -\prod_{n=1}^{\infty} \frac{\mu_n^2 (k + \mathrm{i}k_n)^2}{\lambda_n^2 (k + \mathrm{i}\kappa_n)^2} = -\prod_{n=1}^{\infty} \frac{(k^2 + \kappa_n^2)(k + \mathrm{i}k_n)^2}{(k^2 + k_n^2)(k + \mathrm{i}\kappa_n)^2}, \\ &= -\prod_{n=1}^{\infty} \frac{(k - \mathrm{i}\kappa_n)(k + \mathrm{i}k_n)}{(k - \mathrm{i}k_n)(k + \mathrm{i}\kappa_n)} = -\prod_{n=1}^{\infty} \frac{(\mathrm{i}k/\kappa_n + 1)(-\mathrm{i}k/k_n + 1)}{(\mathrm{i}k/k_n + 1)(-\mathrm{i}k/\kappa_n + 1)}, \end{aligned} \quad (115)$$

which is the same as Eq. (94).

Further Reading

For those interested in mathematical techniques relevant to acoustic waveguides, a few references are provided below. Some, in particular the books of Noble and Mittra & Lee, are at a fairly advanced level. The more basic techniques, such as separation of variables, can be found in standard texts on advanced engineering mathematics.

Argence, E. and Kahan, T. (1967) *Theory of Waveguides and Cavity Resonators.* Blackie & Son.

Billingham, J. and King, A. C. (2000) *Wave Motion.* Cambridge University Press.

Collin, R. E. (1991) *Field Theory of Guided Waves.* IEEE Press.

Coulson, C. A. and Jeffrey, A. (1977) *Waves: A Mathematical Approach to the Common Types of Wave Motion.* Longman.

Courant, R. and Hilbert, D. (1953 & 1962) *Methods of Mathematical Physics, vols. I & II.* Interscience Publishers.

Jones, D. S. (1986) *Acoustic and Electromagnetic Waves.* Clarendon Press.

Leppington, F. G. (1992) Wiener–Hopf Technique, chapter 5 of *Modern Methods in Analytical Acoustics* by Crighton, D. G. *et al.* Springer Verlag.

Mittra, R. and Lee, S. W. (1971) *Analytical Techniques in the Theory of Guided Waves.* Macmillan.

Morse, P. M. and Ingard, K. U. (1986) *Theoretical Acoustics.* Princeton University Press.

Noble, B. (1988) *Methods Based on the Wiener–Hopf Technique for the Solution of Partial Differential Equations.* American Mathematical Society.

Rossing, T. D. and Fletcher, N. H. (1995) *Principles of Vibration and Sound.* Springer Verlag.

Problems

Exercise 6.1 Calculate the modes for a rigid rectangular guide with cross-section

$$-a < z < a, \qquad -b < y < b,$$

the guide being aligned with the x-axis.

(a) Write down the Helmholtz equation in 3-D Cartesian coordinates.
(b) Substitute $\Phi(x, y, z) = X(x)Y(y)Z(z)$ into the Helmholtz equation and re-arrange so as to introduce two separation constants.
(c) Apply the boundary conditions on the guide walls.

(d) Calculate the expression for a general mode in the guide.
(e) What are the cut-off frequencies?

Exercise 6.2 Calculate the modes for a rigid circular cylindrical resonator with cross-section $r < a$ and ends at $x = 0$ and $x = L$.

Exercise 6.3 Solve the boundary-value problem illustrated below.

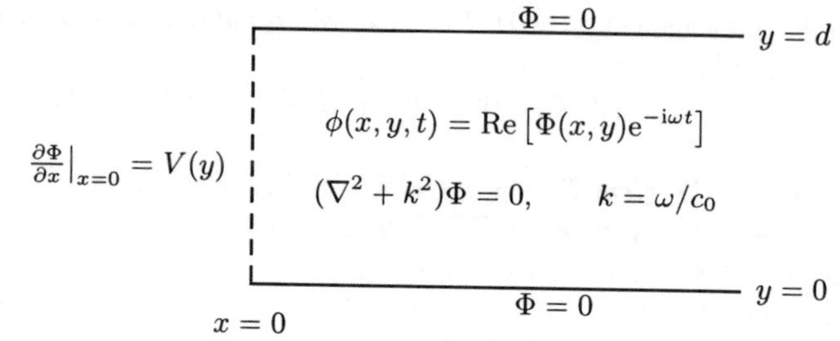

$\Phi = 0$ $y = d$

$\left.\frac{\partial\Phi}{\partial x}\right|_{x=0} = V(y)$

$$\phi(x,y,t) = \mathrm{Re}\left[\Phi(x,y)\mathrm{e}^{-\mathrm{i}\omega t}\right]$$

$$(\nabla^2 + k^2)\Phi = 0, \qquad k = \omega/c_0$$

$\Phi = 0$ $y = 0$

$x = 0$

Exercise 6.4 Set up an infinite system of equations for the problem considered in §6.3 but with the incident wave from the right. Take $\pi/2 < k < \pi$ so that one propagating mode is possible in each of the two regions. It may help to re-label κ_n so that both sets begin at zero, i.e.

$$\psi_n(y) = \epsilon_n^{1/2}\cos\lambda_n y, \qquad k_n = (\lambda_n^2 - k^2)^{1/2}, \qquad \lambda_n = n\pi,$$
$$\Psi_n(y) = 2^{1/2}\sin\mu_n y, \qquad \kappa_n = (\mu_n^2 - k^2)^{1/2}, \qquad \mu_n = (n + \tfrac{1}{2})\pi.$$

Then $k_0 = -\mathrm{i}k$ and $\kappa_0 = -\mathrm{i}\kappa$ are purely imaginary whereas k_n, κ_n, $n \geq 1$ are real and positive. We then have the conditions

$$\Phi \sim T\mathrm{e}^{k_0 x}\psi_0(y) = T\mathrm{e}^{-\mathrm{i}kx} \qquad \text{as } x \to -\infty,$$
$$\Phi \sim (\mathrm{e}^{\kappa_0 x} + R\mathrm{e}^{-\kappa_0 x})\Psi_0(y) = 2^{1/2}\,(\mathrm{e}^{-\mathrm{i}\kappa x} + R\mathrm{e}^{\mathrm{i}\kappa x})\sin\mu_0 y \qquad \text{as } x \to \infty,$$

instead of Eqs. (62) and (63).

Chapter 7

Wavefield Decomposition

M. C. M. Wright
Institute of Sound and Vibration Research
University of Southampton

7.1 Introduction

In Example 3.16 (p. 85) we analysed the field due to a monopole source close to an acoustically hard reflecting boundary. Consider now the more difficult problem of a monopole source with a reflecting boundary that is non-locally reacting. This means that a plane wave incident at angle ψ to the boundary's normal is reflected so that its complex amplitude is modified by a factor $V(\psi)$, which may be complex to allow for a change of phase at the boundary. If the source is a long way from the boundary we could argue that the curvature of wave fronts is negligible and that they can be approximated as plane waves. When the source is near to the boundary, however, it would seem to be necessary to decompose its wavefield into plane wave components so that the reflection condition can be applied to each separately.

7.2 Wavefields

The wavefield generated by a time-harmonic point source at the origin satisfies the inhomogeneous Helmholtz equation

$$(\nabla^2 + k^2)p = \delta(\mathbf{x}), \tag{1}$$

after discarding a time factor of $\exp(-\mathrm{i}\omega t)$ and where $k = \omega/c_0$. The solution that obeys the outgoing wave condition is

$$p(\mathbf{x}) = \frac{\mathrm{e}^{\mathrm{i}k(x^2+y^2+z^2)^{1/2}}}{(x^2+y^2+z^2)^{1/2}} = \frac{\mathrm{e}^{\mathrm{i}kR}}{R}, \tag{2}$$

which corresponds to spherical waves travelling from the origin at speed c_0 whose amplitude is in inverse proportion to their distance from the origin.

On the other hand, plane harmonic waves propagating in the co-ordinate directions have spatial factors proportional to $\mathrm{e}^{\pm\mathrm{i}\lambda x}$, $\mathrm{e}^{\pm\mathrm{i}\mu y}$ and $\mathrm{e}^{\pm\mathrm{i}\nu z}$. Since these functions form a complete set over xyz-space it should be possible to rewrite any other spatial factor of time-harmonic wavefield in terms of these functions (the fact that these functions are mutually orthogonal is not necessary, but it is useful.) This means that any harmonic wavefield can be decomposed into a sum of plane wavefields.

Once we have decomposed our wavefield we shall be able to apply our knowledge of reflection and transmission of plane waves to each of the constituent wavefields, allowing us to solve reflection and transmission problems for the wavefields of simple sources. Of course the reflections and transmissions will have to be from/through plane interfaces, but the problem of the field generated by a simple source near to a plane boundary is common enough to make the solution useful.

7.2.1 *Three-dimensional Fourier transform*

The most obvious way to decompose $p(\mathbf{x})$ is by a triple Fourier transform from the variables $\mathbf{x} = (x, y, z)$ to $\mathbf{k} = (k_x, k_y, k_z)$ by attempting to find a function $P(k_x, k_y, k_z)$ that satisfies

$$p(x,y,z) = \iiint_{-\infty}^{\infty} P(k_x,k_y,k_z)\mathrm{e}^{\mathrm{i}(k_x x+k_y y+k_z z)}\,\mathrm{d}k_x\,\mathrm{d}k_y\,\mathrm{d}k_z, \tag{3}$$

or more compactly

$$p(\mathbf{x}) = \iiint_{-\infty}^{\infty} P(\mathbf{k})\mathrm{e}^{\mathrm{i}(\mathbf{k}\cdot\mathbf{x})}\,\mathrm{d}^2\mathbf{k}. \tag{4}$$

The integrand for any fixed set of (k_x, k_y, k_z) represents a plane wave. This works out particularly simply because standard Fourier transforms decompose functions into integrals of complex exponentials, and plane wave fields are represented by complex exponentials.

It is, in fact, possible to find an equation for $P(\mathbf{k})$ by writing the forward Fourier transform

$$P(k_x,k_y,k_z) = \frac{1}{(2\pi)^3}\iiint_{-\infty}^{\infty}\left(\frac{\mathrm{e}^{\mathrm{i}kR}}{R}\right)\mathrm{e}^{-\mathrm{i}(k_x x+k_y y+k_z z)}\,\mathrm{d}x\,\mathrm{d}y\,\mathrm{d}z, \tag{5}$$

and converting to spherical polar coordinates to obtain the result (Morse & Feshbach, 1953)

$$P(k_x, k_y, k_z) = \frac{1}{(2\pi)^3} \frac{1}{k_x^2 + k_y^2 + k_z^2 - k^2}. \tag{6}$$

This is dominated by its singularities, which occur when $k_x^2 + k_y^2 + k_z^2 = k^2$, at which points the wavefield corresponding to the integrand satisfies the wave equation. Expressions of this form can be very useful, as shown in §4.9 of Lighthill (1978). Here, however, we shall instead use a procedure developed by Weyl (1919), (see also Sommerfeld, 1949, Brekhovskikh & Godin, 1992) to seek a nonsingular expression composed only of such terms, in which case once any two of k_x, k_y, k_z have been specified the value for the third will be automatically determined, so we should only need a double integral.

7.2.1.1 *Sign conventions*

Already we have made two arbitrary choices for the sign of the exponent in terms of the form $\mathrm{e}^{\pm ax}$: once in choosing a time factor of $\mathrm{e}^{-\mathrm{i}\omega t}$ to discard and again when choosing $\mathrm{e}^{+\mathrm{i}k_x x}$ terms for the inverse Fourier transform Eq. (3). We are quite at liberty to reverse either or both of these signs but it is convenient for them to be opposites, since this means that the plane waves into which we are decomposing our wavefield will be propagating in the positive x direction when the transform variable k_x is positive.

7.3 Two-Dimensional Representation

We will ignore k_z for the time being and choose k_x and k_y as the variables we intend to integrate over by considering the wavefield in the plane $z = 0$, where it is given by

$$\frac{\mathrm{e}^{\mathrm{i}k(x^2+y^2)^{1/2}}}{(x^2+y^2)^{1/2}} = \frac{\mathrm{e}^{\mathrm{i}kr}}{r} = \iint_{-\infty}^{\infty} F(k_x, k_y)\mathrm{e}^{\mathrm{i}(k_x x + k_y y)}\, \mathrm{d}k_x\, \mathrm{d}k_y, \tag{7}$$

where

$$F(k_x, k_y) = \frac{1}{(2\pi)^2} \iint_{-\infty}^{\infty} \left(\frac{\mathrm{e}^{\mathrm{i}kr}}{r}\right) \mathrm{e}^{-\mathrm{i}(k_x x + k_y y)}\, \mathrm{d}x\, \mathrm{d}y. \tag{8}$$

It turns out that this integration can be performed by transforming both domains into polar variables through the following substitutions:

$$x = r\cos\phi, \qquad y = r\sin\phi, \qquad r = (x^2+y^2)^{1/2}, \tag{9}$$

and

$$k_x = k_r \cos\gamma, \qquad k_y = k_r \sin\gamma, \qquad k_r = (k_x^2 + k_y^2)^{1/2}. \tag{10}$$

Recall that when substituting multiple variables the new integrand must be multiplied by the absolute[1] value of the determinant of the Jacobian of the transformation, i.e:

$$\begin{aligned} \mathrm{d}x\,\mathrm{d}y &= \left|\det\left(\frac{\partial(x,y)}{\partial(r,\phi)}\right)\right| \mathrm{d}r\,\mathrm{d}\phi \\ &= \left|\det\begin{pmatrix} \cos\phi & r\sin\phi \\ \sin\phi & -r\cos\phi \end{pmatrix}\right| \mathrm{d}r\,\mathrm{d}\phi \\ &= \left|-r(\cos^2\phi + \sin^2\phi)\right| \mathrm{d}r\,\mathrm{d}\phi \\ &= r\,\mathrm{d}r\,\mathrm{d}\phi. \end{aligned} \tag{11}$$

Since x and y run over limits that include the entire plane the new limits of the integral give

$$\begin{aligned} F(k_x, k_y) &= \frac{1}{(2\pi)^2}\int_0^{2\pi}\int_0^{\infty}\left(\frac{\mathrm{e}^{\mathrm{i}kr}}{r}\right)\mathrm{e}^{-\mathrm{i}(qr\cos\phi\cos\gamma + qr\sin\phi\sin\gamma)} r\,\mathrm{d}r\,\mathrm{d}\phi. \\ &= \frac{1}{(2\pi)^2}\int_0^{2\pi}\int_0^{\infty}\mathrm{e}^{\mathrm{i}r[k-q\cos(\gamma-\phi)]}\,\mathrm{d}r\,\mathrm{d}\phi. \\ &= \frac{1}{(2\pi)^2}\int_0^{2\pi}\left[\frac{\mathrm{e}^{\mathrm{i}r[k-q\cos(\gamma-\phi)]}}{\mathrm{i}[k-q\cos(\gamma-\phi)]}\right]_0^{\infty}\mathrm{d}\phi. \end{aligned} \tag{12}$$

For any physically realistic medium there will be some attenuation with distance, which can be modelled by giving the wavenumber k a small positive imaginary part. This means that the integral with respect to r will vanish at the upper limit, and the remaining integral over ϕ can be manipulated into a standard form

[1] Obviously taking the absolute part throws some useful information away and may cause trouble. It would be more correct to allow the sign of the integral to depend on the orientation of the elemental areas but that would require us to identify $\iint f\,\mathrm{d}x\,\mathrm{d}y$ with $-\iint f\,\mathrm{d}y\,\mathrm{d}x$, for which we require the concept of 'differential forms'. Interested readers will find an introduction in Baxandall & Liebeck (1987).

as follows:

$$\begin{aligned}F(k_x, k_y) &= \frac{1}{(2\pi)^2}\int_0^{2\pi}\frac{\mathrm{d}\phi}{-\mathrm{i}[k - q\cos(\gamma-\phi)]}\\ &= \frac{\mathrm{i}}{(2\pi)^2 k}\int_0^{2\pi}\frac{\mathrm{d}\delta}{1-(q/k)\cos\delta}\\ &= \frac{\mathrm{i}}{2\pi}\frac{1}{(k^2-q^2)^{1/2}}\\ &= \frac{\mathrm{i}}{2\pi(k^2-k_x^2-k_y^2)^{1/2}},\end{aligned} \tag{13}$$

which can be substituted into Eq. (7) to give

$$\frac{\mathrm{e}^{\mathrm{i}kr}}{r} = \frac{-\mathrm{i}}{2\pi}\iint_{-\infty}^{\infty}\frac{\mathrm{e}^{\mathrm{i}(k_x x + k_y y)}}{(k^2-k_x^2-k_y^2)^{1/2}}\,\mathrm{d}k_x\,\mathrm{d}k_y. \tag{14}$$

7.4 Extension to Three Dimensions

We wish to extend Eq. (14) into three-dimensional space by including z in the integrand. We would still like the integrand to represent a plane wave that satisfies the wave equation. By the process of analytic continuation (see §2.6 on p. 52) we can verify that these conditions can only be satisfied by applying a factor of $\mathrm{e}^{\pm\mathrm{i}k_z z}$, where

$$k_x^2 + k_y^2 + k_z^2 = k^2, \tag{15}$$

in order to satisfy the wave equation. We can recognise k_z as the denominator of the integrand and write

$$\frac{\mathrm{e}^{\mathrm{i}kR}}{R} = \frac{\mathrm{i}}{2\pi}\iint_{\infty}^{\infty}\frac{\mathrm{e}^{\mathrm{i}(k_x x + k_y y \pm k_z z)}}{k_z}\,\mathrm{d}k_x\,\mathrm{d}k_y, \tag{16}$$

where k_z satisfies Eq. (15). Consider first the case where $k_x^2 + k_y^2 < k^2$ so that k_z is purely real, so that the integrand represents an ordinary plane wave propagating with wavevector $\mathbf{k} = (k_x, k_y, k_z)$. To represent a component of a spherical wave the plane wave should be moving away from the $z = 0$ plane on either side, therefore we should take the positive sign when $z > 0$ and *vice versa*. We can make this explicit by writing separate equations for the two spaces or, more compactly, by writing

$$\frac{\mathrm{e}^{\mathrm{i}kR}}{R} = \frac{\mathrm{i}}{2\pi}\iint_{-\infty}^{\infty}\frac{\mathrm{e}^{\mathrm{i}(k_x x + k_y y + k_z |z|)}}{k_z}\,\mathrm{d}k_x\,\mathrm{d}k_y. \tag{17}$$

As k_x and k_y vary the value of k_z will be either purely real or purely imaginary. A positive imaginary value will attenuate the amplitude of a plane wave propagating in a direction that lies in the $z = 0$ plane by a factor that depends exponentially on the distance from that plane. Waves such as this are referred to as 'evanescent waves'. More properly the wavefield is said to evanesce in the z direction and propagate in the x and y directions. When this happens the apparent wavenumber of the propagating component will be $(k_x^2 + k_y^2)^{1/2}$, so as $k_x^2 + k_y^2$ increases beyond k the wavelength of the propagating wave is reduced. If k_z were negative imaginary then the amplitude would grow exponentially with z, which would be physically unrealistic.

There are four different paths in the complex plane that satisfy $k_z = (k^2 - k_x^2 - k_y^2)^{1/2}$, the alternatives being obtained by reflection in the real and/or imaginary axes. We choose the correct version (shown in Fig. 7.1) on physical grounds, arguing that when real the wavenumber should be positive for the waves to propagate away from the source and that when imaginary it should be positive to ensure exponential decay (or evanescence). We could also reason that in a lossy medium k would have a small positive imaginary part ϵ so that the correct path would be one given by

$$k_z = \lim_{\epsilon \to 0}[(k + \mathrm{i}\epsilon)^2 - k_x^2 - k_y^2]^{1/2}, \tag{18}$$

which follows the path Γ_{k_z} shown in Fig. 7.1. To see that this eliminates two of the four paths consider the locus of the quantity under the square root when ϵ is finite, and recall that taking the square root of a complex number halves its complex angle.

The integrand of Eq. (17) is singular at $k = 0$ so this point on the path of integration would have to be avoided or omitted somehow if this expression were to be evaluated numerically. It turns out that the development that follows avoids this problem.

7.5 Angular Representation

With Eq. (17) we have decomposed a spherical wave into an integral of plane waves as desired. In order to make use of this result, however, it would be useful to express the plane wave that corresponds to the integrand in terms of the angle it makes with some plane, since in our original problem the reflection coefficient was specified to be angle dependent.

In the same way that we converted (x, y) spatial dependence to (r, ϕ) dependence, and converted (k_x, k_y) wavenumber dependence to (k_r, γ) dependence in the plane, we will now convert from (k_x, k_y, k_z) dependence to (k, ψ, γ) dependence where γ is the azimuthal angle as before and ψ is the polar angle, giving

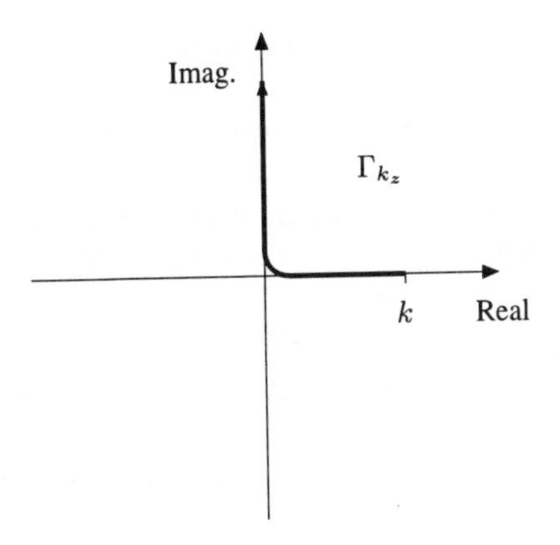

Fig. 7.1 Path of integration for k_z.

the transformation

$$k_x = k \sin\psi \cos\gamma, \tag{19}$$
$$k_y = k \sin\psi \sin\gamma, \tag{20}$$
$$k_z = k \cos\psi. \tag{21}$$

These values can be substituted directly into our integral (although in fact we won't bother to do so just yet), and we can find $\mathrm{d}k_x\,\mathrm{d}k_y$ as before by

$$\begin{aligned}
\mathrm{d}k_x\,\mathrm{d}k_y &= \left|\det\left(\frac{\partial(k_x,k_y)}{\partial(\psi,\gamma)}\right)\right|\,\mathrm{d}\psi\,\mathrm{d}\gamma \\
&= \left|\det\begin{pmatrix} k\cos\psi\cos\gamma & -k\sin\psi\sin\gamma \\ k\cos\psi\sin\gamma & k\sin\psi\cos\gamma \end{pmatrix}\right|\,\mathrm{d}\psi\,\mathrm{d}\gamma \\
&= |k^2\cos\psi\sin\psi|\,\mathrm{d}\psi\,\mathrm{d}\gamma \\
&= kk_z\sin\psi\,\mathrm{d}\psi\,\mathrm{d}\gamma,
\end{aligned} \tag{22}$$

giving us an expression of the form

$$\frac{\mathrm{e}^{\mathrm{i}kR}}{R} = \frac{\mathrm{i}k}{2\pi}\iint_{?} \mathrm{e}^{\mathrm{i}(k_x x + k_y y + k_z|z|)}\sin\psi\,\mathrm{d}\gamma\,\mathrm{d}\psi, \tag{23}$$

for which we must determine the paths of integration. Since the spherical wave we are trying to represent has no particular azimuthal dependence the range for γ can be from 0 to 2π. The polar angle ψ, however, implies dependence on k_z, which, we have seen, can become imaginary. We must choose a path in the complex plane for ψ that corresponds to the path for k_z as determined by Eq. (21) that we have previously chosen. To see how this can be done write $\psi = \alpha + \mathrm{i}\beta$ and recall that

$$k_z/k = \cos\psi = \cos\alpha\cosh\beta - \mathrm{i}\sin\alpha\sinh\beta, \qquad (24)$$

which follows directly from Euler's identity and the definitions of sinh and cosh. Using this expression we can treat the two straight sections of the path separately:

For	$1 \to k_z/k \to 0$	set	$\beta = 0,$	$0 \to \alpha \to \frac{\pi}{2},$
For	$0 \to k_z/k \to \mathrm{i}\infty$	set	$\alpha = \frac{\pi}{2},$	$0 \to \beta \to -\infty,$

giving the path for ψ shown in Fig. 7.2. Imaginary values of ψ will correspond

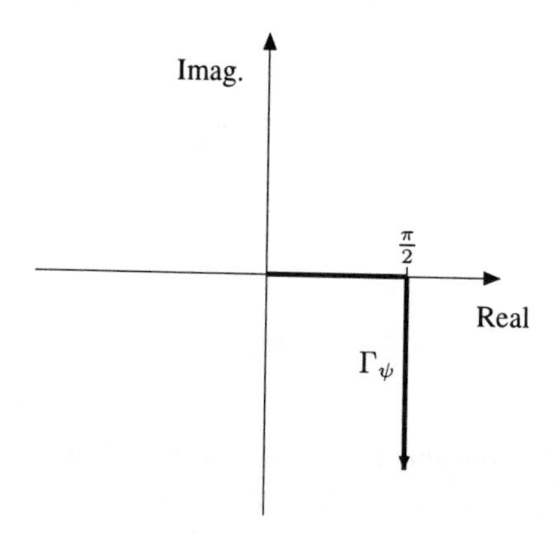

Fig. 7.2 Path of integration for ψ.

to waves that are purely evanescent in the z direction as did imaginary values of k_z. We ought to check that the corresponding values of k_x and k_y remain purely real along the ψ-path, which we can do by recalling that $\sin(\alpha + \mathrm{i}\beta) = \sin\alpha\cosh\beta + \mathrm{i}\cos\alpha\sinh\beta$. If either $\alpha = \frac{\pi}{2}$ and/or $\beta = 0$ then $\sin(\alpha + \mathrm{i}\beta)$ will be purely real as required.

We now have the expression:

$$\frac{\mathrm{e}^{\mathrm{i}kR}}{R} = \frac{\mathrm{i}k}{2\pi}\int_{\Gamma_\psi}\int_0^{2\pi} \mathrm{e}^{\mathrm{i}k(x\sin\psi\cos\gamma + y\sin\psi\sin\gamma + |z|\cos\psi)} \sin\psi \,\mathrm{d}\gamma\,\mathrm{d}\psi, \qquad (25)$$

which decomposes a spherical wave into an integral of plane waves, specified by azimuthal and polar angle of propagation, and whose integral is non-singular for all finite points on the path Γ_ψ.

7.6 Solution to Original Problem

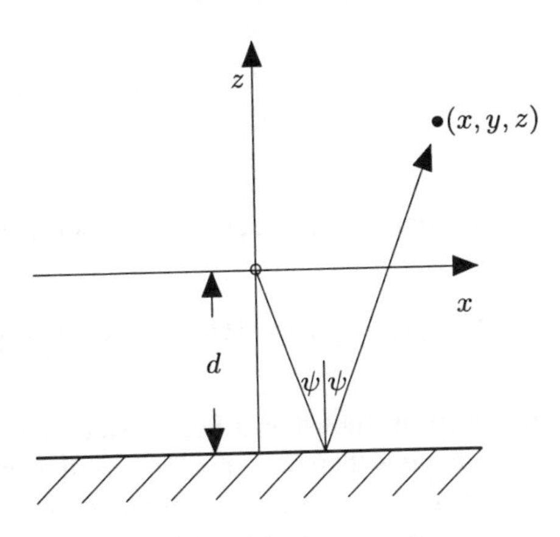

Fig. 7.3 Problem geometry.

Consider a monopole source located at the origin with a reflecting interface parallel to the $z = 0$ plane but a distance d below it. The boundary has the property that a plane wave incident at angle ψ to the normal is reflected so that its complex amplitude is modified by a factor $V(\psi)$ (which may be complex to allow for a change of phase at the boundary). A plane wave component of the reflected field, observed at the point (x, y, z) will have travelled a distance $(x, y, 2d + z)$. The spatial component of the total wavefield at that point will then be

$$p(x,y,z) = \frac{\mathrm{e}^{\mathrm{i}kR}}{R} + p_{\text{refl.}}(x,y,z), \tag{26}$$

where, using Eq. (25)

$$p_{\text{refl.}} = \frac{\mathrm{i}k}{2\pi}\int_{\Gamma_\psi}\int_0^{2\pi} \mathrm{e}^{\mathrm{i}k(x\sin\psi\cos\gamma + y\sin\psi\sin\gamma + (2d+z)\cos\psi)}V(\psi)\sin\psi\,\mathrm{d}\gamma\,\mathrm{d}\psi. \tag{27}$$

This expression can be simplified by writing $x = r\cos\phi$ and $y = r\sin\phi$ and using a convenient identity to evaluate the inner integral:

$$\begin{aligned}\int_0^{2\pi} \mathrm{e}^{\mathrm{i}k(x\cos\gamma+y\sin\gamma)\sin\psi}\,\mathrm{d}\gamma &= \int_0^{2\pi} \mathrm{e}^{\mathrm{i}kr(\cos\phi\cos\gamma+\sin\phi\sin\gamma)\sin\psi}\,\mathrm{d}\gamma \\ &= \int_0^{2\pi} \mathrm{e}^{\mathrm{i}kr\sin\psi\cos(\gamma-\phi)}\,\mathrm{d}\gamma \\ &= 2\pi\,\mathrm{J}_0(kr\sin\psi), \end{aligned} \tag{28}$$

where J_0 is the zeroth order Bessel function. We can now write the reflected wavefield as

$$p_{\text{refl.}} = \mathrm{i}k\int_{\Gamma_\psi} \mathrm{J}_0(kr\sin\psi)\mathrm{e}^{-\mathrm{i}k(2d+z)\cos\psi}V(\psi)\sin\psi\,\mathrm{d}\psi. \tag{29}$$

An expression of this form is known as a Sommerfeld integral and can contain more general transfer functions than the reflection coefficient we have used here. The integrand no longer represents a plane wave field but it does retain explicit dependence on ψ, which allows us to apply the reflection factor. Approximation techniques such as the method of steepest descent (see §4.5, p. 103) can be used to evaluate this integral; alternatively numerical integration is possible.

References

Baxandall, P. and Liebeck, H. (1987) *Vector Calculus.* Oxford University Press.

Brekhovskikh, L. M. and Godin, O. A. (1992) *Acoustics of Layered Media II: Point Sources and Bounded Beams.* Springer Verlag.

Lighthill, J. (1978) *Waves in Fluids.* Cambridge University Press.

Morse, P. M. and Feshbach, H. (1953) *Methods of Theoretical Physics. Vol. II p. 1433* McGraw–Hill.

Sommerfeld, A. (1949) *Partial Differential Equations in Physics.* Academic Press.

Weyl, H. (1919) Ausbreitung elektromagnetischer Wellen über einem ebenen Leiter, *Annals of Physics* **60**, pp. 481–500.

Problems

Exercise 7.1 Write computer programs to verify as many of the equations in this chapter as you can by numerical integration. Wherever possible use symmetry to reduce the dimensionality and size of the region over which you have to solve.

Exercise 7.2 Verify the three-dimensional Fourier transform given in Eq. (6).

Exercise 7.3 Hankel functions are given by $\mathrm{H}_0^{(1)} = \mathrm{J}_0 + \mathrm{i}\,\mathrm{Y}_0$ and $\mathrm{H}_0^{(2)} = \mathrm{J}_0 - \mathrm{i}\,\mathrm{Y}_0$. Draw the path of integration $\Gamma_{\psi,\mathrm{H}_0^{(1)}}$ necessary for the following identity to hold:

$$\frac{\mathrm{e}^{\mathrm{i}kR}}{R} = \mathrm{i}k \int_{\Gamma_{\psi,\mathrm{H}_0^{(1)}}} \mathrm{H}_0^{(1)}(kr\sin\psi)\mathrm{e}^{-\mathrm{i}k|z|\cos\psi}\sin\psi\,\mathrm{d}\psi.$$

Hint: Which functions are odd? Which functions are even?

Exercise 7.4 The Hankel transform pair can be written as

$$\tilde{\phi}(k_r) = \frac{1}{2\pi}\int_0^\infty \mathrm{J}_0(k_r r)\phi(r)\,\mathrm{d}r,$$

and

$$\phi(r) = 2\pi\int_0^\infty \mathrm{J}_0(k_r r)\tilde{\phi}(k_r)k_r\,\mathrm{d}k_r.$$

Show that if $\xi(x, y)$ is azimuthally uniform then $\tilde{\xi}$ is also the two-dimensional Fourier transform of ξ when $k_r^2 = k_x^2 + k_y^2$.

Chapter 8

Acoustics of Rigid–Porous Materials

K. Attenborough & O. Umnova
Department of Engineering
University of Hull

8.1 Introduction

Porous materials are used widely for sound absorption. They may be fibrous, cellular or granular. Fibrous materials may be in the form of mats, boards or preformed elements manufactured of glass, mineral or organic fibers (natural or manmade) and include felts and felted textiles. Cellular materials include polymer foams of varying degrees of rigidity. Increasing use is being made of porous metals, for example aluminium foams, as sound absorbers. In comparison with other materials, glass wool, for example, which gives high absorption over a wide frequency range, metal foams are not very good sound absorbers. However, the high weight-specific stiffness, good crash-energy absorption ability, and fire-resistance of porous metals make them suitable for sound absorption panels in the aircraft and automotive industries.

Granular materials can be regarded as an alternative to fibrous and foam absorbers in many indoor and outdoor applications. Sound absorbing granular materials combine good mechanical strength and very low manufacturing costs. Granular materials may be consolidated through use of some form of binder on the particles as in wood-chip panels, porous concrete and pervious road surfaces. Granular materials may be unconsolidated (loose). There are many naturally-occurring granular materials including sands, gravel, soils and snow. The acoustical properties of such materials are an important factor in outdoor sound propagation.

Six examples of porous materials used for sound absorption are shown in Fig. 8.1. Since the particles are nearly spherical and identical the lead shot serves as a reference material when testing theories for the acoustical properties of rigid-framed porous materials. This contribution investigates some of these theories. In theory rigid–porous materials may be considered to be dissipative fluids since the wave motion is confined to the fluid in the pores. The amplitude of the sound wave in a porous material is attenuated and the phase is changed as a result of viscous drag and thermal exchange in the pores. Ideally these effects should be treated simultaneously but here they are treated separately. This approach has been shown by the authors and others to be justified (see bibliography) and the later comparisons between data and predictions demonstrate its validity.

Fig. 8.1 Seven examples of porous materials.

8.2 Wave Equations

The vector wave equation in an isotropic elastic solid with equilibrium density ρ_0 is given by

$$\rho_0 \frac{\partial^2 \mathbf{u}}{\partial t^2} = (\lambda + \mu)\boldsymbol{\nabla}\boldsymbol{\nabla} \cdot \mathbf{u} + \mu \nabla^2 \mathbf{u} + \mathbf{X}, \tag{1}$$

where $\mathbf{u}$ is the displacement vector, $\mathbf{X}$ is the body force, μ is the rigidity modulus and the bulk modulus is given by $K = \lambda + 2\mu/3$.

In a fluid $\mu = 0$, $\mathbf{X} = 0$; principal stresses $= -p$, where p is the hydrostatic pressure, and $K = \lambda$:

$$\mathbf{u} = \boldsymbol{\nabla}\phi, \qquad K\boldsymbol{\nabla}\boldsymbol{\nabla} \cdot \boldsymbol{\nabla}\phi = \rho_0 \frac{\partial^2}{\partial t^2}\boldsymbol{\nabla}\phi. \tag{2}$$

Using $\nabla \cdot \nabla\phi = \nabla^2\phi$, this leads to

$$\nabla^2\phi = \frac{\rho_0}{K}\frac{\partial^2\phi}{\partial t^2}. \tag{3}$$

A plane harmonic wave solution of this equation is

$$\phi(x,t) = \frac{A}{\rho_0\omega^2}\cos(\omega(t - x/c_0) + \varphi), \tag{4}$$

with $c_0 = (K/\rho_0)^{1/2}$ representing the phase speed of the wave and φ representing the phase. This solution can be written

$$\phi(x,t) = \frac{a}{\rho\omega^2}\mathrm{e}^{\mathrm{i}(kx-\omega t)}, \tag{5}$$

with $k = \omega/c_0$ being the wavenumber and $a = A\mathrm{e}^{-\mathrm{i}\varphi}$. Note the time dependence convention being used here.

Using this expression for ϕ, pressure and velocity are found to be,

$$p(x,t) = -\rho_0\frac{\partial^2\phi}{\partial t^2} = \rho_0\omega^2\phi = a\mathrm{e}^{\mathrm{i}(kx-\omega t)}. \tag{6}$$

$$v(x,t) = \frac{\partial^2\phi}{\partial t\,\partial x} = \frac{a}{\rho_0 c_0}\mathrm{e}^{\mathrm{i}(kx-\omega t)}. \tag{7}$$

8.3 Impedance

The ratio of pressure to velocity in air is the characteristic impedance $p/v = \rho_0 c_0 = Z_c$.

Consider forward and backward-travelling plane harmonic waves at two infinite vertical planes in air separated by distance d (Fig. 8.2). There are both

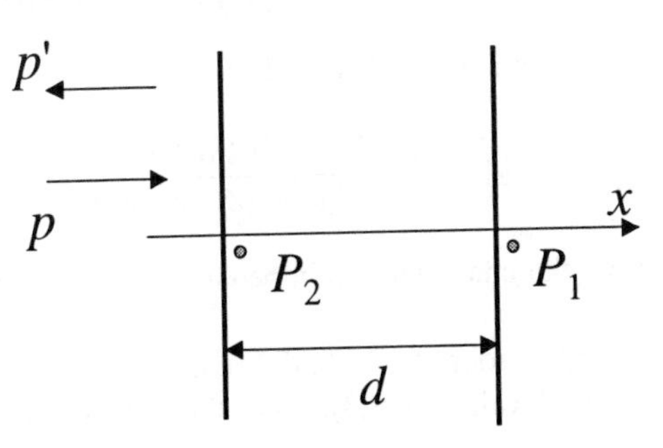

Fig. 8.2 Forward and backward plane waves in air.

positive-going (incident) and negative-going (reflected) waves. If the amplitude of the positive-going wave is a and that of the negative-going wave is a' then the pressure velocity and impedance (i.e. the ratio of pressure to particle velocity) at any point x are given by

$$p(x,t) = a\mathrm{e}^{\mathrm{i}(kx-\omega t)} + a'\mathrm{e}^{\mathrm{i}(-kx-\omega t)}, \tag{8}$$

$$\begin{aligned} v(x,t) &= \frac{a}{\rho_0 c_0}\mathrm{e}^{\mathrm{i}(kx-\omega t)} - \frac{a'}{\rho_0 c_0}\mathrm{e}^{\mathrm{i}(-kx-\omega t)}, \\ &= \frac{a}{Z_c}\mathrm{e}^{\mathrm{i}(kx-\omega t)} - \frac{a'}{Z_c}\mathrm{e}^{\mathrm{i}(-kx-\omega t)}, \end{aligned} \tag{9}$$

$$Z(x,t) = \frac{p(x,t)}{v(x,t)}. \tag{10}$$

At P_1,

$$\begin{aligned} Z(P_1) &= Z_c \frac{a\mathrm{e}^{\mathrm{i}(kx(P_1)-\omega t)} + a'\mathrm{e}^{\mathrm{i}(-kx(P_1)-\omega t)}}{a\mathrm{e}^{\mathrm{i}(kx(P_1)-\omega t)} - a'\mathrm{e}^{\mathrm{i}(-kx(P_1)-\omega t)}}, \\ &= Z_c \frac{a\mathrm{e}^{\mathrm{i}kx(P_1)} + a'\mathrm{e}^{-\mathrm{i}kx(P_1)}}{a\mathrm{e}^{\mathrm{i}kx(P_1)} - a'\mathrm{e}^{-\mathrm{i}kx(P_1)}}. \end{aligned} \tag{11}$$

Hence

$$\frac{a'}{a} = \frac{Z(P_1) - Z_c}{Z(P_1) + Z_c}\mathrm{e}^{2\mathrm{i}kx(P_1)}. \tag{12}$$

At P_2,

$$Z(P_2) = Z_c \frac{a\mathrm{e}^{\mathrm{i}(kx(P_2)-\omega t)} + a'\mathrm{e}^{\mathrm{i}(-kx(P_2)-\omega t)}}{a\mathrm{e}^{\mathrm{i}(kx(P_2)-\omega t)} - a'\mathrm{e}^{\mathrm{i}(-kx(P_2)-\omega t)}}. \tag{13}$$

Using Eqs. (12) and (13), it is possible to write

$$Z(P_2) = Z_c \frac{\mathrm{i}Z(P_1)\cot(kd) + Z_c}{Z(P_1) + \mathrm{i}Z_c\cot(kd)}. \tag{14}$$

Here Z_c and k denote the characteristic impedance and propagation constant in the fluid.

Now consider sound incident from fluid 2 (usually air) on a layer of material (fluid 1) in front of a rigid wall (Fig. 8.3). Since the wall is rigid, $Z(P_1) \to \infty$, so from Eqs. (12) & (13)

$$Z(P_2) = \mathrm{i}Z_c\cot(kd) = Z_c\coth(-\mathrm{i}kd). \tag{15}$$

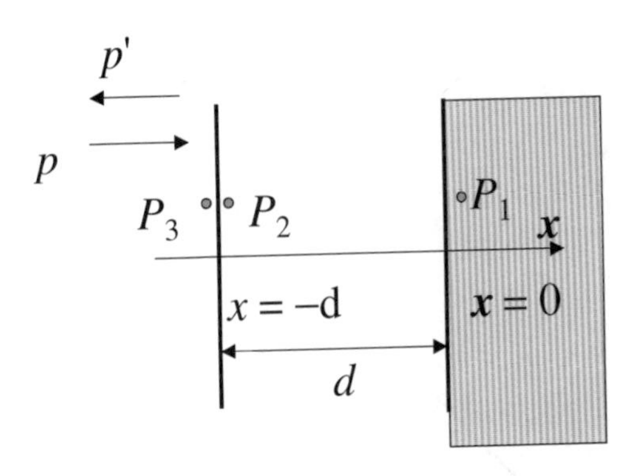

Fig. 8.3 Plane waves incident on a hard-backed layer.

Note that Z_c and k in this equation refer to the characteristic impedance and propagation constant in fluid 1. With the assumption that $Z(P_3) = Z(P_2)$, Eq. (15) gives the *surface impedance of a hard-backed layer* (fluid 1) of thickness d.

The (pressure) reflection coefficient is given by a'/a. The absorption coefficient of the layer follows from $1 - |a'/a|^2$.

As shown later, a consequence of dissipation due to viscous and thermal effects in the pores is that both propagation constant and characteristic impedance in a porous medium are complex:

$$k = k_r + \mathrm{i}k_i, \tag{16}$$

$$Z_c = Z_r + \mathrm{i}Z_i. \tag{17}$$

Also the surface impedance of a porous layer and the plane wave pressure reflection are complex quantities.

8.4 Analysis for a Single Slit

8.4.1 *Viscosity effects in a slit-like pore*

According to Newton's law of viscosity, the viscous stress in a fluid normal to the x_1-direction (Fig. 8.4.1) is given by

$$T_3(x_1) = -\eta \frac{\partial v_3(x_1)}{\partial x_1}, \tag{18}$$

where η is the coefficient of dynamic viscosity in the fluid.

The change in viscous stress across an imaginary fluid element between x_1

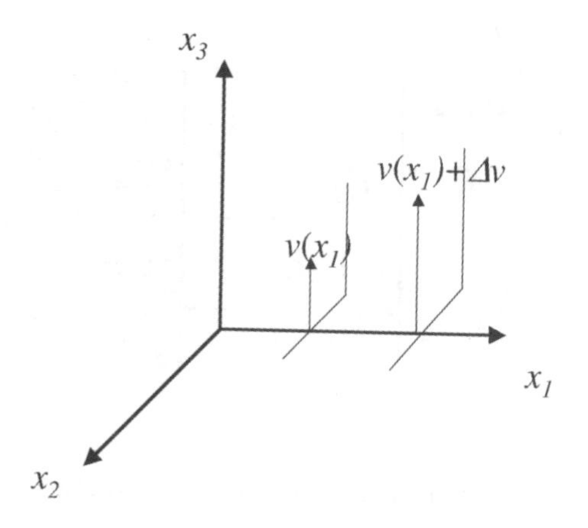

Fig. 8.4 Change in viscous stress across an imaginary fluid element.

and $x_1 + \Delta x_1$ (see Fig. 8.4.1) is given by

$$\Delta F_3 = -\eta \frac{\partial v_3(x_1)}{\partial x_1} + \eta \frac{\partial v_3(x_1 + \Delta x_1)}{\partial x_1}. \tag{19}$$

As the width of the fluid element becomes infinitesimally small, then the viscous force per unit volume, i.e. force per unit length if unit area in the x_1 direction is being considered, is given by

$$X_3 = \lim_{\Delta x_1 \to 0} \frac{\Delta F_3}{\Delta x_1} = \eta \frac{\partial^2 v_3}{\partial x_1^2}. \tag{20}$$

Making use of Eq. (20) in the equation of motion for a sound wave travelling in a parallel-walled slit of width $2a$ (Fig. 8.5),

$$\rho_0 \frac{\partial v_3}{\partial t} = -\mathrm{i}\omega\rho_0 v_3 = -\frac{\partial p}{\partial x_3} + \eta \frac{\partial^2 v_3}{\partial x_1^2}. \tag{21}$$

Applying the 'no-slip' boundary conditions i.e. $v_3 = 0$ at $x_1 = \pm a$, the solution of Eq. (21) is

$$v_3 = \frac{1}{\mathrm{i}\omega\rho_0} \frac{\partial p}{\partial x_3} \left[1 - \frac{\cosh(k_V x_1)}{\cosh(k_V a)}, \right] \tag{22}$$

where $k_V = (-\mathrm{i}\omega\rho_0/\eta)^{1/2}$ is the wavenumber of the viscous wave in a fluid.

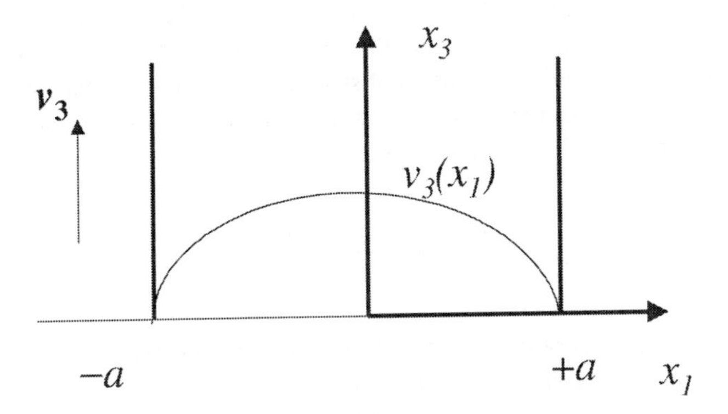

Fig. 8.5 Velocity profile across a parallel-walled slit due to dynamic viscous effects.

The average of the particle velocity over the width of the slit is given by

$$\bar{v}_3 = \frac{\int_{-a}^{a} v_3 \, \mathrm{d}x_1}{2a} = \frac{1}{\mathrm{i}\omega\rho_0}\left[1 - \frac{\tanh(\lambda\sqrt{-\mathrm{i}})}{(\lambda\sqrt{-\mathrm{i}})}\right]\frac{\partial p}{\partial x_3}. \tag{23}$$

Here λ is a dimensionless parameter given by $\lambda = (\omega\rho_0 a^2/\eta)^{1/2} = a(\omega\rho_0/\eta)^{1/2}$.

Equation (23) may be rewritten in the form

$$-\mathrm{i}\omega\rho(\omega)\bar{v}_3 = -\frac{\partial p}{\partial x_3}, \tag{24}$$

where

$$\rho(\omega) = \rho_0/F(\lambda), \tag{25a}$$

$$F(\lambda) = 1 - \frac{\tanh(\lambda\sqrt{-\mathrm{i}})}{\lambda\sqrt{-\mathrm{i}}}, \tag{25b}$$

and $\rho(\omega)$ represents a 'complex density' of air in the pores.

For some developments introduced later, a convenient alternative form for Eq. (24) is

$$-\frac{\partial p}{\partial x_3} = -\mathrm{i}\omega\rho_0\bar{v}_3 - \mathrm{i}\omega\rho_0\bar{v}_3\frac{\tanh(\lambda\sqrt{-\mathrm{i}})/(\lambda\sqrt{-\mathrm{i}})}{1 - \left[\tanh(\lambda\sqrt{-\mathrm{i}})/(\lambda\sqrt{-\mathrm{i}})\right]}. \tag{26}$$

8.4.2 *Thermal effects in a slit-like pore*

As the sound wave propagates in the slit it causes compressions and rarefactions in which the temperature respectively rises above ambient and falls below ambient. If the solid walls of the slit are assumed to be heat conducting there is heat exchange between the walls and the sound wave. After combining the heat conduction equation and the equation of state, it is possible to obtain an equation for the temperature change τ ('acoustical temperature') associated with a plane harmonic sound wave in a slit-like pore:

$$\frac{\partial^2 \tau}{\partial x_1^2} + \mathrm{i}\omega \frac{\gamma}{\upsilon'}\tau = \frac{\mathrm{i}\omega}{\kappa} p, \tag{27}$$

where $\gamma = c_p/c_V$, the ratio of specific heats (constant pressure/constant volume) and $\upsilon' = \kappa/\rho c_V$, κ being the thermal conductivity.

Equation (27) for τ is rather similar to Eq. (21) for v_3. Given the boundary conditions $\tau = 0$ at $x_1 = \pm a$, and, after averaging over the cross-section of the slit, the result for the (complex) bulk modulus of air in the slit, has a similar form to Eq. (24) for complex density, i.e.

$$K = \frac{\gamma P_0}{\gamma - (\gamma - 1)F(\mathrm{Pr}\,\omega)}, \tag{28}$$

where $\mathrm{Pr} = \eta\gamma C_V/\kappa$ is the Prandtl number and γP_0 is the adiabatic bulk modulus of air.

8.4.3 *Viscous and thermal effects in a cylindrical pore*

The equivalent to Eq. (20) for a cylindrical pore may be written

$$X_3 = \eta\left(\frac{\partial^2 v_3}{\partial x_1^2} + \frac{\partial^2 v_3}{\partial x_2^2}\right). \tag{29}$$

After a similar analysis to that for a sound wave in a slit-like pore, Eq. (25b) for a cylindrical pore of radius r may be written

$$F_{\mathrm{cyl}}(\lambda_{\mathrm{cyl}}) = 1 - \frac{2}{\lambda_{\mathrm{cyl}}\sqrt{\mathrm{i}}}\frac{\mathrm{J}_1(\lambda_{\mathrm{cyl}}\sqrt{\mathrm{i}})}{\mathrm{J}_0(\lambda_{\mathrm{cyl}}\sqrt{\mathrm{i}})}, \tag{30}$$

where $\lambda_{\mathrm{cyl}} = (\omega\rho_0 r^2/\eta)^{1/2} = r(\omega\rho_0/\eta)^{1/2}$, and J_1 and J_0 are cylindrical Bessel functions of order one and zero respectively.

Corresponding results have been obtained for triangular pores and rectangular pores. The results for four pore shapes are summarised in Table 8.1.

Table 8.1 Dimensionless parameters and complex density functions for four pore shapes.

Pore shape	λ	$F(\lambda)$
slit (width $2b$)	$b(\omega/\nu)^{1/2}$	$1 - \tanh(\lambda\sqrt{-\mathrm{i}})/(\lambda\sqrt{-\mathrm{i}})$
cylinder (radius a)	$a(\omega/\nu)^{1/2}$	$1 - (2/\lambda\sqrt{\mathrm{i}})\, \mathrm{J}_1(\lambda\sqrt{\mathrm{i}})\, \mathrm{J}_0(\lambda\sqrt{\mathrm{i}})$
equilateral triangle (side d)	$(d\sqrt{3}/4)(\omega/\nu)^{1/2}$	$1 - \coth(\lambda\sqrt{-\mathrm{i}})/(\lambda\sqrt{-\mathrm{i}}) + 3\mathrm{i}/\lambda^2$
rectangle (sides $2a$, $2b$)	$\frac{2a}{\pi\sqrt{a^2+b^2}}(\omega/\nu)^{1/2}$	$\frac{4}{\pi^4}\sum_{m,n=0}^{\infty}[(m+\frac{1}{2})^2(n+\frac{1}{2})^2 \times(1+2\mathrm{i}(m+\frac{1}{2})^2+(n+\frac{1}{2})^2)/\lambda^2]^{-1}$

8.4.4 *Low and high frequency approximations*

The low frequency or small λ approximations of Eq. (26), the equation of particle motion in a slit, and the cylindrical pore equivalent are

$$-\frac{\partial p}{\partial x_3} \approx -\frac{6}{5}\mathrm{i}\omega\rho_0\bar{v}_3 + \frac{3\eta}{a^2}\bar{v}_3, \tag{31}$$

and

$$-\frac{\partial p}{\partial x_3} \approx -\frac{4}{3}\mathrm{i}\omega\rho_0\bar{v}_3 + \frac{8\eta}{r^2}\bar{v}_3. \tag{32}$$

The first terms on the right hand sides of Eqs. (31) and (32) indicate an increase in the effective density of the fluid in the pores that depends on the pore shape. The second terms on the right hand side of Eqs. (31) and (32) represent the effects of viscosity in the pores. For steady flow through a single pore, the ratio of the applied pressure gradient to the average particle velocity is the flow resistivity. Setting $\omega = 0$ gives the 'd.c.' (zero velocity) flow resistivity σ_{pore}, defined as $-(\partial p/\partial x_3)/\bar{v}_3$, so $\sigma_{\text{pore}} = 3\eta/a^2$ in a slit pore of width $2a$ and $\sigma_{\text{pore}} = 8\eta/r^2$ in a cylindrical pore of radius r.

The low frequency or small λ approximations of Eq. (28), the equation for the complex bulk modulus of the fluid in a slit, and the cylindrical pore equivalent are

$$K \approx P_0\left[1 + \frac{1}{3}\frac{\gamma-1}{\gamma}\lambda^2\mathrm{Pr}\right], \tag{33}$$

and

$$K \approx P_0\left[1 + \frac{1}{8}\frac{\gamma-1}{\gamma}\lambda^2\mathrm{Pr}\right]. \tag{34}$$

Since the second terms on the right hand sides of these equations involve the square of λ, they are quite small for small λ. For example in a slit of 20 μm width and at 100 Hz, $\lambda = 0.0064$. So the thermal exchange in pores (of any shape) tends to be isothermal at low frequencies.

For large λ or high frequencies, approximations for a slit are

$$-\frac{\partial p}{\partial x_3} \approx -\mathrm{i}\omega\rho_0\bar{v}_3 + \frac{1-\mathrm{i}}{\sqrt{2}}\frac{\eta}{a}\left(\frac{\omega\rho_0}{\eta}\right)^{1/2}\bar{v}_3, \tag{35}$$

and

$$K \approx \gamma P_0\left[1 - \frac{1+\mathrm{i}}{\sqrt{2}}(\gamma-1)\lambda\sqrt{\mathrm{Pr}}\right]. \tag{36}$$

The first of these approximations indicates that the pore shape effect on the enclosed fluid density is frequency dependent. Even at high frequencies, for typical pore dimensions λ is less than 1, so the high-frequency bulk modulus approximation indicates that the thermal effects tend to become adiabatic at high frequencies.

8.5 Analysis for a Bulk Absorber

8.5.1 *Flow resistivity of a slit absorber*

Consider a hypothetical porous material made from an array of narrow identical slits n per unit length and each of width $2a$ with their axes perpendicular to the surface (Fig. 8.6).

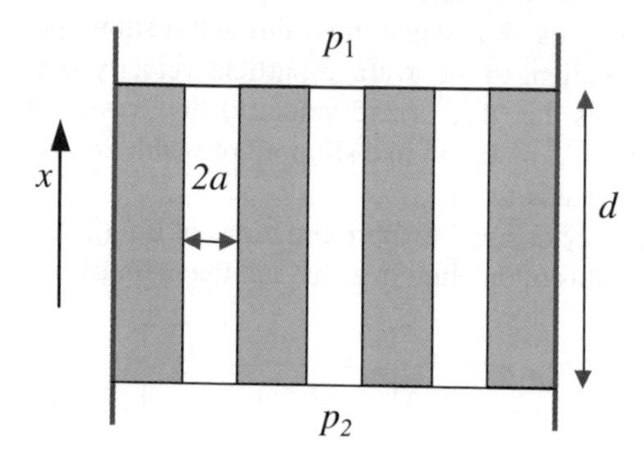

Fig. 8.6 Hypothetical slit pore material exposed to a steady pressure gradient.

The porosity of this material is given by $\Omega = 2na$ and the flow resistivity of the material is given by

$$\sigma = \frac{p_2 - p_1}{2na\bar{v}d} = \frac{p_2 - p_1}{\Omega\bar{v}d}. \tag{37}$$

Since $(p_2 - p_1)/d = 3\eta/a^2$ in each pore, this means that $\sigma = 3\eta/\Omega a^2$. The dimensionless parameter λ can be written in terms of the flow resistivity:

$$\lambda = (\omega \rho_0 a^2/\eta)^{1/2} = (3\omega\rho_0/\Omega\sigma)^{1/2}. \tag{38}$$

8.5.2 *Surface impedance of a slit absorber layer*

Consider a hypothetical region of infinitesimal thickness e above a slit absorber layer (Fig. 8.7). This serves as a transition region between the particle velocity in the slit pores and that outside the material.

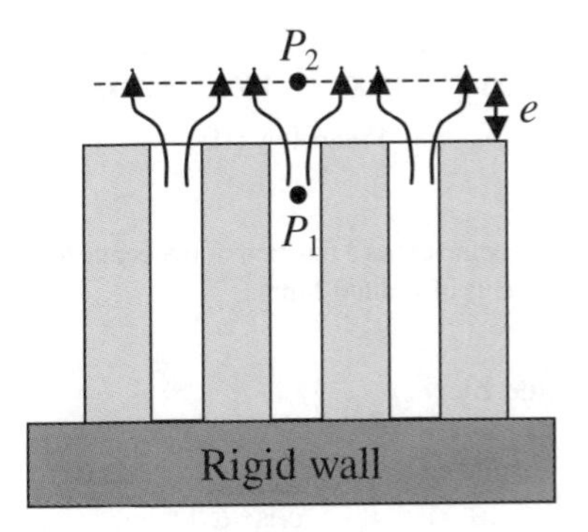

Fig. 8.7 Slit absorber with a hard backing.

Continuity of pressure and velocity at the surface leads to $p(P_2) = p(P_1)$ and $v(P_2) = v(P_1)\Omega$. Hence $Z(P_2) = Z_c \coth(-ikd)/\Omega$ where $Z_c = (K\rho)^{1/2}$, $k = \omega(\rho/K)^{1/2}$, and ρ and k are calculated from Eqs. (25) and (28). Predictions of absorption coefficient as a function of frequency of an absorber consisting of 0.1 m thick identical straight-slit layer with slit width 0.2 mm and surface porosity 0.8 are shown in Fig. 8.8. If such an absorber could be manufactured it should be reasonably useful!

8.5.3 *Tortuosity*

In general, the acoustical properties of a porous material are affected by the fact that sound waves, or more specifically fluid streamlines, are not able to follow straight paths through the material. It is possible to calculate the effects for various idealised microstructures. In the first example (Fig. 8.9a) the flow resistivity is given by

$$\sigma = \frac{p_2 - p_1}{2na\bar{v}d} = \frac{3\eta}{\Omega a^2 \cos^2 \varphi}. \tag{39}$$

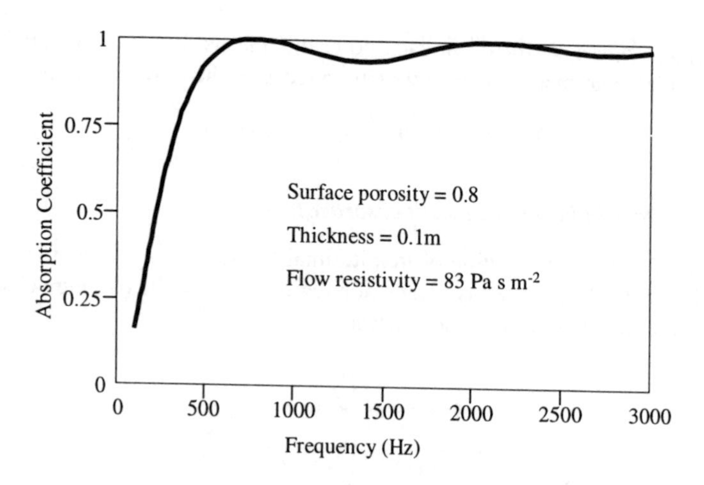

Fig. 8.8 Predicted absorption coefficient as a function of frequency for 0.1 m thick slit absorber layer having surface porosity 0.8 and slits of width 0.2 mm.

If the tortuosity is denoted by

$$T = \frac{1}{\cos^2 \varphi}, \tag{40}$$

then the dimensionless parameter for tortuous slits becomes

$$\lambda = \left(\frac{3\omega\rho_0}{\Omega\sigma T} \right)^{1/2}. \tag{41}$$

In the second case (Fig. 8.9b), assuming constant velocities in each section of the pore, the tortuosity is given by

$$T = \frac{(l_1 A_2 + l_2 A_1)(l_1 A_1 + l_2 A_2)}{(l_1 + l_2)^2 A_1 A_2}. \tag{42}$$

Tortuosity is always greater than 1.

In general, for a porous medium containing tortuous slit pores, the complex density may be written

$$\rho(\omega) = \frac{\rho_0 T}{F(\omega)} = \rho_0 T \left[1 + \frac{\mathrm{i}\sigma\Omega}{\omega\rho_0 T} G(\lambda) \right], \tag{43}$$

where

$$G(\lambda) = \frac{\tanh(\lambda\sqrt{-\mathrm{i}})}{1 - \tanh(\lambda\sqrt{-\mathrm{i}})/(\lambda\sqrt{-\mathrm{i}})}, \tag{44}$$

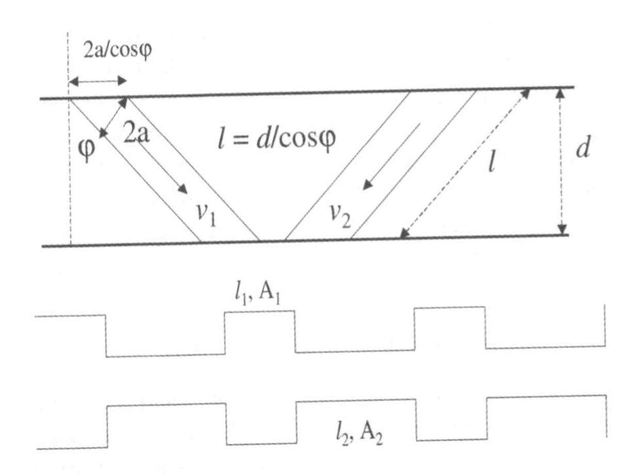

Fig. 8.9 Two idealised examples of tortuosity: due to pores inclined with respect to the surface (upper); due to pores with variable cross-section (lower).

and

$$\lambda = \left(\frac{3\omega\rho_0}{\Omega\sigma T}\right)^{1/2}. \tag{45}$$

The complex bulk modulus is not affected by tortuosity.

8.5.4 *The phenomenological model and structure factor*

A popular model, sometimes called the phenomenological model, for the acoustical properties of porous materials may be represented by an equation of motion and an equation of state respectively as follows:

$$-\frac{\partial p}{\partial x_3} = \frac{\rho_0 k_S}{\Omega}\frac{\partial \bar{v}_3}{\partial t} + \sigma \bar{v}_3, \tag{46}$$

$$\frac{1}{\rho_0 c_0^2}\frac{\partial p}{\partial t} + \frac{\partial v}{\partial x} = 0, \tag{47}$$

where k_S is called a 'structure factor'. Comparison of the first of these with Eqs. (26) and (43) shows that the model corresponds to a low frequency approximation ($G(\lambda) = 1$) of the model for a tortuous slit pore material and that k_S is related to tortuosity. Given that

$$K = -\frac{\partial v/\partial x}{\partial p/\partial t}, \tag{48}$$

then the second equation corresponds to the isothermal approximation. So the phenomenological model ignores frequency dependent viscous and thermal effects. It also neglects effects of pore shape and pore size distribution.

The structure factor has been used as a 'fudge factor' in the past in an attempt to make up for the various deficiencies of the phenomenological model (for example Heckl 1992). However the modern equivalent, *tortuosity*, has a clear physical meaning and there are independently measured or predicted values available for it (see Table 8.2 for example).

8.5.5 *The generalised model of Johnson and Allard*

Eq. (43) can be written down for the various pore shapes listed in Table 8.1 but it has been generalised to deal with pores of arbitrary shape. This has been done at the cost of introducing a further parameter, the **characteristic viscous length**. Hence

$$\rho(\omega) = \rho_0 T \left[1 + \frac{\mathrm{i}\sigma\Omega}{\omega\rho_0 T} G_J(\lambda) \right], \tag{49}$$

where

$$G_J(\omega) = \left(1 - \frac{4\mathrm{i}\omega\rho_0 T^2 \eta}{\sigma^2 \Omega^2 \Lambda^2} \right)^{1/2}, \tag{50}$$

and Λ is the characteristic viscous length defined by

$$\frac{2}{\Lambda} = \frac{\int_S v_S(\mathbf{r})\,\mathrm{d}S}{\int_V v_S(\mathbf{r})\,\mathrm{d}V}. \tag{51}$$

Here v_S represents the particle velocity field on the surface S of the pore in the absence of viscosity. Essentially the viscous characteristic length indicates the size of the narrowest pore sections in the material. These dominate the viscous effects.

Although not set out explicitly here, there is an equivalent expression for complex compressibility. This introduces two additional parameters, the **thermal characteristic length** and the **thermal permeability**. The thermal characteristic length has a simple interpretation in terms of the internal surface area of the material (see Allard 1993).

Equation (49) may be written

$$\frac{\rho_J(\omega)}{\rho_0} = T + \frac{\sigma\Omega}{-\mathrm{i}\omega\rho_0} \sqrt{1 + \frac{4T^2\eta\rho_0(-\mathrm{i}\omega)}{\Omega^2\Lambda^2\sigma^2}}, \tag{52}$$

which can be approximated for low frequencies such that

$$\omega < \frac{\Omega^2 \Lambda^2 \sigma^2}{4T^2 \eta \rho_0}, \tag{53}$$

so that the second term in the square root can be neglected, to give

$$\frac{\rho_s(\omega)}{\rho_0} = T + \frac{\sigma\Omega}{-\mathrm{i}\omega\rho_0}. \tag{54}$$

This three-parameter model is rather similar to the phenomenological model introduced earlier (see Eq. (46)).

8.5.6 *Acoustical properties of stacked spheres*

An idealisation appropriate for predicting the acoustical properties of granular materials is that of a random packing of identical spherical particles. In a granular medium the pores are formed by the interstices between the solid particles. By solving for the dynamic viscous flow around an individual sphere of radius R and then using a 'cell model' approach to account or the influence of neighbouring spheres, the following expression for the flow resistivity and viscous characteristic dimension have been derived in terms of the sphere radius and porosity of the packing:

$$\sigma = \frac{\eta}{k_0}, \qquad k_0 = \frac{2\Omega^2 R^2}{9(1-\Omega)(1-\varphi)W_k}, \tag{55}$$

where

$$W_k = \frac{5}{5 - 9\varphi^{1/3} + 5\varphi - \varphi^2}, \tag{56}$$

$$\varphi = \frac{3}{\sqrt{2\pi}}(1-\Omega) \approx 0.675(1-\Omega), \tag{57}$$

$$\Lambda = \frac{4(1-\varphi)T}{9(1-\Omega)}R. \tag{58}$$

8.6 Comparisons with Data and Nonlinear Aspects

8.6.1 *Comparisons with data for three materials*

Fig. 8.10 shows comparisons between predictions of the Johnson–Allard model and impedance tube data for the reflection coefficient and surface impedance of a 15 cm thick hard-backed layer of 1.89 mm radius lead shot. Note that the frequency range of the data shown is around the later resonance frequency i.e. the frequency at which the reflection coefficient has a minimum. For 15 cm thick layer of 1.89 mm radius lead shot this frequency is 415 Hz.

Table 8.2 Parameters for three rigid–porous materials.

	1.89 mm radius lead shot	Rigid aluminium	Rigid concrete
Porosity ϕ	0.385	0.93	0.3
Flow resistivity σ_0 (Pa s m^{-2})	1373	204.6	3619
Characteristic viscous length Λ (m)	5.5×10^{-4}	7.7×10^{-4}	2.2×10^{-4}
Tortuosity α_∞	1.6	1.07	1.8

Equations (49) & (50) have been used to calculate the complex density and the model for stacked spheres has been used to calculate the complex compressibility (see Johnson *et al.* 1987). The parameters required for the acoustical characterisation of lead shot are listed in Table 8.2. The flow resistivity and porosity values have been measured non-acoustically and the viscous characteristic length has been calculated from the stacked sphere model (result given above). The tortuosity has been calculated from an empirical relationship with porosity. Figures 8.11 and 8.12 compare data and predictions for porous aluminium and porous concrete. For these materials, the flow resistivity and porosity have been measured non-acoustically. However the viscous characteristic lengths and tortuosity have been adjusted to fit the data at the layer resonance.

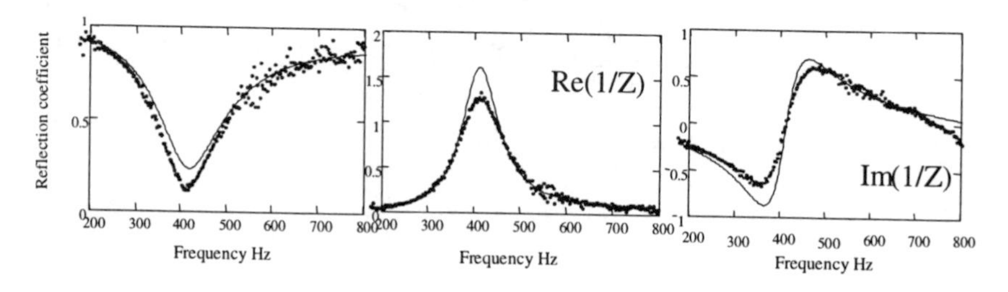

Fig. 8.10 Data (dots) and predictions (lines) for the acoustical properties of a 15 cm Layer of 2 mm lead shot for a frequency range around the layer resonance frequency (415 Hz). In this and the two subsequent figures the middle and right hand graphs show real and imaginary parts of the normalised admittance (inverse of impedance).

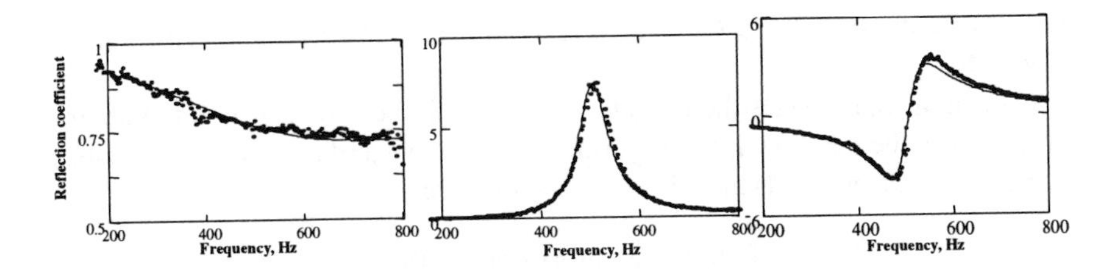

Fig. 8.11 Data (dots) and predictions (lines) for the acoustical properties of a 15 cm layer of porous aluminium for a frequency range around the layer resonance (600–700 Hz).

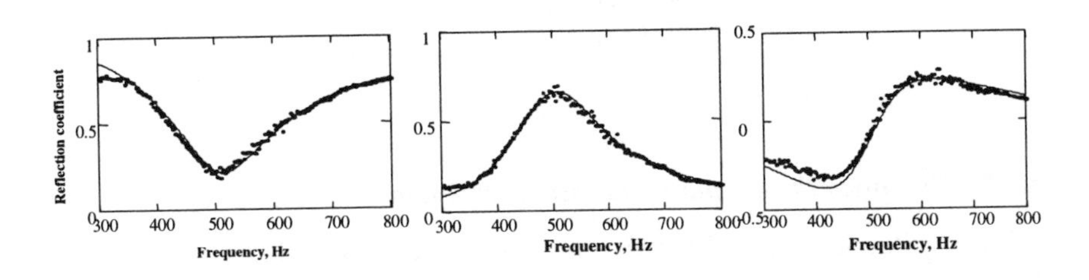

Fig. 8.12 Data (dots) and predictions (lines) for acoustical properties of 15 cm layer of porous concrete for a frequency range around the layer resonance at approximately 500 Hz.

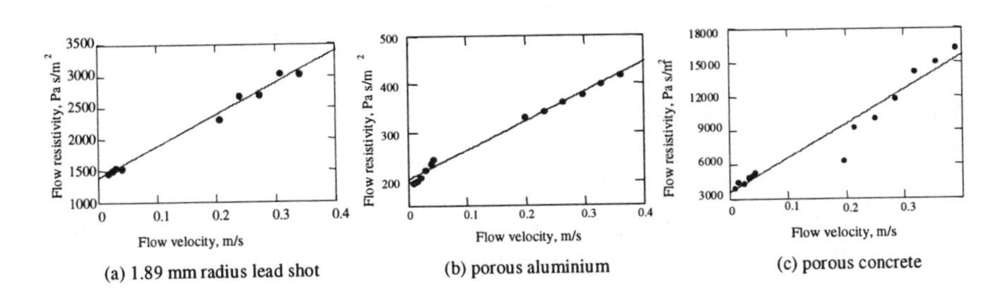

Fig. 8.13 Measured dependence of flow resistivity on flow velocity and straight line approximations to the data.

In view of the good straight-line fits (see Fig. 8.13), flow resistivity can be expressed as a linear function of flow velocity:

$$\sigma = \sigma_0(1 + \xi\Omega\,|v|), \tag{59}$$

where σ_0 is the d.c. flow resistivity and ξ is known as Forchheimer's nonlinearity parameter. The factor of porosity is required to match the flow velocity outside the material with the particle velocity inside the material (see Fig. 8.7). The nonlinearity parameter is calculated from the gradient of the straight line fit to the flow resistivity data. Table 8.3 gives values of the nonlinearity parameter for three materials.

For high sound levels the particle velocities will be such that it is not appropriate to use the d.c. flow resistivity values and the linear theory for the acoustics of rigid–porous materials must be developed further.

Table 8.3 Measured values of Forchheimer's nonlinearity parameter ξ for three materials.

Material	ξ (s/m)
1.89 mm radius lead shot	3.7
Porous aluminium	2.9
Porous Concrete	8.3

8.6.2 *Nonlinear theory and comparisons with data*

After substituting Eq. (59) for flow resistivity into Eq. (49), the latter can be rewritten

$$-\mathrm{i}\omega\rho(\omega)v = -\frac{\partial p}{\partial x}, \tag{60}$$

$$\rho(\omega) = \rho_0\left(T + \frac{\sigma_0\Omega(1+\xi\Omega\,|v|)}{-\mathrm{i}\omega\rho_0}\sqrt{1-\mathrm{i}\omega\frac{4T^2\rho_0\eta}{\sigma_0^2\Omega^2\Lambda^2(1+\xi\Omega\,|v|)^2}}\right). \tag{61}$$

When Eq. (61) is used in Eq. (60), it gives rise to nonlinear terms in v, i.e. the resulting differential equation is nonlinear.

The solution by means of the method of slowly varying amplitudes and the mean field approximation (see Gibbs 1985) gives rise to a transcendental equation for the surface impedance of a rigid–porous layer of thickness d as a function of incident sound pressure P:

$$\begin{aligned} Z(\omega,P) =& \frac{1}{\Omega}\sqrt{\rho(\omega)K(\omega)}\left[1-\frac{M(\omega)}{\mathrm{i}k(\omega)}G(U(\omega))\right] \\ &\times \coth\left\{-\mathrm{i}k(\omega)d\left[1-\frac{M(\omega)}{2\mathrm{i}k(\omega)}G(U(\omega))\right]\right\}, \end{aligned} \tag{62}$$

$$U(\omega) = \frac{2P}{\Omega[\rho_0 c_0 + Z(\omega,P)]} \tag{63}$$

where

$$G(x) = \sqrt{1+\mathrm{i}\frac{\omega_c}{\omega}(1+\xi\Omega x)^2} - \sqrt{1+\mathrm{i}\frac{\omega_c}{\omega}}, \qquad \omega_c = \frac{\sigma_0^2\Omega^2\Lambda^2}{4T^2\rho_0\eta}, \tag{64}$$

$$M(\omega) = (1 - \mathrm{i})\frac{T\omega^2 \rho_0}{k(\omega)K(\omega)}\frac{\delta(\omega)}{2\Lambda}, \qquad \delta(\omega) = \sqrt{\frac{2\eta}{\omega\rho_0}}. \tag{65}$$

This nonlinear equation has to be solved numerically. In Fig. 8.14 some predictions for the pressure dependence of the reflection coefficient are compared with data obtained using pure tones in an impedance tube.

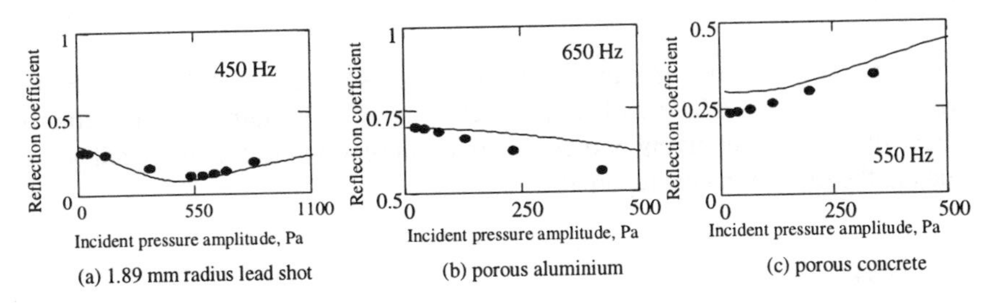

Fig. 8.14 Measured and predicted variation in reflection coefficient magnitude with incident sound pressure for 15 cm layers of three materials.

The measured data for lead shot and porous aluminium show a significant initial decrease in reflection coefficient with increasing pressure whereas the data for porous concrete show a significant increase in reflection coefficient as pressure increases. Note that 500 Pa is equivalent to 148 dB re 20 μPa. The predictions reproduce these trends in the data. For porous aluminium, it is predicted (numerically) that the reflection coefficient has a minimum at a pressure of between 4 and 5 kPa but this level of (continuous) sound pressure was beyond the capability of the experiment.

8.7 Conclusions

Aspects of theories for the acoustical properties of rigid–porous materials have been explored. The theories involve solution of the wave equation for a fluid including viscous dissipation terms and assuming various idealised forms of the pores. For example it has been shown possible to predict the sound absorbing properties of a material containing identical tortuous slits. Theory that allows for an arbitrary pore structure is available but requires six parameters. Consideration of an idealised material made from stacked spheres shows how the parameters depend only on sphere radius and packing porosity.

The variation of flow resistivity with flow velocity means that the acoustical behaviour of porous materials depends on the incident sound pressure. The resulting theory involves solution of nonlinear wave equations. It has been found to obtain reasonable agreement between data and predictions using the theories outlined. In particular it has been found that some materials will have a reflection coefficient that increases with incident sound pressure whereas others will have a reflection coefficient that decreases with incident sound pressure initially but then increases with further increase in incident sound pressure.

Acknowledgements

The work was supported in part by USARSDG (UK), contract R7D 8901-EN-01 with funds from the US Army ERDC BT-25 Program. The authors are grateful to Marshalls plc. for supplying porous concrete samples and to Edwin Standley for making the measurements reported.

References

Allard, J. F. (1993) *Propagation of Sound in Porous Media.* Elsevier Science.

Attenborough, K. (1993) Models for the acoustical characteristics of air filled granular materials, *Acta Acustica* **1**, pp. 213–226.

Gibbs, H. M. (1985) *Optical Bistability: Controlling Light With Light.* Academic Press.

Heckl, M. (1992) Reverberation, chapter 22 of *Modern Methods in Analytical Acoustics* by Crighton, D. G. *et al.* Springer Verlag.

Johnson, D. L., Koplik, J. and Dashen, R. (1987) Theory of dynamic permeability and tortuosity in fluid-saturated porous media, *Journal of Fluid Mechanics* **176**, pp. 379–402.

Umnova, O., Attenborough, K. and Li, K. M. (2000) Cell model calculations of the dynamic drag parameters in packings of spheres, *Journal of the Acoustical Society of America* **107**, pp. 3113–3119.

Umnova, O., Attenborough, K. and Li, K. M. (2001) A cell model for the acoustical properties of packings of spheres, *Acta Acustica* **87**, pp. 226–235.

Umnova, O., Attenborough, K., Standley, E. and Cummings, A. (2003) Behavior of rigid–porous layers at high levels of continuous acoustic excitation: theory and experiment, *Journal of the Acoustical Society of America* **114**, pp. 1346–1356.

PART III
Aeroacoustics

Chapter 9

Generalised Functions in Aeroacoustics

N. Peake

Department of Applied Mathematics and Theoretical Physics
University of Cambridge

9.1 Introduction

9.1.1 *Introductory remarks and basic concept*

In fluid mechanics in general, and acoustics in particular, one is often interested in dealing theoretically with objects that have either discontinuities or even singularities in space or time. Obvious examples are the point source of sound, localised at a single point in space, the shock wave, described in inviscid theory by a discontinuity in the fluid properties, and even a solid body immersed in a fluid, in which the body surface provides a distinct edge to the fluid-mechanical domain of interest. Conventional functions, by which we mean here continuous and differentiable functions defined over all space, cannot describe any of these structures, and to do this we need to introduce a new sort of function, the **generalised function**. Definitions and properties will be described in §9.1. It will become clear that generalised functions can play an important role in acoustics, and a particular aim of these notes is to describe a very compact derivation of the fundamental equation of aeroacoustics using generalised functions (in §9.2). The use of Fourier transforms and generalised functions will be introduced in §9.3.

9.1.2 *Definitions*

There are two ways of defining generalised functions. One way is to define the generalised function by its effect on ordinary, so-called 'test', functions. For instance, consider the famous **Dirac delta function** in one dimension, denoted $\delta(x)$.

For any continuous test function, $f(x)$, the delta function has the property that

$$\int_{-\infty}^{\infty} \delta(x) f(x)\, \mathrm{d}x = f(0). \tag{1}$$

In other words, $\delta(x)$ 'projects out' the value of $f(x)$ at $x = 0$. A second way of thinking about generalised functions is as the limit of ordinary functions. For instance, consider the ordinary function

$$\delta_n(x) \equiv \sqrt{\frac{n}{\pi}} \exp\left(-nx^2\right). \tag{2}$$

The delta function can be defined to be

$$\delta(x) \equiv \lim_{n\to\infty} \delta_n(x). \tag{3}$$

This idea is shown in Fig. 9.1—note how as n increases the functions become higher and thinner, but always have unit area. The definition Eqs. (2,3) is by no means unique (see exercises).

Much of what we have to say about generalised functions will be about the delta function, so we can now set down the key result as follows, which is just a slight modification of Eq. (1) to account for a shifted origin:

$$\int_a^b f(x)\delta(x-c)\, \mathrm{d}x = \begin{cases} f(c) & \text{if } c \in (a,b), \\ 0 & \text{if } c \notin [a,b]. \end{cases} \tag{4}$$

Generalised functions can be multiplied by ordinary functions. Given the way that $\delta(x)$ projects out the value at $x = 0$ in Eq. (1) it is easy to see that as generalised functions

$$g(x)\delta(x) = g(0)\delta(x), \tag{5}$$

i.e. it is only the value of $g(x)$ at $x = 0$ that has a bearing on the function $g(x)\delta(x)$. This means that

$$x\,\delta(x) = 0. \tag{6}$$

Therefore, the solution of the equation $xf(x) = 0$ is not $f(x) \equiv 0$, as it would be for ordinary continuous functions, but we now know that

$$xf(x) = 0 \quad \text{for all } x, \quad \text{so} \quad f(x) = C\delta(x) \quad \text{for any constant } C. \tag{7}$$

This is an important example, and in fact the basis of a fundamental idea, which is used to determine the stability of shear flows, and will also be used in §9.3 when we think about Fourier transforms of generalised functions.

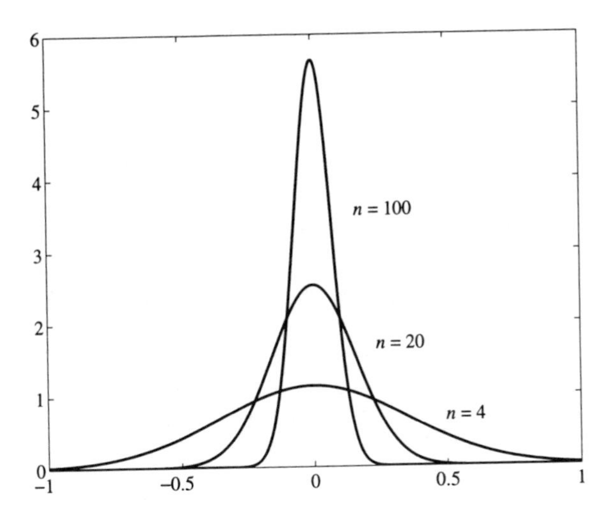

Fig. 9.1 Limit of ordinary functions tending to $\delta(x)$—see Eq. (2).

9.1.3 *Scaling and general argument*

We will be interested not just in $\delta(x)$ but also $\delta(f(x))$, i.e. a delta function with a more complicated argument. By applying the substitution $x = 2y$ in Eq. (1), and then by trying $x = -y$ in Eq. (1), we can see that as generalised functions

$$\delta(2x) = \tfrac{1}{2}\delta(x), \qquad \delta(-x) = \delta(x). \tag{8}$$

If we now suppose that $f(x) = 0$ at the single point $x = x_0$, then we can generalise Eq. (8) to yield

$$\delta(f(x)) = \frac{1}{|f'(x_0)|}\delta(x - x_0). \tag{9}$$

We can see straight from Eq. (9) that x_0 has to be a simple zero of $f(x)$, i.e. $f'(x_0) \neq 0$, so that for instance $\delta(x^2)$ is not meaningful.

A further generalisation of Eq. (9) is possible if $f(x)$ has multiple simple zeros, say at $x = x_0, x_1, \ldots, x_n$, in which case

$$\delta(f(x)) = \sum_{i=0}^{n} \frac{1}{|f'(x_i)|}\delta(x - x_i), \tag{10}$$

i.e. $\delta(f(x))$ is simply equivalent to a series of delta functions located at each of the roots of $f(x) = 0$, scaled by the modulus of the derivative at those points.

Example 9.1 Evaluate $\int_{-1}^{2} \delta(x^2 + 2x - 3) \cos x \, \mathrm{d}x$.

Solution Here we have

$$f(x) = x^2 + 2x - 3 = (x + 3)(x - 1), \tag{11}$$

so that the zeros of $f(x)$ are $x = -3, 1$, while

$$f'(x) = 2x + 2 \qquad \text{implying that} \qquad |f'(-3)| = |f'(1)| = 4. \tag{12}$$

From Eq. (10) we therefore see that

$$\delta(x^2 + 2x - 3) = \frac{\delta(x+3)}{4} + \frac{\delta(x-1)}{4}, \tag{13}$$

so that the integral becomes

$$\int_{-1}^{2} \frac{\delta(x+3)}{4} \cos x \, \mathrm{d}x + \int_{-1}^{2} \frac{\delta(x-1)}{4} \cos x \, \mathrm{d}x. \tag{14}$$

The first term in the above is zero, since -3 lies outside the range of integration, while the second term yields $(\cos 1)/4$.

9.1.4 *Differentiation*

The derivative of a generalised function can be defined as follows. Consider the generalised function $G(x)$, and write its derivative as $G'(x)$. Then we define $G'(x)$ via the formula

$$\int_{-\infty}^{\infty} G'(x) f(x) \, \mathrm{d}x = -\int_{-\infty}^{\infty} G(x) f'(x) \, \mathrm{d}x, \tag{15}$$

where $f(x)$ is an ordinary test function. Equation (15) is just integration by parts, in which the end-point contribution at $\pm\infty$ have been set to zero provided $G(x) = 0$ for sufficiently large x. Hence, the derivative of the delta function is described by the formula

$$\int_{-\infty}^{\infty} \delta'(x) f(x) \, \mathrm{d}x = -\int_{-\infty}^{\infty} \delta(x) f'(x) \, \mathrm{d}x = -f'(0), \tag{16}$$

i.e. $\delta'(x)$ projects out the value of $-f'(x)$ at $x = 0$.

By repeated application of Eq. (16) we can also define the nth derivative of $\delta(x)$, denoted $\delta^{(n)}(x)$:

$$\int_{-\infty}^{\infty} \delta^{(n)}(x) f(x)\,\mathrm{d}x = -\int_{-\infty}^{\infty} \delta^{(n-1)}(x) f'(x)\,\mathrm{d}x = \cdots$$
$$= (-1)^n \int_{-\infty}^{\infty} \delta(x) f^{(n)}(x)\,\mathrm{d}x = (-1)^n f^{(n)}(0). \quad (17)$$

At this point we can introduce another generalised function, the familiar **Heaviside step function**, $\mathrm{H}(x)$, defined such that

$$\mathrm{H}(x) = \begin{cases} 0 & \text{if } x < 0, \\ 1 & \text{if } x > 0. \end{cases} \quad (18)$$

If we take Eq. (4) with $a \to -\infty$, $b = x$, $c = 0$ then we can see that

$$\int_{-\infty}^{x} \delta(x)\,\mathrm{d}x = \mathrm{H}(x), \quad (19)$$

leading to the important result

$$\mathrm{H}'(x) = \delta(x). \quad (20)$$

Having defined the step function, we can introduce the 'sign' function, $\mathrm{sgn}(x)$, to be

$$\mathrm{sgn}(x) = \begin{cases} -1 & \text{if } x < 0, \\ 1 & \text{if } x > 0. \end{cases} \quad (21)$$

We can easily see that

$$\mathrm{sgn}(x) = 2\,\mathrm{H}(x) - 1. \quad (22)$$

Example 9.2 Differentiate the discontinuous function

$$f^+(x)\,\mathrm{H}(x) + f^-(x)\,\mathrm{H}(-x),$$

where $f^{\pm}(x)$ are differentiable.

Solution

$$\frac{\mathrm{d}}{\mathrm{d}x}\left\{f^+(x)\,\mathrm{H}(x) + f^-(x)\,\mathrm{H}(-x)\right\} =$$
$$\mathrm{H}(x)\frac{\mathrm{d}f^+}{\mathrm{d}x} + \mathrm{H}(-x)\frac{\mathrm{d}f^-}{\mathrm{d}x} + f^+(x)\delta(x) - f^-(x)\delta(-x), \quad (23)$$

where the final term has been evaluated using Eq. (8). Using Eq. (8) and Eq. (5), the right hand side can then finally be reduced to

$$\mathrm{H}(x)\frac{\mathrm{d}f^+}{\mathrm{d}x} + \mathrm{H}(-x)\frac{\mathrm{d}f^-}{\mathrm{d}x} + (f^+(0) - f^-(0))\delta(x). \tag{24}$$

9.1.5 *Extension to higher dimensions*

Much of what we have said already extends to three dimensions. So for instance, we can consider $\delta(\mathbf{x} - \mathbf{x_0})$, which has the property that

$$\iiint_\Sigma \delta(\mathbf{x} - \mathbf{x}_0)\,\mathrm{d}V(\mathbf{x}) = \begin{cases} 1 & \text{if } \mathbf{x}_0 \in \Sigma, \\ 0 & \text{if } \mathbf{x}_0 \notin \Sigma. \end{cases} \tag{25}$$

Another idea we will need soon is the concept of the step function across a surface. Consider the (moving) surface described by the equation $f(\mathbf{x}, t) = 0$; on one side of the surface f is positive and on the other side f is negative. As a simple example, consider the sphere of radius r and centre located at $(t, 0, 0)$. Define $f(x, y, z, t) = (x - t)^2 + y^2 + z^2 - r^2$, then the sphere surface is $f = 0$, $f > 0$ is outside the sphere and $f < 0$ is inside the sphere. The step function $\mathrm{H}(f(\mathbf{x}, t))$ now takes the value $+1$ on one side (where $f > 0$) and 0 on the other side (where $f < 0$). We will want to differentiate this step function with respect to both space and time, and the results are

$$\boldsymbol{\nabla}(\mathrm{H}(F)) = (\boldsymbol{\nabla}F)\delta(F), \qquad \frac{\partial}{\partial t}\mathrm{H}(F) = \frac{\partial F}{\partial t}\delta(F), \tag{26}$$

which are simply the result of differentiating the argument F and then multiplying it by the derivative of the step function, i.e. the delta function.

9.2 Application in Aeroacoustics

9.2.1 *A basic equation*

We will now use the material from the previous section to derive the governing equation for the generation of sound waves by a moving body. This equation was first derived by Ffowcs Williams and Hawkings and bears their names, and has very significant practical impact in aeroacoustics prediction.

We start with a moving rigid body, whose surface is given by the equation $F(\mathbf{x}, t) = 0$, with $F < 0$ inside the body and $F > 0$ outside the body—see Fig. 9.2. The body moves through the surrounding fluid (typically air or water in applications), and because the surface is rigid it follows that no fluid can enter the

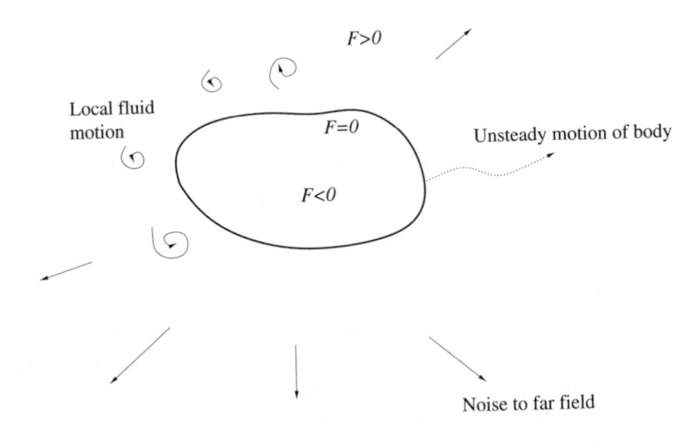

Fig. 9.2 Rigid body in arbitrary unsteady motion, generating far-field sound.

body, so that the surface $F = 0$ is a material surface in the fluid, i.e.

$$\frac{DF}{Dt} \equiv \frac{\partial F}{\partial t} + \mathbf{u} \cdot \boldsymbol{\nabla} F = 0 \qquad \text{on } F = 0, \tag{27}$$

where $\mathbf{u}(\mathbf{x}, t)$ is the fluid velocity. Without loss of generality we will suppose that

$$|\boldsymbol{\nabla} F| = 1 \qquad \text{on } F = 0. \tag{28}$$

Now consider the equation of mass conservation in the fluid, which states that

$$\frac{\partial \rho}{\partial t} + \boldsymbol{\nabla} \cdot (\rho \mathbf{u}) = 0, \tag{29}$$

where $\rho(\mathbf{x}, t)$ is the fluid density. This equation is only valid in the fluid, i.e. in $F > 0$, but we can make it valid throughout all space, including inside the body, by the simple trick of multiplying Eq. (29) by $\mathrm{H}(F)$. The new equation is then Eq. (29) in $F > 0$ and is $0 = 0$ in $F < 0$. We now proceed to rearrange this extended version of Eq. (29), by taking the $\mathrm{H}(F)$ inside the operators, using the expressions Eq. (26) for the space and time derivatives of $\mathrm{H}(F)$, and then finally applying Eq. (27), to end up with

$$\frac{\partial}{\partial t}\{\rho \mathrm{H}(F)\} + \boldsymbol{\nabla} \cdot (\rho \mathbf{u} \mathrm{H}(F)) = 0. \tag{30}$$

Now, in acoustics we are really interested in the density fluctuation at infinity, so if we suppose that the undisturbed fluid has uniform density ρ_0 then the acoustic fluctuating density will be $\rho - \rho_0$. Putting this into Eq. (30), and applying Eq. (27)

once again, leads us to

$$\frac{\partial}{\partial t}\left\{(\rho-\rho_0)\,\mathrm{H}(F)\right\}+\boldsymbol{\nabla}\cdot(\rho\mathbf{u}\mathrm{H}(F))=\rho_0(\mathbf{u}\cdot\boldsymbol{\nabla}F)\delta(F). \tag{31}$$

Now we turn our attention to the momentum equation. We will write this, using suffix notation, in the slightly unfamiliar form

$$\frac{\partial}{\partial t}(\rho u_i)+\frac{\partial}{\partial x_j}(\rho u_i u_j+p\delta_{ij}-\sigma_{ij})=0, \tag{32}$$

where $p(\mathbf{x},t)$ is the pressure and σ_{ij} is the viscous part of the stress tensor. Rearrangement of this equation, and use of mass conservation Eq. (29) leads exactly to the usual form of the Navier–Stokes equation. Once again Eq. (32) is only valid in the fluid, $F>0$, so we render it valid everywhere by multiplying by $\mathrm{H}(F)$, and then taking the $\mathrm{H}(F)$ inside the derivatives and using Eq. (27) we find that

$$\frac{\partial}{\partial t}\{\rho u_i\mathrm{H}(F)\}+\frac{\partial}{\partial x_j}\{(\rho u_i u_j+p\delta_{ij}-\sigma_{ij})\mathrm{H}(F)\}=\{p\delta_{ij}-\sigma_{ij}\}\frac{\partial F}{\partial x_j}\delta(F). \tag{33}$$

Our aim is now to combine Eqs. (31) and (33) in a particular way to produce an equation that looks like

$$\left\{\frac{\partial^2}{\partial t^2}-c_0^2\nabla^2\right\}(\rho-\rho_0)\,\mathrm{H}(F)=Q(\mathbf{x},t), \tag{34}$$

where c_0 is the speed of sound of the undisturbed fluid at infinity. Equation (34) is the wave equation for propagation through a uniform quiescent medium, with the right hand side, $Q(\mathbf{x},t)$, being the so-called 'source term'. The idea is that if Q can be approximated or determined in some other way then Eq. (34) can be solved so as to predict the sound far away from the body.

The algebra required to find Eq. (34) is slightly cumbersome, but not technically difficult. First, we form the following equation:

$$\frac{\partial}{\partial t}\left\{\text{Eq. (31)}\right\}-\boldsymbol{\nabla}\cdot\left\{\text{Eq. (33)}\right\}-c_0^2\nabla^2\{(\rho-\rho_0)\,\mathrm{H}(F)\}, \tag{35}$$

and then, again making use of the result Eq. (26), finally find that

$$\left\{\frac{\partial^2}{\partial t^2}-c_0^2\nabla^2\right\}\{(\rho-\rho_0)\mathrm{H}(F)\}=$$
$$\frac{\partial^2}{\partial x_i\partial x_j}\{T_{ij}\,\mathrm{H}(F)\}+\frac{\partial}{\partial t}\{\rho_0V_n\delta(F)\}-\frac{\partial}{\partial x_i}\{\mathcal{R}_i\delta(F)\}. \tag{36}$$

This is the famous Ffowcs Williams–Hawkings equation (in differential form).

We can now identify the terms on the right hand side of Eq. (36) as noise source terms:

(1) The term containing T_{ij} is located throughout the fluid (due to the presence of $H(F)$), although in many practical applications it can be localised close to the body. Here

$$T_{ij} = \rho u_i u_j - \sigma_{ij} + (p - \rho c_0^2)\delta_{ij} \tag{37}$$

is the **Lighthill stress tensor**, and arises from the unsteady motion of the fluid around the body produced by the body motion. This term tells us that noise is generated by unsteady stress in the fluid, including the Reynolds stress, the viscous stress and the isotropic term $(p - c_0^2\rho)\delta_{ij}$. The term containing T_{ij} in Eq. (36) is referred to as a **quadrupole** source, and is characterised by its double spatial derivatives.

(2) The term involving V_n is located only on the body surface, thanks to the presence of $\delta(F)$). Here $V_n = \mathbf{u} \cdot \boldsymbol{\nabla} F$ is the body normal velocity into the fluid (recall that $\boldsymbol{\nabla} F$ is the *unit* vector normal to the body surface pointing into the fluid). This term therefore corresponds to noise being generated by the unsteady injection of volume into the fluid by the body motion (note the appearance of ρ_0, the density at infinity, rather than ρ, the local fluid density on the surface, telling us that this is volume, rather than mass, injection). This term in Eq. (36), with no spatial derivatives, is referred to as a **monopole**.

(3) The remaining term is also localised on the body surface, and contains the quantity

$$\mathcal{R}_i = (p\delta_{ij} - \sigma_{ij})\frac{\partial F}{\partial x_j}, \tag{38}$$

which is the normal force exerted by the body on the fluid. Noise is therefore also generated by the (unsteady) injection of momentum into the fluid by the body motion. This term, with a single spatial derivative, is referred to as a **dipole**.

9.2.2 *Shock noise*

As a second example, suppose that we remove the body from the flow completely, so that the surface terms disappear and leaving the equation

$$\left\{\frac{\partial^2}{\partial t^2} - c_0^2\nabla^2\right\}(\rho - \rho_0) = \frac{\partial^2}{\partial x_i \partial x_j}\{T_{ij}\}. \tag{39}$$

This equation predates the Ffowcs Williams–Hawkings equation, and was derived by Lighthill to describe noise generated by turbulence in jets. It is universally acknowledged as being the foundation of modern aeroacoustics.

Suppose that a shock is present in the flow, so that, at least according to the Euler equations, the flow variables will jump discontinuously across the shock.

A practical example of this would be a transonic helicopter rotor blade with an attached shock on the upper surface. Let's suppose that the shock surface is described by the equation $S(\mathbf{x},t) = 0$, and following Example 9.2 we will write the Lighthill quadrupole in the form

$$T_{ij}(\mathbf{x},t) = T^+_{ij}(\mathbf{x},t)\,\mathrm{H}(S) + T^-_{ij}(\mathbf{x},t)\,\mathrm{H}(-S), \tag{40}$$

where $T^{\pm}_{ij}$ are smooth functions.

We will require the second derivative of T_{ij}. The first derivative is

$$\begin{aligned}\frac{\partial T_{ij}}{\partial x_j} &= \frac{\partial T^+_{ij}}{\partial x_j}\,\mathrm{H}(S) + \frac{\partial T^-_{ij}}{\partial x_j}\,\mathrm{H}(-S) + T^+_{ij}\frac{\partial S}{\partial x_j}\delta(S) - T^-_{ij}\frac{\partial S}{\partial x_j}\delta(S)\\ &= \frac{\partial T^+_{ij}}{\partial x_j}\,\mathrm{H}(S) + \frac{\partial T^-_{ij}}{\partial x_j}\,\mathrm{H}(-S) + [T_{ij}]n_j\delta(S),\end{aligned} \tag{41}$$

where $\mathbf{n} = \boldsymbol{\nabla} S$ is the normal to the shock and the notation $[\cdot]$ denotes the jump across the shock. Note that in this last step we have used the property Eq. (5) to replace $T^{\pm}_{ij}(\mathbf{x},t)$ by their values on the shock.

Finally, we can calculate the second derivative of T_{ij}, leading to

$$\begin{aligned}\frac{\partial^2 T_{ij}}{\partial x_i \partial x_j} &= \frac{\partial^2 T^+_{ij}}{\partial x_i \partial x_j}\,\mathrm{H}(S) + \frac{\partial^2 T^-_{ij}}{\partial x_i \partial x_j}\,\mathrm{H}(-S)\\ &\quad + \left[\frac{\partial T_{ij}}{\partial x_j} n_i\right]\delta(S) + \frac{\partial}{\partial x_i}\left\{[T_{ij}]n_j\delta(S)\right\}.\end{aligned} \tag{42}$$

The first two terms in this equation correspond to the quadrupole source terms on either side of the shock, with the same double-derivative structure as in Eq. (39). However, the third and fourth terms are new, and arise from the presence of the shock itself. The third term, with no spatial derivative on the delta function, is akin to a monopole term, while the fourth term, with its single spatial derivative, is like the dipole term.

9.3 Fourier Transforms

Generalised functions can be Fourier transformed in much the same way as ordinary functions, but some special care is needed. This is a very big topic, and only a few brief examples will be given here.

9.3.1 *Definitions and basics*

Define the Fourier transform of $f(x)$ as $\overline{f}(k)$, with the transform definition and the inversion theorem given by

$$\overline{f}(k) = \int_{-\infty}^{\infty} f(x)\exp(-\mathrm{i}kx)\,\mathrm{d}x, \tag{43}$$

$$f(x) = \frac{1}{2\pi}\int_{-\infty}^{\infty} \overline{f}(k)\exp(\mathrm{i}kx)\,\mathrm{d}k \tag{44}$$

respectively. Two standard results that we will require later are the transform of $\mathrm{d}f/\mathrm{d}x$,

$$\overline{\frac{\mathrm{d}f}{\mathrm{d}x}}(k) = \mathrm{i}k\overline{f}(k), \tag{45}$$

and the transform of $x^n f(x)$,

$$\overline{x^n f}(k) = \int_{-\infty}^{\infty} x^n f(x)\exp(-\mathrm{i}kx)\,\mathrm{d}x = \mathrm{i}^n\frac{\mathrm{d}^n\overline{f}}{\mathrm{d}k^n}. \tag{46}$$

9.3.2 *Some examples*

The easiest starting point is the Fourier transform of the delta function: from Eq. (43)

$$\overline{\delta}(k) = \int_{-\infty}^{\infty} \delta(x)\exp(-\mathrm{i}kx)\,\mathrm{d}x = 1, \tag{47}$$

where the last step has been completed using property Eq. (1) of the delta function. Putting Eq. (47) into Eq. (44) gives the further result

$$\int_{-\infty}^{\infty} \exp(\mathrm{i}kx)\,\mathrm{d}k = 2\pi\delta(x). \tag{48}$$

It follows that

$$\overline{1} = 2\pi\delta(k). \tag{49}$$

Now consider $\mathrm{sgn}(x)$. We know that $\mathrm{sgn}'(x) = 2\delta(x)$, so that using Eq. (47)

$$\overline{\mathrm{sgn}'}(k) = 2. \tag{50}$$

We now apply result Eq. (45), so that

$$\mathrm{i}k\overline{\mathrm{sgn}}(k) = 2. \tag{51}$$

The solution of Eq. (51) is

$$\overline{\operatorname{sgn}}(k) = \frac{2}{\mathrm{i}k} + C\delta(k) \tag{52}$$

for any constant C (remember that this follows from Eq. (7)). However, we can argue that $C = 0$; this is because $\operatorname{sgn}(x)$ is an odd function of x so that its Fourier transform is an odd function of k, and the only way that the right hand side of Eq. (52) is odd in k is if $C = 0$. This gives us

$$\overline{\operatorname{sgn}}(k) = \frac{2}{\mathrm{i}k}. \tag{53}$$

We can now use Eq. (22) to work out the Fourier transform of the step function,

$$\overline{\mathrm{H}}(k) = \frac{1}{\mathrm{i}k} + \pi\delta(k). \tag{54}$$

In Eqs. (53) and (54) we must be careful to say exactly what $1/k$ means. In fact, in this context $1/k$ is also a generalised function, which must be interpreted as being the Cauchy principal value when integrated.

We can now use properties (45) and (46) to generate some more results:

Example 9.3 The Fourier transform of x^n is obtained by applying Eq. (46) with $f(x) = 1$. We have:

$$\begin{aligned}
\overline{x^n} &= \overline{x^n.1} \\
&= \mathrm{i}^n \frac{\mathrm{d}^n}{\mathrm{d}k^n}\overline{1} && \text{from Eq. (46)} \\
&= 2\pi \mathrm{i}^n \delta^{(n)}(k) && \text{from Eq. (49).}
\end{aligned} \tag{55}$$

Example 9.4 We can use property (46) to calculate the Fourier transforms of $x^n \operatorname{sgn} x$ and $x^n\,\mathrm{H}(x)$ simply by differentiating n times the results (53) and (54) respectively:

$$\overline{x^n \operatorname{sgn} x}(k) = \frac{2\mathrm{i}^{n-1}(-1)^n n!}{k^{n+1}} \tag{56}$$

$$\overline{x^n\,\mathrm{H}(x)}(k) = \frac{\mathrm{i}^{n-1}(-1)^n n!}{k^{n+1}} + \mathrm{i}^n \pi \delta^{(n)}(k). \tag{57}$$

Example 9.5 The Fourier transform of $\mathrm{H}(x)x^\alpha$ for $\alpha > -1$ not an integer is a bit trickier (assume x^α is positive for real $x > 0$, with branch cut along negative real axis). Consider first the ordinary function $\exp(-tx)\,\mathrm{H}(x)x^\alpha$ with $t > 0$. The Fourier transform of this function is

$$\overline{\exp(-tx)\,\mathrm{H}(x)x^\alpha}(k) = \int_0^\infty \exp(-tx - \mathrm{i}kx)x^\alpha\,\mathrm{d}x = \frac{\alpha!}{(t + \mathrm{i}k)^{\alpha+1}}, \tag{58}$$

where the last step has used the definition $\alpha! \equiv \int_0^\infty y^\alpha \exp(-y)\,\mathrm{d}y$. We now take the limit $t \to 0$, to find that as a generalised function

$$\overline{\mathrm{H}(x)x^\alpha}(k) = \frac{\alpha!}{|k|^{\alpha+1}} \exp(-\mathrm{i}\pi \operatorname{sgn}(k)(\alpha+1)/2), \tag{59}$$

where the final exponential has arisen from taking the fractional power of $\mathrm{i}k$.

Example 9.6 The Fourier transform of $\log|x|$ is

$$-\frac{\pi \operatorname{sgn}(k)}{k}.$$

Further Reading

Two excellent books on the theory of generalised functions from a practical viewpoint are:

Strichartz, R. S. (2003) *A Guide to Distribution Theory and Fourier Transforms.* World Scientific.

Woyczynski, W. A. and Saichev, A. I. (1997) *Distributions in the Physical and Engineering Sciences, Vol.1: Distributional and Fractal Calculus, Integral Transforms and Wavelets (Applied and Numerical Harmonic Analysis).* Birkhauser.

An authoritative review of generalised functions in aeroacoustics is provided by the following two NASA reports:

Farassat, F. (1994) Introduction to generalized functions with applications in aerodynamics and aeroacoustics, NASA Technical Paper 3428, (corrected April 1996).
http://techreports.larc.nasa.gov/ltrs/PDF/tp3428.pdf

Farassat, F. (1996) The Kirchhoff formulas for moving surfaces in aeroacoustics—the subsonic and supersonic cases, NASA Technical Memo. 110285.
http://techreports.larc.nasa.gov/ltrs/PDF/NASA-96-tm110285.pdf

A short but informative introduction to generalised functions and Fourier transforms, which also introduces the important idea of how the results of §9.3 above can be used for both approximating Fourier integrals and approximately inverting Fourier series, is given by:

Lighthill, M. J. (1958) *An Introduction to Fourier Analysis and Generalised Functions.* Cambridge University Press.

Problems

Exercise 9.1 Alternative definitions of $\delta(x)$.

(a) Use the limiting sequence for $\delta(x)$ given in Eq. (3) to find, and sketch, a limiting sequence for $\delta'(x)$.

(b) By differentiating the identity

$$\operatorname{sgn} x = \lim_{\epsilon \to 0} \frac{|x|^{\epsilon+1}}{x},$$

show that

$$\delta(x) = \lim_{\epsilon \to 0} \frac{\epsilon}{2} |x|^{\epsilon - 1}.$$

(c) By Fourier transforming the identity

$$1 = \lim_{n \to \infty} [\mathrm{H}(x+n) - \mathrm{H}(x-n)],$$

i.e. the increasingly wide step, show that

$$\delta(x) = \lim_{\epsilon \to 0} \frac{\sin nx}{\pi x}.$$

(d) Sketch graphs of the limiting sequences in (b) and (c).

Exercise 9.2 Evaluation of integrals involving generalised functions.

(a) Find

$$\int_{-1}^{3} \delta(x^2 + x - 6)\, \mathrm{d}x.$$

(b) Find

$$\int_{-\infty}^{\infty} \frac{\delta(\sin \pi x)}{1 + x^2}\, \mathrm{d}x.$$

(c) Show that

$$\int_{0}^{\infty} \delta'(x^2 - 1) x^2\, \mathrm{d}x = -\tfrac{1}{4}.$$

Hint: Write $x = \sqrt{u}$ first.

Exercise 9.3 Poisson's summation formula.
Starting from the famous result

$$\sum_{n=-\infty}^{\infty} f(n/\Delta) = \Delta \sum_{m=-\infty}^{\infty} \overline{f}(2m\pi\Delta),$$

where $\overline{f}(k)$ is the Fourier transform of $f(x)$, calculate:

(i) $\displaystyle\sum_{n=-\infty}^{\infty} \delta(L-n),$

(ii) $\displaystyle\sum_{n=-\infty}^{\infty} \exp(-\lambda^2 n^2),$

(iii) $\displaystyle\sum_{n=1}^{\infty} \frac{\cos(n\lambda)}{1+n^2}, \qquad 0<\lambda<2\pi.$

Hint: Take the Fourier transform of $\exp(\mathrm{i}x\lambda)/(1+x^2)$ using the residue theorem.

Exercise 9.4 Shock noise.
A one-dimensional shock wave moves at speed c_s into an otherwise still fluid. Ahead of and behind the shock the density and pressure are the (constant) values $\rho_{1,2}$, $p_{1,2}$ respectively, and the velocity behind the shock is u_2 (also constant). Simplify the noise sources for this problem, i.e. the right hand side of equation (36), as much as you can.

Exercise 9.5 Inversion.
Check that the Fourier inversion theorem works for the Fourier transforms of

(i) $\mathrm{sgn}(x)$,
(ii) $\mathrm{H}(x)$.

Hint: Complete using the residue theorem, but note that

$$\int_{-\infty}^{\infty} \frac{\exp(\mathrm{i}kx)}{k}\, \mathrm{d}k$$

is to be interpreted as

$$\lim_{\epsilon\to 0}\left\{\int_{-\infty}^{-\epsilon} + \int_{\epsilon}^{\infty}\right\}.$$

Exercise 9.6 Fourier transform of $\log|x|$**.**

(a) Use Eq. (53) to show that

$$\overline{\frac{1}{x}}(k) = -\mathrm{i}\pi \,\mathrm{sgn}(k).$$

(b) Starting from

$$\frac{\mathrm{d}}{\mathrm{d}x} \log|x| = \frac{1}{x},$$

determine the Fourier Transform of $\log|x|$.

Chapter 10

Monopoles, Dipoles, and Quadrupoles

C. J. Chapman
Department of Mathematics
University of Keele

10.1 Introduction

This chapter gives solutions of the forced wave equation of acoustics, including solutions for monopole, dipole, and quadrupole sources. The following three-way classification is adopted.

(i) The number of space dimensions is three or two. Although most real-life problems are three-dimensional, the acoustic field in parts of the near-field may be approximately two-dimensional; thus two-dimensional theory is of value.

(ii) Results are given both for the usual wave equation and for the convected wave equation. The latter incorporates the effect of a uniform mean flow. Admittedly, the frame of reference may be chosen to move with the mean flow, in which case the convected wave equation is transformed to the usual wave equation; but this passes the convection velocity onto the sources. In many problems, it is simpler and more lucid to work with the convected wave equation from the start.

(iii) The source is taken to be an arbitrary function of time, or to contain only a single frequency.

Thus there are eight cases altogether, and we shall consider all of them. To give examples of monopole, dipole, and quadrupole sources for all eight cases would make the chapter too long. Accordingly, the examples chosen are for the three-dimensional usual wave equation. However, the general solutions presented for the eight cases reduce any specific problem to evaluation of an integral; thus the chapter may be useful both as a tutorial and a reference. The way in which the values of the general integrals depend on the parameters of a problem is of considerable interest, both mathematically and physically.

10.2 The Three-Dimensional Wave Equation

10.2.1 *Arbitrary forcing*

The acoustic pressure p in a uniform stationary medium with density ρ_0 and sound speed c_0 is taken to satisfy the equation

$$\left(\frac{1}{c_0^2}\frac{\partial^2}{\partial t^2} - \nabla^2\right) p = \rho_0 \frac{\partial q}{\partial t} - \boldsymbol{\nabla}\cdot\mathbf{F} + \frac{\partial^2 T_{ij}}{\partial x_i \partial x_j}, \tag{1}$$

where t denotes time and $\mathbf{x} = (x_i)$ denotes position. The summation convention is used for repeated indices. The scalar $q = q(\mathbf{x}, t)$ is a volume source per unit volume per unit time, giving a monopole source strength $\rho_0 \partial q/\partial t$; the vector $\mathbf{F} = \mathbf{F}(\mathbf{x}, t)$ is an applied force per unit volume, giving a dipole source strength $-\boldsymbol{\nabla}\cdot\mathbf{F}$; and the second-order tensor $T_{ij} = T_{ij}(\mathbf{x}, t)$ is the Lighthill stress tensor, with dimensions of force per unit area, giving a quadrupole source strength $\partial^2 T_{ij}/\partial x_i \partial x_j$. For a derivation of Eq. (1) see the section on the Ffowcs Williams–Hawkings equation in Chapter 9: 'Generalised Functions in Aeroacoustics'.

The source terms on the right-hand side of Eq. (1) provide the 'forcing' for the acoustic pressure field. If they are collectively denoted $f(\mathbf{x}, t)$, then Eq. (1) may be written

$$\left(\frac{1}{c_0^2}\frac{\partial^2}{\partial t^2} - \nabla^2\right) p = f(\mathbf{x}, t). \tag{2}$$

The free-space solution of Eq. (2) is

$$p = \frac{1}{4\pi}\iiint_{\mathbb{R}^3} \frac{f(\mathbf{x}', t - |\mathbf{x}-\mathbf{x}'|/c_0)}{|\mathbf{x}-\mathbf{x}'|}\,\mathrm{d}^3\mathbf{x}'. \tag{3}$$

The derivation of Eq. (3) may be found in many textbooks and monographs, e.g. Dowling & Ffowcs Williams (1983) and Howe (1998, 2003), and will not be repeated here. The physical explanation of Eq. (3) is that each source at $\mathbf{x}'$ produces a pressure field that decays with distance like $1/\{4\pi|\mathbf{x}-\mathbf{x}'|\}$. The retarded time $t - |\mathbf{x}-\mathbf{x}'|/c_0$ accounts for the propagation speed c_0 of a signal travelling in a straight line from source point $\mathbf{x}'$ to observation point $\mathbf{x}$.

An alternative way of writing the solution of Eq. (2) is

$$p = \frac{1}{4\pi}\int_{-\infty}^{\infty}\iiint_{\mathbb{R}^3} f(\mathbf{x}', t')\frac{\delta(t - t' - |\mathbf{x}-\mathbf{x}'|/c_0)}{|\mathbf{x}-\mathbf{x}'|}\,\mathrm{d}^3\mathbf{x}'\,\mathrm{d}t'. \tag{4}$$

The physical explanation of Eq. (4) is that a source at $\mathbf{x}'$ acting only at the instant t' produces a pressure field proportional to $\delta(t - t' - |\mathbf{x} - \mathbf{x}'|/c_0)/\{4\pi|\mathbf{x} - \mathbf{x}'|\}$. The previous result (3) is obtained from Eq. (4) by performing the integration with respect to t', using the defining property of the delta function. In some problems, it is easier to perform the $\mathbf{x}'$ integration first; then (4) is preferable to (3).

When the general solution (3) is applied to (1), the result is

$$\begin{aligned} p = \frac{\rho_0}{4\pi}\frac{\partial}{\partial t}\iiint_{\mathbb{R}^3} \frac{q(\mathbf{x}', t - |\mathbf{x} - \mathbf{x}'|/c_0)}{|\mathbf{x} - \mathbf{x}'|}\, \mathrm{d}^3\mathbf{x}' \\ - \frac{1}{4\pi}\boldsymbol{\nabla}\cdot\iiint_{\mathbb{R}^3} \frac{\mathbf{F}(\mathbf{x}', t - |\mathbf{x} - \mathbf{x}'|/c_0)}{|\mathbf{x} - \mathbf{x}'|}\, \mathrm{d}^3\mathbf{x}' \\ + \frac{1}{4\pi}\frac{\partial^2}{\partial x_i \partial x_j}\iiint_{\mathbb{R}^3} \frac{T_{ij}(\mathbf{x}', t - |\mathbf{x} - \mathbf{x}'|/c_0)}{|\mathbf{x} - \mathbf{x}'|}\, \mathrm{d}^3\mathbf{x}'. \end{aligned} \tag{5}$$

To obtain Eq. (5), consider 'sources' q, $\mathbf{F}$, and T_{ij}, use the general solution (3), and at the end apply $\rho_0\partial/\partial t$, $-\boldsymbol{\nabla}\cdot$, and $\partial^2/\partial x_i \partial x_j$.

10.2.2 *Single-frequency forcing*

If the sources in (2) contain only the single frequency ω, the forcing function may be written

$$f(\mathbf{x}, t) = f_1(\mathbf{x})\mathrm{e}^{-\mathrm{i}\omega t}. \tag{6}$$

Throughout this chapter, the subscript 1 indicates the function of position that remains on extracting a time-harmonic factor $\mathrm{e}^{-\mathrm{i}\omega t}$. Expression (6) gives

$$f(\mathbf{x}', t - |\mathbf{x} - \mathbf{x}'|/c_0) = f_1(\mathbf{x}')\mathrm{e}^{\mathrm{i}\omega|\mathbf{x} - \mathbf{x}'|/c_0}\mathrm{e}^{-\mathrm{i}\omega t}. \tag{7}$$

Hence the acoustic field (3) becomes

$$p = p_1(\mathbf{x})\mathrm{e}^{-\mathrm{i}\omega t} \tag{8}$$

where

$$p_1(\mathbf{x}) = \frac{1}{4\pi}\iiint_{\mathbb{R}^3} f_1(\mathbf{x}')\frac{\mathrm{e}^{\mathrm{i}\omega|\mathbf{x} - \mathbf{x}'|/c_0}}{|\mathbf{x} - \mathbf{x}'|}\, \mathrm{d}^3\mathbf{x}'. \tag{9}$$

Equivalently, expressions (6) and (8) may be substituted into the original wave equation (2). Cancellation of $\mathrm{e}^{-\mathrm{i}\omega t}$ gives

$$(\nabla^2 + (\omega/c_0)^2)p_1 = -f_1(\mathbf{x}). \tag{10}$$

This is the reduced wave equation, or the Helmholtz equation. Hence we have solved this equation: the solution is Eq. (9). Alternatively, Eq. (10) may be solved directly, e.g. by Fourier transforms. Evaluation of a Fourier inversion integral gives Eq. (9).

10.2.3 *Examples*

10.2.3.1 *Stationary point monopole*

In the wave equation (1), let the source term $q(\mathbf{x}, t)$ have the form

$$q = Q(t)\delta(\mathbf{x}). \tag{11}$$

Since q has the dimension $1/\text{time}$, and $\delta(\mathbf{x})$ has the dimension $1/\text{volume}$, it follows that $Q(t)$ has the dimensions volume/time. A conventional but unfortunate name for $Q(t)$ is the volume velocity. Corresponding to Eq. (11), the forcing is

$$f(\mathbf{x}, t) = \rho_0 \frac{\partial q}{\partial t} = \rho_0 \dot{Q}(t)\delta(\mathbf{x}), \tag{12}$$

where a dot denotes differentiation with respect to the argument. Thus the acoustic field (3) is

$$p = \frac{\rho_0}{4\pi} \iiint_{\mathbb{R}^3} \frac{\dot{Q}(t - |\mathbf{x} - \mathbf{x}'|/c_0)}{|\mathbf{x} - \mathbf{x}'|} \delta(\mathbf{x}') \, \mathrm{d}^3\mathbf{x}' = \rho_0 \frac{\dot{Q}(t - r/c_0)}{4\pi r}. \tag{13}$$

We use the notation $r = |\mathbf{x}|$. For a single-frequency volume velocity we take $Q(t) = Q_0 \mathrm{e}^{-\mathrm{i}\omega t}$, where Q_0 is a constant with dimensions volume/time. Then Eq. (13) gives

$$p = -\mathrm{i}\rho_0 \omega Q_0 \frac{\mathrm{e}^{-\mathrm{i}\omega(t - r/c_0)}}{4\pi r}, \qquad p_1 = -\mathrm{i}\rho_0 \omega Q_0 \frac{\mathrm{e}^{\mathrm{i}\omega r/c_0}}{4\pi r}. \tag{14}$$

10.2.3.2 *Stationary point dipole*

In Eq. (1) let the source term $\mathbf{F}(\mathbf{x}, t)$ have the form

$$\mathbf{F} = \mathbf{f}(t)\delta(\mathbf{x}). \tag{15}$$

Since $\mathbf{F}$ has the dimensions force/volume, it follows that $\mathbf{f}(t)$ has the dimensions of a force. Corresponding to Eq. (15), the right-hand side of Eq. (2) is

$$f(\mathbf{x}, t) = -\boldsymbol{\nabla} \cdot \mathbf{F} = -\boldsymbol{\nabla} \cdot (\mathbf{f}(t)\delta(\mathbf{x})). \tag{16}$$

To solve Eq. (2) with right-hand side (16), note that if the vector function $\mathbf{A}(\mathbf{x}, t)$ satisfies

$$\left(\frac{1}{c_0^2} \frac{\partial^2}{\partial t^2} - \nabla^2 \right) \mathbf{A} = -\mathbf{F} \tag{17}$$

then $\nabla \cdot \mathbf{A}$ satisfies

$$\left(\frac{1}{c_0^2}\frac{\partial^2}{\partial t^2} - \nabla^2\right)(\nabla \cdot \mathbf{A}) = -\nabla \cdot \mathbf{F}. \tag{18}$$

Hence we may solve Eq. (17) for $\mathbf{A}$, using a vector form of Eq. (3), and take the divergence of the solution to obtain the solution of Eq. (18) for $\nabla \cdot \mathbf{A}$, i.e. the acoustic pressure p. Thus

$$\mathbf{A} = -\frac{1}{4\pi}\iiint_{\mathbb{R}^3} \frac{\mathbf{f}(t - |\mathbf{x} - \mathbf{x}'|/c_0)\delta(\mathbf{x}')}{|\mathbf{x} - \mathbf{x}'|}\,\mathrm{d}^3\mathbf{x}' = -\frac{\mathbf{f}(t - r/c_0)}{4\pi r}. \tag{19}$$

Therefore

$$p = -\frac{1}{4\pi}\nabla \cdot \left(\frac{\mathbf{f}(t - r/c_0)}{r}\right). \tag{20}$$

This is a basic formula of acoustics. To expand the divergence, write $\mathbf{x} = (x_1, x_2, x_3)$, so that $r = (x_1^2 + x_2^2 + x_3^2)^{1/2}$. Then $\partial r/\partial x_i = x_i/r \equiv \hat{x}_i$, where $\hat{\mathbf{x}} \equiv (\hat{x}_1, \hat{x}_2, \hat{x}_3)$ is a unit vector in the $\mathbf{x}$ direction. Thus

$$p = -\frac{\partial}{\partial x_i}\left\{\frac{f_i(t - r/c_0)}{4\pi r}\right\} = \frac{1}{4\pi c_0 r}\,\hat{x}_i\{\dot{f}_i(t - r/c_0) + \frac{c_0}{r} f_i(t - r/c_0)\}. \tag{21}$$

Here the dot denotes differentiation with respect to the argument. The second term in Eq. (21) is important only in the near field. The far field is

$$p \simeq \frac{1}{4\pi c_0 r}\hat{\mathbf{x}} \cdot \dot{\mathbf{f}}(t - r/c_0). \tag{22}$$

Given $\mathbf{x}$ and t, let θ be the angle between $\dot{\mathbf{f}}(t - r/c_0)$ and the radius vector $\mathbf{x}$, and let $\dot{f} \equiv |\dot{\mathbf{f}}|$. Then $\mathbf{x} \cdot \dot{\mathbf{f}} = r\dot{f}\cos\theta$ and $\hat{\mathbf{x}} \cdot \dot{\mathbf{f}} = \dot{f}\cos\theta$, so that the far field is

$$p \simeq \frac{1}{4\pi c_0}\frac{\cos\theta}{r}\dot{f}(t - r/c_0). \tag{23}$$

Therefore the far field is proportional to $\cos\theta$, where the direction $\theta = 0$ is the direction of $\dot{\mathbf{f}}(t - r/c_0)$. This statement must be interpreted with caution, because for given t the direction $\theta = 0$ depends on r. If $\mathbf{f}$ is always parallel to a fixed direction then the main beams are centred on the directions $\theta = 0$ and $\theta = \pi$, and no sound is radiated in the plane $\theta = \frac{1}{2}\pi$.

10.2.3.3 *Stationary point quadrupole*

In Eq. (1), let the source term $T_{ij}(\mathbf{x}, t)$ have the form

$$T_{ij} = \tilde{T}_{ij}(t)\delta(\mathbf{x}). \tag{24}$$

The dimensions of $\tilde{T}_{ij}(t)$ are force $\times$ length. Corresponding to Eq. (24), expression (5) for the acoustic pressure is

$$p = \frac{\partial^2}{\partial x_i \partial x_j} \iiint_{\mathbb{R}^3} \frac{\tilde{T}_{ij}(t - |\mathbf{x} - \mathbf{x}'|/c_0)}{4\pi|\mathbf{x} - \mathbf{x}'|} \delta(\mathbf{x}')\, \mathrm{d}^3\mathbf{x}'$$

$$= \frac{\partial^2}{\partial x_i \partial x_j} \left\{ \frac{\tilde{T}_{ij}(t - r/c_0)}{4\pi r} \right\}. \tag{25}$$

The far-field approximation, obtained by evaluating the partial derivatives in (25) and keeping the dominant terms, is

$$p \sim \frac{1}{4\pi c_0^2 r} \hat{x}_i \hat{x}_j \ddot{\tilde{T}}_{ij}(t - r/c_0). \tag{26}$$

The quadrupole directivity $\hat{x}_i \hat{x}_j$ is readily interpreted in spherical polar coordinates, say (r, θ, ϕ), where $x_1 = r \sin\theta \cos\phi$, $x_2 = r \sin\theta \sin\phi$, $x_3 = r\cos\theta$. Thus x_3 is the polar axis. Then $\hat{x}_1 = \sin\theta\cos\phi$, $\hat{x}_2 = \sin\theta \sin\phi$, $\hat{x}_3 = \cos\theta$, and typical products $\hat{x}_i \hat{x}_j$ are

$$\hat{x}_1\hat{x}_1 = \tfrac{1}{4}(1 - \cos 2\theta)(1 + \cos 2\phi), \qquad \hat{x}_1 \hat{x}_2 = \tfrac{1}{4}(1 - \cos 2\theta)\sin 2\phi,$$
$$\hat{x}_1\hat{x}_3 = \tfrac{1}{2}\sin 2\theta \cos\phi. \tag{27}$$

If $T_{11} = T_{22}$, for example, the directivity contains a term

$$\hat{x}_1\hat{x}_1 + \hat{x}_2\hat{x}_2 = \tfrac{1}{2}(1 - \cos 2\theta), \tag{28}$$

and if $T_{33} = -T_{11} = -T_{22}$, the directivity contains a term

$$\hat{x}_3\hat{x}_3 - \hat{x}_2\hat{x}_2 - \hat{x}_1\hat{x}_1 = \cos 2\theta. \tag{29}$$

The expression $\cos 2\theta$ corresponds to a four-lobed directivity pattern in a meridional plane, i.e. in a plane containing the axis of the polar coordinate system, namely the x_3 axis. Other four-lobed patterns are readily obtained.

10.2.3.4 *Moving point monopole*

In the wave equation (1), let the source term $q(\mathbf{x}, t)$ correspond to a time-varying point source of volume velocity $Q(t)$, in which the point source moves on an arbitrary path $\mathbf{x} = \mathbf{X}(t)$. Thus $Q(t)$ and $\mathbf{X}(t)$ are specified functions, and

$$q = Q(t)\delta(\mathbf{x} - \mathbf{X}(t)). \tag{30}$$

From (4) and (5), the acoustic field is

$$p = \frac{\rho_0}{4\pi}\frac{\partial}{\partial t}\int_{-\infty}^{\infty}\iiint_{\mathbb{R}^3} Q(t')\delta(\mathbf{x}' - \mathbf{X}(t'))\frac{\delta(t - t' - |\mathbf{x} - \mathbf{x}'|/c_0)}{|\mathbf{x} - \mathbf{x}'|}\,\mathrm{d}^3\mathbf{x}'\,\mathrm{d}t'$$
$$= \frac{\rho_0}{4\pi}\frac{\partial}{\partial t}\int_{-\infty}^{\infty} Q(t')\frac{\delta(t - t' - |\mathbf{x} - \mathbf{X}(t')|/c_0)}{|\mathbf{x} - \mathbf{X}(t')|}\,\mathrm{d}t'. \tag{31}$$

Hence the acoustic field at position $\mathbf{x}$ and time t involves $Q(t')$ only at the values of t' satisfying

$$|\mathbf{x} - \mathbf{X}(t')| = c_0(t - t'). \tag{32}$$

These values of t' are the retarded times corresponding to $(\mathbf{x}, t)$. By a basic rule for delta functions (see Chapter 9), evaluation of the integral in Eq. (31) involves division by the factor

$$\left|\frac{\mathrm{d}}{\mathrm{d}t'}\left(t - t' - |\mathbf{x} - \mathbf{X}(t')|/c_0\right)\right|, \tag{33}$$

i.e. by

$$\left|-1 + \frac{\mathbf{x} - \mathbf{X}(t')}{|\mathbf{x} - \mathbf{X}(t')|}\cdot\frac{\dot{\mathbf{X}}(t')}{c_0}\right|. \tag{34}$$

This factor is $|1 - \hat{\mathbf{e}}\cdot\dot{\mathbf{X}}(t')/c_0|$, where $\hat{\mathbf{e}}$ is a unit vector from a retarded position $\mathbf{X}(t')$ to the observation point $\mathbf{x}$. Since $\hat{\mathbf{e}}\cdot\dot{\mathbf{X}}(t')/c_0$ is a Mach-vector component in the observer direction, we write it M_r. Hence the factor is $|1 - M_r|$. Thus Eq. (31) becomes

$$p = \frac{\rho_0}{4\pi}\frac{\partial}{\partial t}\sum_{t'}\frac{Q(t')}{r\,|1 - M_r|}, \tag{35}$$

where $r = |\mathbf{x} - \mathbf{X}(t')|$ and the sum is over the roots t' of Eq. (32), i.e. over the retarded times. For a subsonic source velocity, i.e. when $|\dot{\mathbf{X}}(t)| < c_0$ for all t, there is only one retarded time t' for any given $(\mathbf{x}, t)$.

In deriving Eq. (35), we have assumed that the source is a moving injection of volume, at a rate $Q(t)$ per unit time. In general, there may be an associated injection of momentum, and thus an associated moving point dipole. Thus in many problems Eq. (35) would give only part of the acoustic field.

10.3 The Three-Dimensional Convected Wave Equation

The rest of this chapter is a collection of general results. It may be found useful for reference purposes.

10.3.1 *Arbitrary forcing*

If there is a uniform mean flow of speed U in the positive x direction, the acoustic pressure p satisfies the convected wave equation,

$$\left\{\frac{1}{c_0^2}\left(\frac{\partial}{\partial t}+U\frac{\partial}{\partial x}\right)^2-\nabla^2\right\}p=f(\mathbf{x},t). \tag{36}$$

The Mach number M is defined by $M = U/c_0$. We shall assume that the mean flow is subsonic, i.e. $M < 1$. The free-space solution of (36) may then be expressed in terms of the Doppler factor $\beta = (1 - M^2)^{1/2}$ and the Doppler-transformed coordinates

$$\begin{aligned}\bar{\mathbf{x}} &= (\bar{x},\bar{y},\bar{z}) = (x/\beta^2, y/\beta, z/\beta),\\ \bar{\mathbf{x}}' &= (\bar{x}',\bar{y}',\bar{z}') = (x'/\beta^2, y'/\beta, z'/\beta)\end{aligned} \tag{37}$$

as

$$p=\frac{1}{4\pi\beta^2}\iiint_{\mathbb{R}^3}\frac{f(\mathbf{x}',t+M(\bar{x}-\bar{x}')/c_0-|\bar{\mathbf{x}}-\bar{\mathbf{x}}'|/c_0)}{|\bar{\mathbf{x}}-\bar{\mathbf{x}}'|}\,\mathrm{d}^3\mathbf{x}'. \tag{38}$$

In Eq. (38), there is no bar on the space argument of f or on $\mathrm{d}^3\mathbf{x}'$. One method of deriving (38) is to apply a Lorentz transformation to the usual wave equation (2), as described in Chapman (2000). Another method is to take the Fourier transform of the convected wave equation (36), and evaluate the resulting inversion integral.

10.3.2 *Single-frequency forcing*

If the sources in Eq. (36) contain only the single frequency ω, then as in §10.2.2 we may write

$$f(\mathbf{x},t)=f_1(\mathbf{x})\mathrm{e}^{-\mathrm{i}\omega t},\qquad p(\mathbf{x},t)=p_1(\mathbf{x})\mathrm{e}^{-\mathrm{i}\omega t}, \tag{39}$$

and then Eq. (38) gives

$$p_1(\mathbf{x})=\frac{1}{4\pi\beta^2}\iiint_{\mathbb{R}^3}f_1(\mathbf{x}')\frac{\mathrm{e}^{-\mathrm{i}(\omega/c_0)M(\bar{x}-\bar{x}')+\mathrm{i}(\omega/c_0)|\bar{\mathbf{x}}-\bar{\mathbf{x}}'|}}{|\bar{\mathbf{x}}-\bar{\mathbf{x}}'|}\,\mathrm{d}^3\mathbf{x}'. \tag{40}$$

Equivalently, substitution of Eq. (39) into Eq. (36) shows that expression (40) solves the convected Helmholtz equation

$$\left\{\nabla^2+\left(\frac{\omega}{c_0}+\mathrm{i}M\frac{\partial}{\partial x}\right)^2\right\}p_1=-f_1(\mathbf{x}). \tag{41}$$

10.4 The Two-Dimensional Wave Equation

10.4.1 *Arbitrary forcing*

We now assume that no quantity depends on z. The wave equation (2) still applies, but with $\mathbf{x} = (x, y)$ and $\nabla^2 = \partial^2/\partial x^2 + \partial^2/\partial y^2$. The free-space solution corresponding to Eq. (4) is

$$p = \frac{1}{2\pi} \int_{-\infty}^{t} \iint \frac{f(\mathbf{x}', t')}{\{(t-t')^2 - |\mathbf{x} - \mathbf{x}'|^2/c_0^2\}^{1/2}} \, \mathrm{d}^2\mathbf{x}' \, \mathrm{d}t', \tag{42}$$

in which the spatial integration region comprises all $\mathbf{x}'$ for which $|\mathbf{x} - \mathbf{x}'|^2/c_0^2 \leq (t - t')^2$. Expression (42) may be obtained by performing the integration with respect to z in Eq. (4), or directly from the wave equation by Fourier transforms.

10.4.2 *Single-frequency forcing*

If $f(\mathbf{x}, t)$ contains only the single frequency ω, then substitution of Eqs. (39) into Eq. (42) gives a t' integral that can be evaluated analytically in terms of the Hankel function of order zero of the first kind, i.e. $\mathrm{H}_0^{(1)} \equiv \mathrm{J}_0 + \mathrm{i}\,\mathrm{Y}_0$, where J_0 and Y_0 are the Bessel functions of order zero of the first and second kinds. The result is

$$p_1(\mathbf{x}) = \tfrac{1}{4}\mathrm{i} \iint_{\mathbb{R}^2} f_1(\mathbf{x}') \, \mathrm{H}_0^{(1)}(\omega|\mathbf{x} - \mathbf{x}'|/c_0) \, \mathrm{d}^2\mathbf{x}'. \tag{43}$$

This function $p_1(\mathbf{x})$ solves the two-dimensional Helmholtz equation

$$(\nabla^2 + (\omega/c_0)^2)p_1 = -f_1. \tag{44}$$

10.5 The Two-Dimensional Convected Wave Equation

10.5.1 *Arbitrary forcing*

The solution of the two-dimensional version of the convected wave equation (36) at subsonic Mach number $M = U/c_0$ involves, as before, Doppler-transformed coordinates. These are now $\bar{\mathbf{x}} = (\bar{x}, \bar{y}) = (x/\beta^2, y/\beta)$ and $\bar{\mathbf{x}}' = (\bar{x}', \bar{y}') = (x'/\beta^2, y'/\beta)$, with $\beta = (1 - M^2)^{1/2}$, and the two-dimensional solution is

$$p = \frac{1}{2\pi\beta} \int_{-\infty}^{t} \iint \frac{f(\mathbf{x}', t' + M(\bar{x} - \bar{x}')/c_0)}{\{(t-t')^2 - |\bar{\mathbf{x}} - \bar{\mathbf{x}}'|^2/c_0^2\}^{1/2}} \, \mathrm{d}^2\mathbf{x}' \, \mathrm{d}t'. \tag{45}$$

The spatial integration region is as for Eq. (42)

10.5.2 *Single-frequency forcing*

At frequency ω, substitution of Eqs. (39) into Eq. (45), followed by evaluation of the t' integral, gives

$$p_1(\mathbf{x}) = \frac{\mathrm{i}}{4\beta} \iint_{\mathbb{R}^2} f_1(\mathbf{x}') \mathrm{e}^{-\mathrm{i}(\omega/c_0)M(\bar{x}-\bar{x}')} \, \mathrm{H}_0^{(1)}(\omega|\bar{\mathbf{x}} - \bar{\mathbf{x}}'|/c_0) \, \mathrm{d}^2\mathbf{x}'. \quad (46)$$

The function $p_1(\mathbf{x})$ solves the two-dimensional version of (41), i.e. the two-dimensional convected Helmholtz equation.

10.6 Conclusion

The results in this chapter are widely used in acoustics, and have many extensions, for example to the effects of boundaries and to realistic source modelling, i.e. the determination of a suitable function $f(\mathbf{x}, t)$ in a given practical situation. It is hoped that the earlier sections of the chapter, with the examples chosen, will be found useful as a tutorial, and the later sections, containing general formulae, will be found useful as a reference for describing some effects of mean flow and of variation in the number of space dimensions.

References

Chapman, C. J. (2000) Similarity variables for sound radiation in a uniform flow, *Journal of Sound & Vibration* **233**, pp. 157–164.

Dowling, A. P. and Ffowcs Williams, J. E. (1983) *Sound and Sources of Sound.* Ellis Horwood.

Howe, M. S. (1998) *Acoustics of Fluid–Structure Interactions.* Cambridge University Press.

Howe, M. S. (2003) *Theory of Vortex Sound.* Cambridge University Press.

Problems

Exercise 10.1 For the monopole source $q = Q(t)\delta(\mathbf{x})$, some simple forms of $Q(t)$, involving parameters Q_0, ω, and τ, are

$$Q(t) = Q_0\mathrm{e}^{-\mathrm{i}\omega t}, \;\; Q_0\delta(t/\tau), \;\; Q_0, \;\; Q_0\mathrm{e}^{-(1/2)(t/\tau)^2}, \;\; Q_0\,\mathrm{H}(t/\tau). \quad (47)$$

For a selection of these, calculate the acoustic pressure when the governing equation is (a) the usual wave equation, and (b) the convected wave equation, both for three space dimensions and for two space dimensions. (In practice, interest lies not only in general solutions but also in the dependence of a solution on physically meaningful parameters, e.g. frequency, time scale of forcing, size of forcing region, flow speed, etc.)

Exercise 10.2 Repeat Ex. 10.1 for the dipole source $\mathbf{F} = \mathbf{f}(t)\delta(\mathbf{x})$, where the force $\mathbf{f}$, depending on a constant vector $\mathbf{f}_0$, takes the forms

$$\mathbf{f}(t) = \mathbf{f}_0 \mathrm{e}^{-\mathrm{i}\omega t}, \quad \mathbf{f}_0\delta(t/\tau), \quad \mathbf{f}_0, \quad \mathbf{f}_0 \mathrm{e}^{-(1/2)(t/\tau)^2}, \quad \mathbf{f}_0\, \mathrm{H}(t/\tau). \tag{48}$$

Exercise 10.3 A simple model for rotor noise is to take the sources as axial point forces acting at points that move in a circle. Analyse the sound field as far as you can.

Exercise 10.4 This chapter contains eight forced wave equations, classified as in §10.1. Solve a selection of these by Fourier transforms, evaluating the inversion integrals as in Chapter 3: 'Integral Transforms'.

Exercise 10.5 When a monopole point source translates at uniform subsonic speed, the acoustic pressure field is symmetric fore-and-aft. Verify this fact in two ways: (a) solve the convected wave equation with a stationary point source; (b) solve the usual wave equation with a moving point source, and change the coordinates in the answer to a coordinate system that moves with the source. (This result is far from obvious, because in method (b) the denominator $r\,|1 - M_r|$ arising in Eq. (35) does not appear to have fore-and-aft symmetry; one might at first think that the field beams forward.)

Chapter 11

Corrugated Pipe Flow

J. W. Elliott
Department of Mathematics
University of Hull

11.1 The Problem

Here we are concerned with the phenomenon of tone, or multi-tone, sound generation in a pipe with a corrugated wall, due to air flowing through it. This phenomenon can be demonstrated surprisingly easily by blowing through a section of corrugated hose. Indeed a popular toy in the 1970s was the *Hummer* or *Whirl-a-sound*, in which a corrugated tube, of 1 m in length and 2.5 cm in diameter, was swung around one's head. Here centrifugal force pumped air from the inner to the outer end of the pipe. For a sufficiently fast rotation a loud, clear, discrete tone is heard. Moreover the faster the pipe is swung, the higher the pitch of the note emitted.

In addition to the benign manifestation in toys, tone generation in corrugated pipes can constitute a severe problem in other contexts, such as in domestic and industrial appliances. For example, in HVAC control systems where compressed air is fed through small-diameter corrugated tubes, it is found that high intensity, high frequency whistling can occur. Also flexible corrugated air pipes, used in the control systems of heating ducts in buildings, can generate high noise levels.

11.1.1 *The experimental set-up*

In order to investigate this phenomenon some preliminary experimental research has taken place at the University of Hull, in conjunction with Prof. A. Cummings. Additionally there have been some limited experiments by Prof. V. F. Kopiev, at

TsAGI (Moscow).

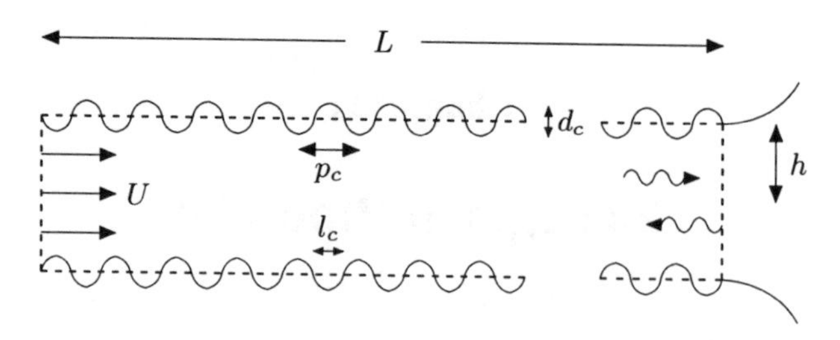

Fig. 11.1 Experimental geometry.

Consider a corrugated pipe of length L, mean radius h and a corrugation pitch p_c, with a uniform cavity length l_c and cavity depth d_c, as shown in Fig. 11.1. Typical dimensions of the pipes investigated at Hull are given in Table 11.1, where approximately $h = R + \frac{1}{2}d_c = D - \frac{1}{2}d_c$, if the thickness of the pipe wall is neglected.

Table 11.1 Dimensions of pipes used in experiments (mm).

	Pipe 1	Pipe 2	Pipe 3	Pipe 4
Length L	1031	1020	1350	931
Internal diameter $2R$	10.65	13.75	16.66	19.23
External diameter $2D$	14.35	18.85	21.86	25.53
Corrugation pitch p_c	2.13	2.72	2.83	3.08
Cavity depth d_c	1.85	2.55	2.6	3.15
Cavity length l_c	1.45	1.9	2.05	1.65

11.1.2 *Experimental results*

When there is air flow of a sufficient mean speed, U, along the pipe, a tone, or multi-tone, noise is generated. The corrugations are essential, for in their absence no sound is heard. Moreover it is clear that the tones heard are associated with the phenomenon of resonance, for if a horn is added to the pipe exit then no sound is heard. The effect of the horn is to reduce the strength of the acoustic reflection at the pipe exit, so effectively only downstream travelling waves are present.

The tones heard are the harmonics of the pipe, whose frequencies, f_m, are all multiples of some fundamental tone f_1. Indeed in the absence of any mean flow the resonant frequencies for a corresponding smooth hard-walled tube, are

$$f_m = mf_1 = \frac{mc_0}{2L}(1 - M^2) \qquad \text{for } m = 1, 2, 3, \ldots, \tag{1}$$

where c_0 is the speed of sound and $M = U/c_0$ is the Mach number of the flow. The effect of the mean flow and corrugations being to modify these resonant frequencies reducing them so that are approximately 10% smaller than those predicted for the smooth hard-walled pipe. From a simple model of the corrugations we were able to obtain an improved prediction

$$f_m = \frac{mc_0}{2L_e} \frac{(1 - M^2)}{\left[1 + \frac{d_c}{R}\left(\frac{l_c}{p_c}\right)\left(1 + \frac{d_c}{2R}\right)\right]} \qquad \text{for } m = 1, 2, 3, \ldots, \tag{2}$$

where L_e is the *effective* length of the pipe.

In general it is almost impossible to make the fundamental sing. Indeed, we tried to excite these fundamental modes by introducing external forcing using a loudspeaker. Table 11.2 compares the frequencies obtained with the above predictions.

Table 11.2 Predicted and experimental frequencies (Hz).

Axial mode number	Frequency of noise fed to loudspeaker	Frequency of self-excited oscillation	Frequency of predicted oscillation	Frequency of equivalent smooth tube
3	492.5	—	495.3	543
4	660	—	660.4	724
5	825	—	825.4	905
6	990	—	990.6	1087
7	1159	—	1156	1268
8	1325	—	1321	1449
9	1491	1488	1486	1630
10	1655	1650	1651	1811
11	1823	1812	1816	1992
12	1992	1975	1981	2173

It has been asserted that the level of turbulence within the pipe is important for the production of sound. We, however, failed to confirm this. Indeed when a smooth-contoured bell-mouth inlet was machined to smoothly fit one of our pipes, we found the results were very little different from those obtained without a bell-mouth inlet. This would imply that any flow disturbance in the immediate inlet region of the corrugated tube is of little importance in the sound generation process.

When we measure the sound spectrum we observe that is a collection of closely located distinct peaks distributed in a narrow band region about a main, or central, frequency $f = F_v$, with each peak, or each radiating frequency, being a resonant mode of the pipe lying within the domain of radiation. That is $F_v = f_m$, for some integer m.

Moreover this narrow band of closely located excited frequencies is continu-

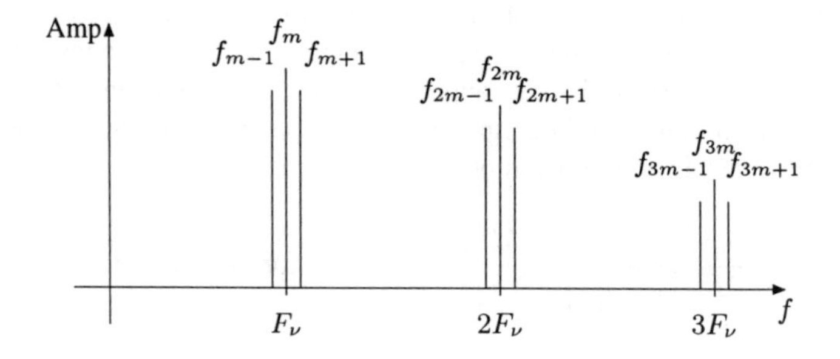

Fig. 11.2 Spectrum of resonances.

ally repeated at higher and higher frequencies, as shown in Fig. 11.2, with the main frequency of each radiating domain being approximately $f = nF_v$, for $n = 1, 2, \ldots$. The noise mechanism is very stable and acoustic output levels at the pipe exit are very high.

At first sight it appears that the sound generated does not depend on the length of the pipe L. For if we increase the pipe length from (say) half a metre to one of several metres, we find that F_v remains fixed.

For each excited mode, we can identify a maximum intensity associated with a given critical speed. As the flow velocity increases the frequency F_v increases, realized as a discrete jump in the frequency of the most noisy mode, $F_v = f_m$, to the neighbouring radiating mode, $F_v = f_{m+1}$. The region of radiation also drifts to a higher frequency domain in line with the increase in the main frequency. This mechanism is repeated for all the radiating frequency ranges, as shown in Fig. 11.3.

The frequency F_v appears to increase proportionally with the mean flow velocity U, or $F_v \propto U$. Indeed if f is the frequency, then over the entire first domain of radiation the Strouhal number, the dimensionless number defined by

$$\mathrm{St} = \frac{f l_c}{U_c}, \tag{3}$$

lies within the domain $S_{\min} \leq \mathrm{St} \leq S_{\max}$. If U_0 is the average fluid speed, then $U_c \approx \frac{1}{2} U$ is the grazing speed directly above the cavity. The maximum radiation is given by $\mathrm{St} \approx 0.5$.

An analysis of the experimental data shows that the spectrum of generated noise is characterised by two factors: (i) the resonance ability of the pipe and (ii) the frequency instability region(s) where there is a positive energy flux from the mean flow into the oscillations. The dominant frequency (and instability region) is characterised by the flow velocity rather than the pipe length. Of course

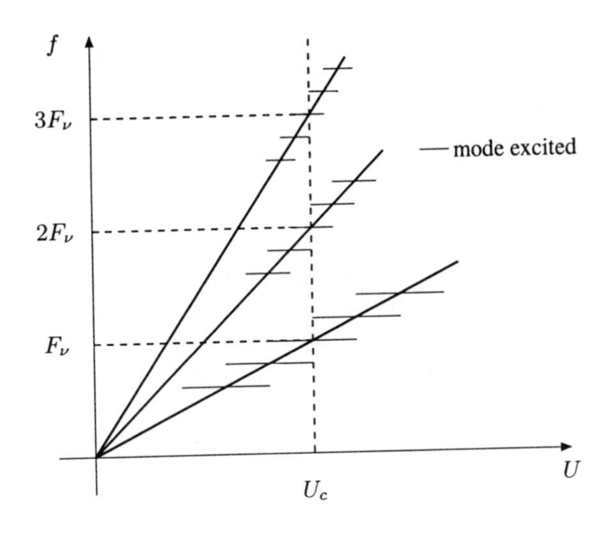

Fig. 11.3 Variation in resonance frequency with velocity.

each excited resonance has a frequency that depends on the pipe length L.

We should note that the selection and final amplitudes of the excited modes, is determined by some nonlinear process.

11.1.3 *Proposed mechanism*

A pipe with axial corrugations can be thought of as consisting of many cylindrical cavities all placed together. There is a great deal of literature covering noise due to cavity flow and the associated phenomena of self-sustaining cavity oscillations. Cavity oscillations are classified as being either deep, if $d_c/l_c > 1$, or shallow, if $d_c/l_c < 1$.

It is suggested that the oscillations are caused by the selective amplification of vortical fluctuations in the layer of fluid passing directly over the cavity, due to the following feedback loop. A sound pressure fluctuation induces an annular vortex in the free shear layer at the cavity's leading edge. The vortex is then convected at a speed $U_c \approx 0.5U$ where it subsequently hits the downstream edge of the cavity. The impingement of the vortex produces a fluctuating axial force that acts as an acoustic dipole. This dipole drives a resonant acoustic mode of the tube, amplifying the sound pressure fluctuation. These pressure fluctuations then propagate back to the leading edge of the cavity, where they induce more vortices.

This mechanism is very similar to that proposed for edge-tones, for jet speeds that are sufficiently large, except that the latter does not involve the presence of a resonator. Instead there is a direct acoustic feedback from the edge to the jet mouth. Both mechanisms involve an acoustic dipole source, which causes an unsteady pressure differential across a free shear layer, giving rise to vortices that are convected with the mean flow and interact with a solid surface.

11.1.4 *Other experiments*

An additional series of experiments was carried out to investigate whether the location of the cavities was itself important. In particular we were concerned with whether the sound was generated principally by the corrugations near the inlet. To this end 10 corrugations of tube 1 were placed between two lengths of a plain (smooth) tube of equivalent diameter. The total length of the combined tube was held fixed, $L = 1021$ mm, but the distance between the corrugations and the inlet, L_{ent}, as shown in Fig. 11.4, was varied.

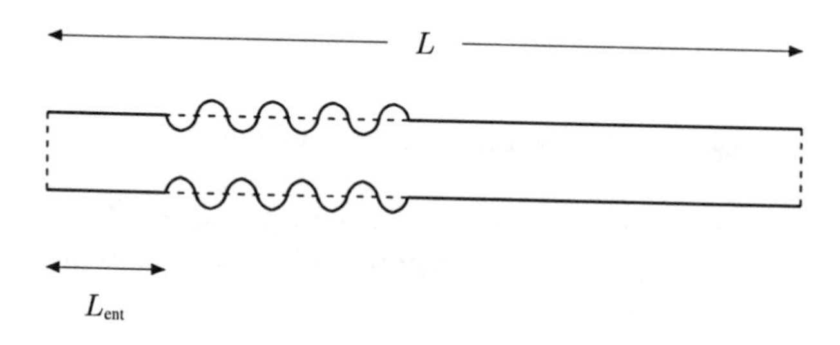

Fig. 11.4 Variation of entry conditions.

It was found that for $L_{\mathrm{ent}} = 0$, that is with the corrugated section at the inlet, self-excited oscillations were easily produced. For $L_{\mathrm{ent}} = 100$ mm the oscillations were still evident, but a higher flow speed was required to initiate them. For $L_{\mathrm{ent}} > 100$ mm no oscillations could be produced. This possibly suggests that the noise sources are located principally in the vicinity of the tube inlet, and that the source mechanism may require a thin boundary layer and fairly low levels of turbulence.

11.2 Pipe Flow

Consider a smooth circular pipe, of length L and radius h. Denoting the pressure by p and the fluid velocity by $\mathbf{u} = u\hat{\mathbf{i}} + v\hat{\mathbf{r}}$, where u, v are the axial and radial components respectively, we linearise about a uniform mean state $p = p_0$, $\mathbf{u} = U\hat{\mathbf{\imath}}$. The unsteady pressure and velocity fluctuations, p' and $\mathbf{u}'$, then satisfy

$$\left[\frac{1}{c_0^2}\left(\frac{\partial}{\partial t}+U\frac{\partial}{\partial x}\right)^2-\nabla^2\right]p'=0, \qquad \rho_0\left(\frac{\partial}{\partial t}+U\frac{\partial}{\partial x}\right)\mathbf{u}'=-\boldsymbol{\nabla}p', \quad (4)$$

for $0 \le r \le h$, $0 \le x \le L$, where r and x denote radial and axial distances. If we look for a harmonic response with angular frequency ω by taking

$$p'(x,r,t)=\mathrm{Re}\left[\bar{p}(x,r)\mathrm{e}^{-\mathrm{i}\omega t}\right], \qquad \mathbf{u}'=\mathrm{Re}\left\{\left[\bar{u}(x,y),\bar{v}(x,r)\right]\mathrm{e}^{-\mathrm{i}\omega t}\right\}, \quad (5)$$

then the convected wave equation for the acoustic pressure reduces to

$$-\left[\nabla^2+\left(\kappa+\mathrm{i}M\frac{\partial}{\partial x}\right)^2\right]\bar{p}=0. \quad (6)$$

Here $\kappa = \omega/c_0$ is the acoustic wavenumber and $M = U/c_0$ is the Mach number. Having solved for the pressure the velocity components are then given by

$$-\mathrm{i}\rho_0 c_0\left(\kappa+\mathrm{i}M\frac{\partial}{\partial x}\right)\bar{u}=-\frac{\partial\bar{p}}{\partial x}, \qquad -\mathrm{i}\rho_0 c_0\left(\kappa+\mathrm{i}M\frac{\partial}{\partial x}\right)\bar{v}=-\frac{\partial\bar{p}}{\partial r}. \quad (7)$$

For total reflection at the pipe inlet/exit, we require

$$\bar{p}=0 \qquad \text{at } x=0 \text{ and } x=L. \quad (8)$$

Further if Z is the wall impedance and $\beta = Z^{-1}$ is the wall admittance, then

$$\frac{\partial\bar{p}}{\partial r}=\mathrm{i}\rho_0 c_0\left(\kappa+\mathrm{i}M\frac{\partial}{\partial x}\right)(\beta p) \quad \text{at } r=h, \quad \text{where } Z=\frac{1}{\beta}=\left.\frac{\bar{p}}{\bar{v}}\right|_{r=h}. \quad (9)$$

Thus for $Z = \infty$, $\beta = 0$, we have a rigid wall, with zero normal flow, $\bar{v} = 0$.

11.2.1 *The admittance*

In the case of zero mean flow, where $M = 0$, the problem for $\bar{p}$ reduces to Helmholtz's equation, which in cylindrical polar coordinates, is given by

$$-\left(\nabla^2+\kappa^2\right)\bar{p}\equiv-\left[\frac{1}{r}\frac{\partial}{\partial r}\left(r\frac{\partial}{\partial r}\right)+\frac{\partial^2}{\partial x^2}+\kappa^2\right]\bar{p}=0, \quad (10)$$

subject to $\partial\bar{p}/\partial r = \mathrm{i}\rho_0 c_0 \kappa\beta p$ at $r = h$. If we look for a separable solution, then

$$\bar{p}(x,r)=\sum_{n=0}^{\infty}\left[A_n\mathrm{e}^{\mathrm{i}k_n x}+B_n\mathrm{e}^{-\mathrm{i}k_n x}\right]\mathrm{J}_0(\alpha_n r/h), \quad (11)$$

where $k_n^2 = \kappa^2 - (\alpha_n/h)^2$ and J_0 is the zeroth order Bessel function, which satisfies

$$t\,\mathrm{J}_0'' + \mathrm{J}_0' + t\,\mathrm{J}_0 = 0, \qquad \mathrm{J}_0(0) = 1. \tag{12}$$

Here $\alpha = \alpha_n$ are the discrete solutions to the transcendental equation

$$\alpha_n\,\mathrm{J}_0'(\alpha_n) - \mathrm{i}\rho_0 c_0 \kappa h \beta\,\mathrm{J}_0(\alpha_n) = 0. \tag{13}$$

For a rigid wall, $\beta = 0$, we have real roots to the equation $\mathrm{J}_0'(\alpha_n) = -\,\mathrm{J}_1(\alpha_n) = 0$, of which the first, $\alpha_0 = 0$, corresponds to planar flow. For $\beta \ll 1$ we have an expansion of the form

$$k_n = k_{n0} + \mathrm{i}\left(\frac{\rho_0 c_0}{h}\right)\left(\frac{\kappa}{k_{n0}}\right)\beta + O(\beta^2). \tag{14}$$

Here k_{n0} are the real wavenumbers for a rigid wall, with $k_{00} = \kappa$. This gives

$$\mathrm{e}^{\mathrm{i}k_n x} = \mathrm{e}^{\mathrm{i}k_{n0} x}\exp\left(-\left(\frac{\rho_0 c_0}{h}\right)\left(\frac{\kappa}{k_{n0}}\right)\beta x\right). \tag{15}$$

The dissipative case, where the wall absorbs the pipe's wave energy, corresponds to $\mathrm{Re}(\beta) > 0$ and exponential decay of the pressure. In contrast $\mathrm{Re}(\beta) < 0$ corresponds to energy input by the wall and exponential pressure growth.

Alternatively consider a small section of pipe, of length δx, cross-sectional area $A = \pi h^2$ and perimeter $P = 2\pi h$. Suppose a plane wave, of acoustic intensity $I_x(x)A$, enters the section and a plane wave of intensity

$$I_x(x + \delta x)A \approx \left[I_x(x) + I'(x)\,\delta x\right]A, \tag{16}$$

flows out of the volume. Now for $T = 2\pi/\omega$ we have the time-average

$$\langle I_x(x)\rangle = \frac{1}{T}\int_0^T p'u'\,\mathrm{d}t = \tfrac{1}{4}(\bar{p}\bar{u}^* + \bar{p}^*\bar{u}) = \frac{1}{2\rho_0 c_0}|\bar{p}|^2, \tag{17}$$

since $\bar{p} = \pm\rho_0 c_0 \bar{u}$ for a plane wave. In addition we have acoustic energy $I_y(x)P\,\delta x = p'v'P\,\delta x$ flowing out of across the pipe walls, where

$$\langle I_y(x)\rangle = \tfrac{1}{4}(\bar{p}\bar{v}^* + \bar{p}^*\bar{v}) = \tfrac{1}{4}\left(\beta + \beta^*\right)|\bar{p}|^2 = \tfrac{1}{2}\,\mathrm{Re}(\beta)|\bar{p}|^2, \tag{18}$$

since $\bar{v} = \beta\bar{p}$. By conservation of energy we would expect

$$\frac{\mathrm{d}}{\mathrm{d}x}\left(\frac{1}{2\rho_0 c_0}|\bar{p}|^2 A\,\delta x\right) + \tfrac{1}{2}\,\mathrm{Re}(\beta)|\bar{p}|^2 P\,\delta x = 0, \tag{19}$$

showing that $\mathrm{Re}(\beta) > 0$ leads to exponential decay of the pressure amplitude, with

$$|\bar{p}| \propto \exp\left(-\frac{\rho_0 c_0}{h}\,\mathrm{Re}(\beta)x\right). \tag{20}$$

11.2.2 *Planar pipe resonance*

11.2.2.1 *Low speed flow*

Taking $\bar{v} = \partial/\partial r = 0$ and $M \ll 1$ the problem for $\bar{p}$ reduces to Helmholtz's equation

$$-\left(\nabla^2 + \kappa^2\right)\bar{p} \equiv -\left[\frac{\partial^2}{\partial x^2} + \kappa^2\right]\bar{p} = 0, \tag{21}$$

whose solution is given by

$$\bar{p} = A\left[e^{i\kappa x} + Re^{-i\kappa x}\right], \qquad \bar{u} = \frac{A}{\rho_0 c_0}\left[e^{i\kappa x} - Re^{-i\kappa x}\right], \tag{22}$$

where A is the pressure amplitude and R is the reflection coefficient. Note that the *specific* acoustic impedance

$$\bar{Z} = \frac{Z}{\rho_0 c_0} = \frac{1}{\rho_0 c_0}\frac{\bar{p}}{\bar{u}} = \frac{(1+R) + i(1-R)\tan(\kappa x)}{(1-R) + i(1+R)\tan(\kappa x)} = \frac{Z_0 + i\tan(\kappa x)}{1 + iZ_0\tan(\kappa x)}, \tag{23}$$

where $Z_0 = (1+R)/(1-R)$. Thus, in general, $\bar{Z}$ is complex with

$$\bar{Z} = Z_r + iZ_i = \frac{Z_0\left[1 + Z_0^2\tan(\kappa x)\right] + i\tan(\kappa x)\left[1 - Z_0^2\tan(\kappa x)\right]}{1 + Z_0^2\tan^2(\kappa x)}. \tag{24}$$

Here Z_r is the specific acoustic *resistance* and Z_i is the specific acoustic *reactance*. The inlet condition, $\bar{p} = 0$ at $x = 0$, then yields $R = -1$ and

$$\bar{p} = 2iA\sin(\kappa x), \qquad \bar{u} = \frac{2A}{\rho_0 c_0}\cos(\kappa x). \tag{25}$$

Hence the impedance is purely imaginary (reactive), as

$$Z = \frac{1}{\beta} = \frac{\tilde{p}}{\tilde{u}} = \frac{\bar{p}}{\bar{u}} = i\rho_0 c_0\tan(\kappa x). \tag{26}$$

To satisfy the exit condition, $\bar{p} = 0$ at $x = L$, we must have

$$\sin(\kappa L) = 0 \quad \text{so} \quad \kappa L = n\pi \quad \text{or} \quad \omega = c_0\kappa = \frac{n\pi c}{L}, \quad \text{for } n = 1, 2, 3, \ldots. \tag{27}$$

The resonance frequencies of the smooth hard-walled pipe are the discrete values

$$f_n = \frac{\omega}{2\pi} = \frac{nc_0}{2L}, \qquad \text{for } n = 1, 2, 3, \ldots. \tag{28}$$

11.2.2.2 *Higher speed flow*

For $M = O(1)$ the pressure satisfies the convected wave equation

$$-\left[\frac{\partial^2}{\partial x^2} + \left(\kappa + \mathrm{i}M\frac{\partial}{\partial x}\right)^2\right]\bar{p} = 0. \tag{29}$$

If we look for a solution of the form $\mathrm{e}^{\mathrm{i}\kappa k x}$, where $\kappa = \omega/c_0$ and k is the dimensionless acoustic wavenumber, then we obtain a dispersion relation

$$(1 - kM)^2 = k^2 \qquad \text{so} \qquad k = k_+, \qquad k = -k_-, \tag{30}$$

where the upstream and downstream axial wavenumbers are given by

$$k_+ = \frac{1}{1+M}, \qquad k_- = \frac{1}{1-M} \qquad \text{so} \qquad k_+ + k_- = \frac{2}{1-M^2}. \tag{31}$$

The phase speed of the upstream and downstream travelling waves are then

$$c_\pm = \frac{\omega}{\kappa k_\pm} = \frac{c_0}{k_\pm} = (1 \pm M)c_0. \tag{32}$$

We again have a solution

$$\bar{p} = A\left[\mathrm{e}^{\mathrm{i}\kappa k_+ x} + R\mathrm{e}^{-\mathrm{i}\kappa k_- x}\right], \qquad \bar{u} = \frac{A}{\rho_0 c_0}\left[\mathrm{e}^{\mathrm{i}\kappa k_+ x} - R\mathrm{e}^{-\mathrm{i}\kappa k_- x}\right], \tag{33}$$

and again the entrance condition, $\bar{p} = 0$ at $x = 0$, yields $R = -1$. Furthermore, in order to satisfy the exit condition

$$\bar{p} \equiv A\left[\mathrm{e}^{\mathrm{i}\kappa k_+ L} - \mathrm{e}^{-\mathrm{i}\kappa k_- L}\right] = A\mathrm{e}^{-\mathrm{i}\kappa k_- L}\left[\mathrm{e}^{\mathrm{i}\kappa(k_+ + k_-)L} - 1\right] = 0, \tag{34}$$

we must have

$$\kappa(k_+ + k_-)L = 2n\pi \qquad \text{so} \qquad \frac{2\kappa L}{1-M^2} = 2n\pi, \tag{35}$$

and therefore

$$\omega = c_0\kappa = \frac{n\pi c_0}{L}\left(1 - M^2\right), \tag{36}$$

for integer n. Thus the resonance frequencies of the smooth hard-walled pipe are the discrete values

$$f_n = \frac{\omega}{2\pi} = \frac{nc_0}{2L}\left(1 - M^2\right), \qquad \text{for } n = 1, 2, 3, \ldots. \tag{37}$$

11.3 Cummings's Corrugated Model

A simple model of corrugated flow, presented by Prof. A. Cummings, is as follows. Let the corrugations be rectangular cylinders, having a cavity length l_c and cavity depth d_c, and a corrugation pitch p_c as shown in Fig. 11.5.

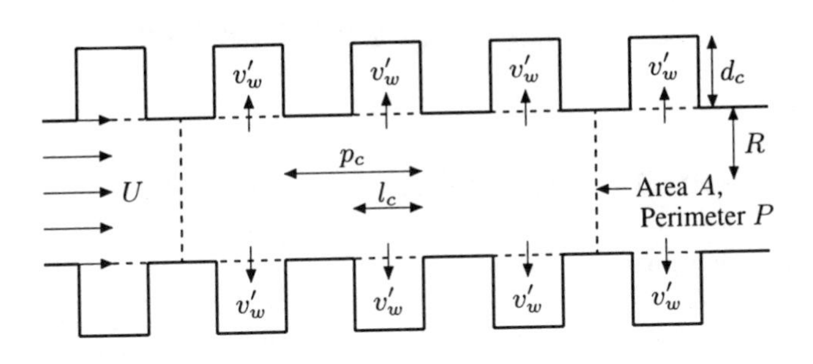

Fig. 11.5 Rectangular cylinder model.

Let a basic uniform flow, for $0 < r < R$, be separated from one at rest, for $R < r < R + d_c$, by a shear layer of zero thickness. We shall then assume that the fluctuations do not vary across the pipe cross-section, but allow a small radial particle velocity at the shear layer.

By considering a control volume, indicated by the dotted lines, whose cross-section has a perimeter $P = 2\pi R$ and an area $A = \pi R^2$, it can be shown that in the interior of the pipe, the Euler equations have a quasi 1-D form

$$\rho_0 \left(\frac{\partial}{\partial t} + U \frac{\partial}{\partial x} \right) \mathbf{u}' = -\frac{\partial p}{\partial x}, \tag{38}$$

$$\frac{1}{c_0^2} \left(\frac{\partial}{\partial t} + U \frac{\partial}{\partial x} \right) p' + \rho_0 \frac{\partial u'}{\partial x} = -\rho_0 \left(\frac{P}{A} \right) \langle v' \rangle, \tag{39}$$

where $\langle v' \rangle$ is the space-averaged radial velocity. Combining these two equations results in a single equation of the form

$$\left[\frac{1}{c_0^2} \left(\frac{\partial}{\partial t} + U \frac{\partial}{\partial x} \right)^2 - \frac{\partial^2}{\partial x^2} \right] p' = -\rho_0 \frac{P}{A} \left(\frac{\partial}{\partial t} + U \frac{\partial}{\partial x} \right) \langle v' \rangle. \tag{40}$$

Taking ξ' to be the particle displacement and assuming the pressure and radial particle displacement to be continuous across the shear layer, gives

$$\langle v' \rangle = \left(\frac{\partial}{\partial t} + U \frac{\partial}{\partial x} \right) \langle \xi' \rangle, \qquad \langle v'_w \rangle = \frac{\partial}{\partial t} \left(\langle \xi' \rangle \right) \qquad \text{at } r = h, \tag{41}$$

where v'_w is the radial velocity within the corrugations. Taking

$$(p', u', \langle v' \rangle, \langle \xi' \rangle) = \left(\bar{p}, \bar{u}, \langle \bar{v} \rangle, \langle \bar{\xi} \rangle \right) \mathrm{e}^{\mathrm{i}(\kappa k - \omega t)}, \tag{42}$$

results in

$$-\kappa^2 \left[(1 - kM)^2 - k^2 \right] \bar{p} = \mathrm{i} \frac{P}{A} \rho_0 c_0 \kappa \, (1 - kM) \, \langle \bar{v} \rangle. \tag{43}$$

Moreover since we have

$$\langle \bar{v} \rangle = -\mathrm{i} c_0 \kappa \, (1 - kM) \, \langle \bar{\xi} \rangle, \qquad \langle \bar{v}_w \rangle = -\mathrm{i} c_0 \kappa \langle \bar{\xi} \rangle, \tag{44}$$

then for a prescribed space-averaged wall admittance, given by

$$\langle \beta_w \rangle = \rho_0 c_0 \frac{\langle \bar{v}_w \rangle}{\bar{p}} = \rho_0 c_0 \, (1 - kM)^2 \, \frac{\langle \bar{v} \rangle}{\bar{p}}, \tag{45}$$

we have the dispersion relation

$$-\kappa^2 \left[(1 - kM)^2 - k^2 \right] = \mathrm{i} \frac{P}{A} \kappa \langle \beta_w \rangle (1 - kM)^2. \tag{46}$$

Alternatively we can write this as the quadratic

$$\left[1 + \left(1 - \mathrm{i} \frac{P}{A} \frac{\langle \beta_w \rangle}{\kappa} \right) \right] M^2 k^2 - 2M \left(1 - \mathrm{i} \frac{P}{A} \frac{\langle \beta_w \rangle}{\kappa} \right) k + \left(1 - \mathrm{i} \frac{P}{A} \frac{\langle \beta_w \rangle}{\kappa} \right) = 0, \tag{47}$$

which has solutions $k = k_+$ and $k = -k_-$

$$k_+ = \frac{\left(1 - \mathrm{i} \frac{P}{A} \frac{\langle \beta_w \rangle}{\kappa} \right)^{1/2}}{1 + M \left(1 - \mathrm{i} \frac{P}{A} \frac{\langle \beta_w \rangle}{\kappa} \right)^{1/2}}, \qquad k_- = \frac{\left(1 - \mathrm{i} \frac{P}{A} \frac{\langle \beta_w \rangle}{\kappa} \right)^{1/2}}{1 - M \left(1 - \mathrm{i} \frac{P}{A} \frac{\langle \beta_w \rangle}{\kappa} \right)^{1/2}}. \tag{48}$$

For resonance we must again have

$$\omega = c_0 \kappa = \frac{2 n \pi c_0}{(k_+ + k_-) L_e} = \frac{n \pi c_0}{L} \frac{\left[1 - M^2 \left(1 - \mathrm{i} \frac{P}{A} \frac{\langle \beta_w \rangle}{\kappa} \right)^{1/2} \right]}{\left(1 - \mathrm{i} \frac{P}{A} \frac{\langle \beta_w \rangle}{\kappa} \right)^{1/2}} \tag{49}$$

for integer n, where here L_e is the effective length of the pipe. Thus the resonance frequencies of the smooth hard-walled pipe are the discrete values

$$f_n = \frac{\omega}{2\pi} = \frac{nc_0}{2L} \frac{\left[1 - M^2\left(1 - \mathrm{i}\frac{P}{A}\frac{\langle\beta_w\rangle}{\kappa}\right)^{1/2}\right]}{\left(1 - \mathrm{i}\frac{P}{A}\frac{\langle\beta_w\rangle}{\kappa}\right)^{1/2}}, \qquad \text{for } n = 1, 2, 3, \ldots \tag{50}$$

For low frequencies, Cummings assumes that the walls have a spring-like, purely imaginary, impedance (involving no energy loss/gain) and sets

$$\beta_w = \begin{cases} \mathrm{i}\kappa V/S & \text{over the cavities,} \\ 0 & \text{over the rigid wall,} \end{cases} \tag{51}$$

where $S = 2\pi R l_c$ is the area presented by the cavity to the tube and V is the cavity volume given by

$$V = \pi l_c\left[(R+d)^2 - R^2\right] = 2\pi R l_c d\left(1 + \frac{d}{2R}\right) = Sd\left(1 + \frac{d}{2R}\right), \tag{52}$$

then we have

$$\langle\beta_w\rangle = \mathrm{i}\kappa\left(\frac{l_c}{p_c}\right)\frac{V}{S} = \mathrm{i}\kappa\left(\frac{l_c}{p_c}\right)d\left(1 + \frac{d}{2R}\right). \tag{53}$$

Upon substitution, and recalling that $P/A = 2/R$, we obtain the result

$$k_\pm = \frac{\left[1 + \frac{2d}{R}\left(\frac{l_c}{p_c}\right)\left(1 + \frac{d}{2R}\right)\right]^{1/2}}{1 \pm M\left[1 + \frac{2d}{R}\left(\frac{l_c}{p_c}\right)\left(1 + \frac{d}{2R}\right)\right]^{1/2}} \approx \frac{1 + \frac{d}{R}\left(\frac{l_c}{p_c}\right)\left(1 + \frac{d}{2R}\right)}{1 \pm M}, \tag{54}$$

if we assume that

$$\frac{2d}{R}\left(\frac{l_c}{p_c}\right)\left(1 + \frac{d}{2R}\right) \ll 1. \tag{55}$$

Since the phase speed of the upstream and downstream travelling waves is $c_\pm = c_0/k_\pm$ we see the effect of the corrugations is to slow down the wave. Finally, in such a limit the resonant frequencies of the pipe are given by

$$f_n = \frac{nc_0}{2L}\frac{(1 - M^2)}{\left[1 + \frac{d}{R}\left(\frac{l_c}{p_c}\right)\left(1 + \frac{d}{2R}\right)\right]}, \qquad \text{for } n = 1, 2, \ldots \tag{56}$$

11.4 Further Thoughts

Clearly the phenomenon of tone generation is closely connected with the flow within the corrugation cavity. The main problem is to identify the main mechanism for the energy flux. One recent approach follows from the theory of vortex sheet/sound interactions, where we regard the effect of corrugations in terms of the acoustic wall admittance of an equivalent smooth cylindrical pipe. The vortex sheet, which forms across each cavity, changes the effective acoustic admittance of the pipe wall so that it can now have a real part.

The radiating frequencies are those for which the real part of admittance is negative, corresponding to an input to the pipe wave energy. Those eigenmodes whose frequencies lie within these radiating domains are then excited. The region of radiation, which in general must be obtained by experiment can, for the simplest geometries, be obtained analytically.

The simplest corrugation configuration is one of length l_c and infinite depth, the basic flow being that of a uniform stream of speed U separated from a fluid at rest by two semi-infinite flat plates located at $y = 0$ for $x < 0$ and $x > l_c$ respectively, as shown in Fig. 11.6.

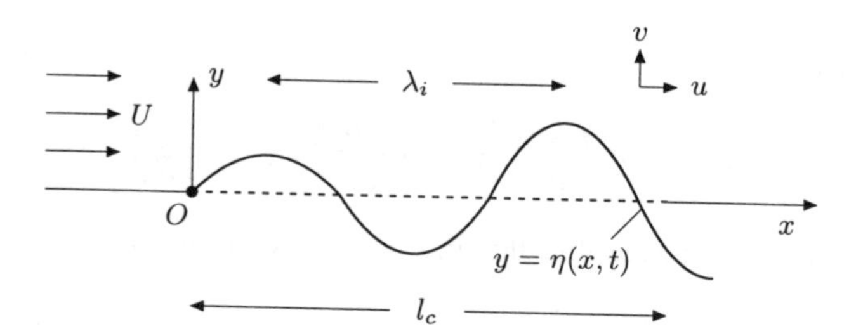

Fig. 11.6 The vortex sheet.

Consequently there is a vortex sheet across the cavity at $y = 0$ emanating from the leading edge of the cavity. We therefore consider the scattering problem due to a wave of frequency $\omega = c_0\kappa$. For small disturbances the vortex sheet is displaced to $y = \eta(x, t) = \bar{\eta}(x)\mathrm{e}^{-\mathrm{i}\omega t}$. On physical grounds the vortex sheet must vanish at the trailing edge of the cavity.

For simplicity let us consider the scattering problem for the trailing edge of a single semi-infinite flat plate. It can be shown that the vortex sheet leads to growing and decaying Kelvin–Helmholtz instability waves of the form

$$\mathrm{e}^{\mathrm{i}(\alpha x - \omega t)} \qquad \text{where } \alpha = \frac{\kappa}{M}(1 \pm \mathrm{i}) = \frac{\omega}{U}(1 \pm \mathrm{i}). \tag{57}$$

These waves are hydrodynamic, rather than acoustic, with a wave-speed $c_{ph} = \omega/\alpha$ proportional to U rather than c_0. Consequently the instability wavelength $\lambda_i = 2\pi/\alpha_r$, at least for $M \ll 1$, is much shorter than the acoustic wavelength.

Some of these instability waves would indeed vanish at trailing edge of the cavity and some would not. Clearly to find F_v, the frequency of the most excited mode, we need to integrate $\bar{v}_0 = \bar{v}(x, 0+)$, the reduced normal velocity across the cavity, over the cavity, for various values of λ_i, to find the one that gives the largest negative value for the real part of the wall admittance β. However, given that the wave also grows, the largest contribution to this integral will come from the vicinity of the trailing edge. Indeed we would expect the most excited mode $f = F_v$ to correspond to

$$\lambda_i \sim \frac{U_c}{f} = O(l_c), \quad \text{so} \quad \mathrm{St} = \frac{f l_c}{U} = O(1), \quad \text{giving} \quad F_v \propto \frac{U}{l_c}, \tag{58}$$

implying that the centre of the first radiating region has a frequency that is proportional to the mean flow velocity U. This radiating region will be repeated at higher frequencies, when more than one instability wavelength is contained within the cavity length. Thus the most excited modes will have frequencies $nF_v \propto nU/l_c$.

In order to derive the scattering coefficients the Fourier transform of the problem is taken, and the Wiener–Hopf equations are formulated. Crucial to the outcome is the application of the correct Kutta condition. The quantity of interest, the acoustic impedance (admittance) is related to the phase of the reflection coefficients.

Integrating the wall pressure over the cavity length $l_c \ll L$, yields

$$\bar{p}(t) = \int_0^{l_c} p'(x, 0+, t)\,\mathrm{d}x = \int_0^{l_c} \bar{p}_0\left(\frac{x}{L}\right)\mathrm{e}^{-\mathrm{i}\omega t}\,\mathrm{d}x = \bar{p}_0 l_c \mathrm{e}^{-\mathrm{i}\omega t}, \tag{59}$$

since $\bar{p}_0$ varies over the scale of the pipe length. However the vortex sheet, and hence the normal velocity v', will vary over the shorter scale of the cavity. Consequently upon integration over the cavity we have

$$\bar{v}(t) = \int_0^{l_c} v'(x, 0+, t)\,\mathrm{d}x = \int_0^{l_c} \bar{v}_0\left(\frac{x}{l_c}, \frac{\lambda_i}{l_c}\right)\mathrm{e}^{-\mathrm{i}\omega t}\,\mathrm{d}x = v_0^*\left(\frac{\lambda_i}{l_c}\right)\mathrm{e}^{-\mathrm{i}\omega t}, \tag{60}$$

where λ_i is the wavelength of the instability wave and v_0^* is a complex constant. This implies that the wall admittance over the slot, $\beta = \bar{v}/\bar{p}$, is a complex constant. Indeed analysis of the trailing edge problem suggests that β is of the form

$$\beta = -\frac{\bar{v}}{\bar{p}} = -\frac{v_0^*}{\bar{p}_0 l_c} \sim -\beta_0 \mathrm{e}^{\mathrm{i}(\kappa l_c/M - 2\pi\theta)}, \tag{61}$$

where $\beta_0 > 0$ is real. In order for the real part of β to be negative we require

$$-\frac{\pi}{2} \leq \frac{\kappa l_c}{M} - 2\pi\theta \leq \frac{\pi}{2} \qquad \text{so} \qquad 2\pi(\theta - \tfrac{1}{4}) \leq \frac{\kappa l_c}{M} \leq 2\pi(\theta + \tfrac{1}{4}). \tag{62}$$

Now since

$$\mathrm{St} = \frac{f l_c}{U} = \frac{\omega l_c}{2\pi U} = \frac{\kappa l_c}{2\pi M}, \tag{63}$$

the frequencies for the first domain of radiation lie in the range

$$\theta - \tfrac{1}{4} \leq \mathrm{St} \leq \theta + \tfrac{1}{4}. \tag{64}$$

Indeed we would expect the most excited mode, $f = F_v$, to correspond to $\mathrm{St} = \theta$, where the imaginary part of β, is zero. Analysis suggests that

$$\theta \approx \tfrac{9}{16} \approx 0.562\,. \tag{65}$$

Consider the following experimental configuration, namely $L = 0.8$ m, $l_c = 0.02$ m, $c_0 = 340$ ms^{-1}, $U = 10$ ms^{-1}. Then the frequency $f = F_v$ of the most excited mode in the first domain of radiation will be located near the value

$$F_v = \theta\frac{U}{l_c} = 2810 \text{ Hz}, \tag{66}$$

and the distance between radiating peaks will be

$$\Delta f \approx \frac{c_0}{2L} = 212 \text{ Hz}. \tag{67}$$

We see that the model is in broad agreement with the experimental findings.

PART IV
Signal Processing

Chapter 12

Digital Filters

P. J. Duncan

Salford Acoustics Audio and Video
University of Salford

12.1 Mathematical Overview

The charts shown in Figs. 12.1 and 12.2 show the main areas of mathematics used to describe and analyze discrete and continuous systems and signals in both the time and frequency domain.

Physical systems can often be modelled on the basis that their overall behaviour is linear and time-invariant. With real systems, this is always an approximation, but in many cases the nonlinear contributions are insignificant and may be ignored.

A linear system conforms to the principle of superposition. This means that if an input x_1 produces an output y_1, and another input x_2 produces an output y_2, then the sum of these inputs will produce an output $y_1 + y_2$. This may be formally expressed as

$$ax_1 + bx_2 \Rightarrow ay_1 + by_2 \tag{1}$$

A system is said to be time-invariant if a time shifted input signal produces an equally time shifted output signal.

$$\text{If} \qquad x(t) \Rightarrow y(t) \qquad \text{then} \qquad x(t+\tau) \Rightarrow y(t+\tau). \tag{2}$$

The system in Fig. 12.3 can be described in the time domain by the impulse response $h(t)$, or the frequency domain by its frequency response $H(\omega)$, and if we know what the output is for a given input it is possible to estimate the physical properties of the system. This is the basis of acoustic analyzers such as MLSSA.

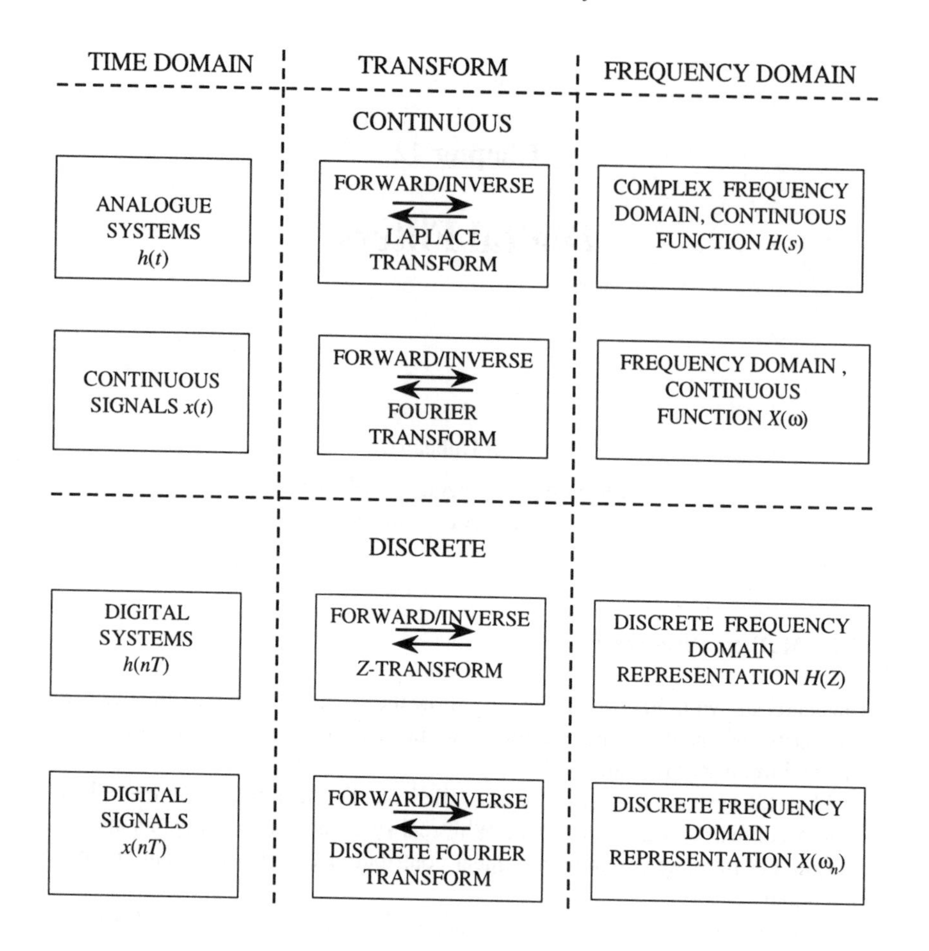

Fig. 12.1 Overview of mathematics in signal processing.

What we are interested in specifically are the time and frequency domain relationships between the input and output and how these can be exploited to extract the physical parameters of the system under analysis.

12.2 Fourier Transform

Time domain and frequency domain variables are related via the Fourier transform which allows any function in time to be represented as an infinite series of complex exponential 'cyssoid' functions in frequency. The relationship between the two domains is as follows. For the forward Fourier transform, going from time domain

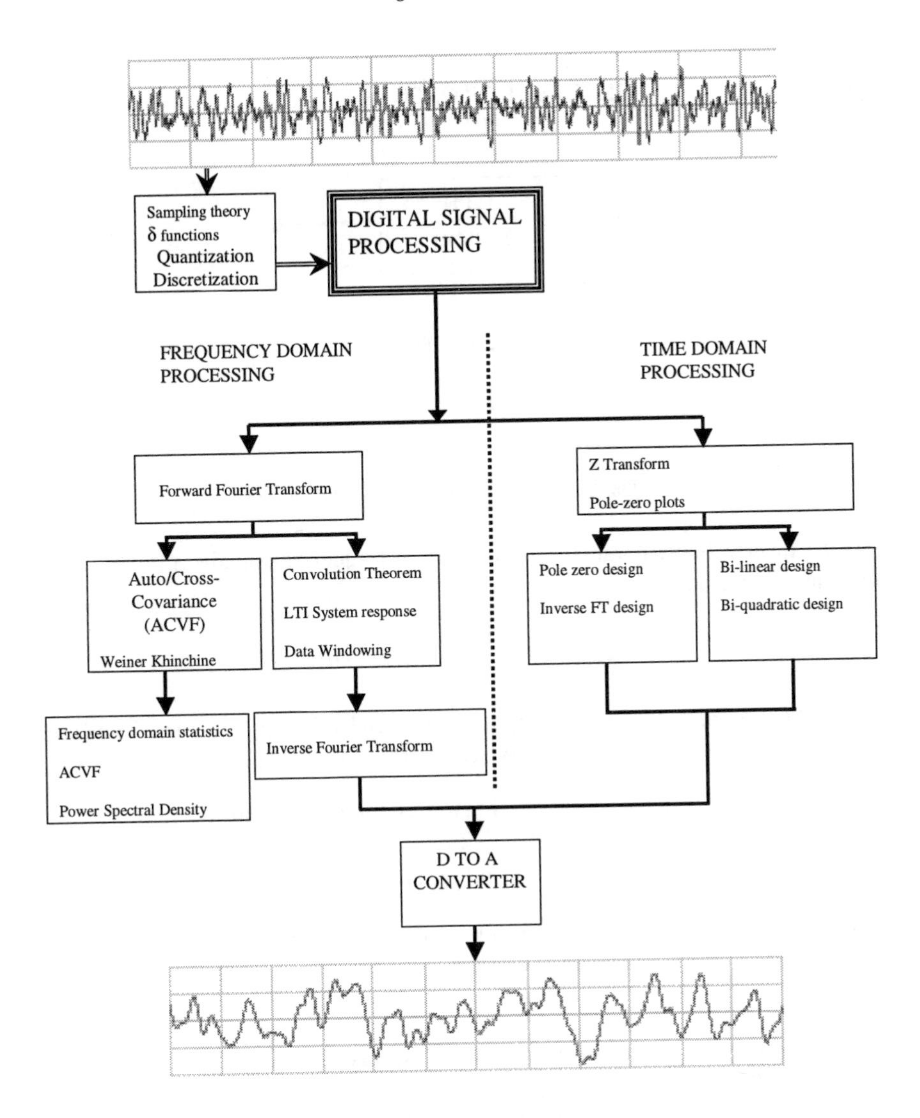

Fig. 12.2 Signal processing flowchart.

$x(t)$ to frequency domain $X(\omega)$ we have

$$X(\omega) = \int_{-\infty}^{\infty} x(t)\mathrm{e}^{-\mathrm{i}\omega t}\,\mathrm{d}t. \tag{3}$$

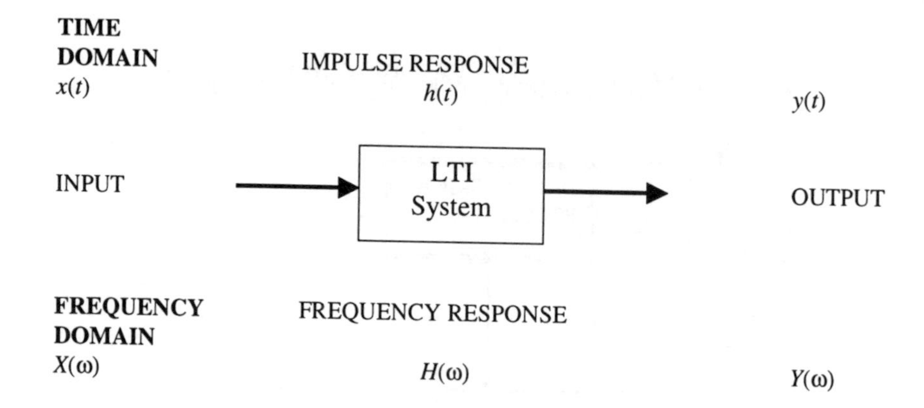

Fig. 12.3 A linear, time-invariant system.

For the inverse Fourier transform, going from frequency domain $X(\omega)$ to time domain $x(t)$ we have

$$x(t) = \frac{1}{2\pi}\int_{-\infty}^{\infty} X(\omega)\mathrm{e}^{\mathrm{i}\omega t}\,\mathrm{d}t. \tag{4}$$

In the LTI system, the inputs and outputs are related by the equations above and are said to be Fourier Transform pairs. We can introduce a notation for this purpose:

$$\mathcal{F}[x(t)] = X(\omega) \qquad \text{Forward Fourier Transform,} \tag{5}$$

$$\mathcal{F}[X(\omega)] = x(t) \qquad \text{Inverse Fourier Transform.} \tag{6}$$

The notation in this and the following chapters is always upper case for frequency variables and lower case for time variables

12.3 Impulse Response and Frequency Response

The LTI system response may be specified in the time or frequency domain. For the kind of systems that we are interested in (filters, rooms, musical instruments etc) it makes sense to talk about the frequency domain behaviour, or frequency response. If we envisage a signal $x(t)$ which may be represented in the frequency domain as a distribution of frequency components $X(\omega)$ being input to the LTI system, the output $Y(\omega)$ will be the *product* of the frequency response $H(\omega)$ and $X(\omega)$. An illustrative example might involve a white noise input signal passed through a low-pass filter. The output will consist of only low frequency components of the noise.

12.4 Convolution Principle

The way that the LTI system interacts with an input signal in the time domain is less straightforward to understand and involves a principle of fundamental importance in signal processing, the principle of convolution.

We have mentioned previously that our LTI system behaviour in time can be specified by the time domain impulse response. This describes the system output response to a unit impulse at the input. If we consider an input signal in time as consisting of an infinite sequence of weighted unit impulses, the output will then be a corresponding sequence of weighted impulse responses. This is formally mathematically expressed by the convolution equation:

$$y(t) = \int_0^\infty h(\tau)x(t-\tau)\,\mathrm{d}\tau, \tag{7}$$

where $h(\tau)$ is the system impulse response and τ is the duration of the impulse response. The output is then the product of the signal $x(t)$ and the impulse response $h(\tau)$ integrated over the system response time. The following notation is often used to denote convolution:

$$y(t) = h(t) * x(t). \tag{8}$$

The symbol $\otimes$ is sometimes used instead of $*$. In order to get a handle on convolution, consider a perfect amplifier where the output is an exact copy of the input, multiplied by a gain factor. For this to be the case, the system response time must be zero and since the impulse response and frequency response are reciprocally related, this implies that the amplifier must have infinite bandwidth. Conversely a band limited amplifier will have a finite response time over which the input signal is ‘smeared’.

We have seen that in the frequency domain, the output is obtained by simple multiplication of the input frequency distribution $X(\omega)$ and the system frequency response $H(\omega)$. In the time domain the relationship between input and output is determined by the convolution integral. To summarize, a multiplication in the frequency domain is equivalent to a convolution in the time domain.

$$y(t) = h(t) * x(t), \tag{9}$$

$$Y(\omega) = H(\omega)X(\omega). \tag{10}$$

This also works the other way round, multiplication in the time domain is equivalent to convolution in the frequency domain (applies to FFT windowing and consequent spectral smearing).

12.5 Dirac Delta Functions and Sifting Property

The Dirac delta function $\delta(t - a)$ is fundamental to digital sampling theory. It is a mathematical construction representing a perfect impulse having infinite amplitude and zero width at a point in time $t = a$. At this point, the contents of the bracket sum to zero and this defines the position of the delta function in time (see Fig. 12.4).

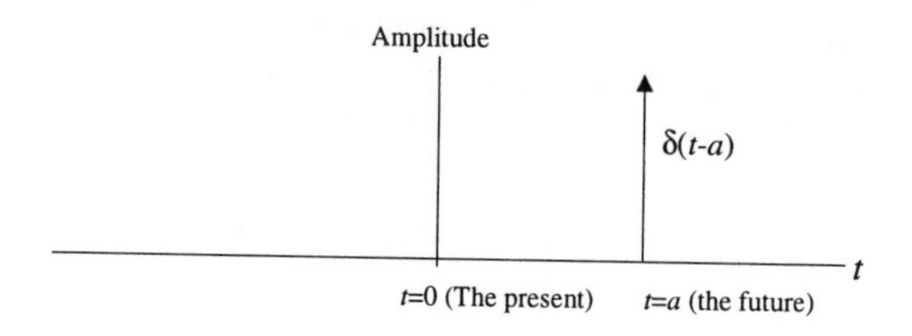

Fig. 12.4 A delta function.

If a single delta function is integrated as a product with a continuous function in time (a signal for example), the outcome will be the value of the signal at the time the delta function occurred (see Fig. 12.5.) This is the **sifting property** of delta functions,.

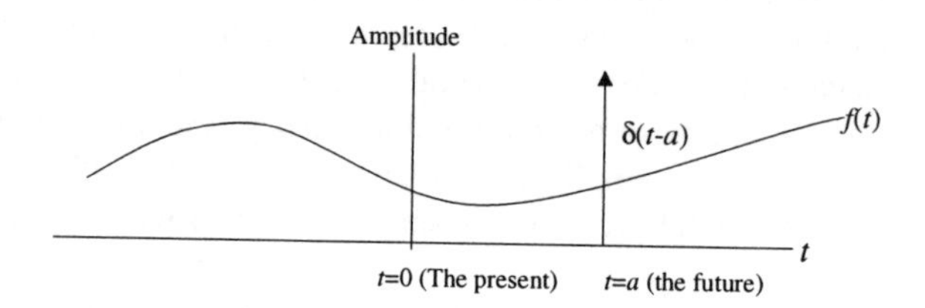

Fig. 12.5 A delta function multiplied by an ordinary function.

$$\int_{-\infty}^{\infty} f(t)\delta(t - a)\,\mathrm{d}t = f(a). \tag{11}$$

The expression above is a formal statement of the sifting property. The only point in time where the integral can have any meaningful value is at $t = a$, and the value of the integral is the value of the function $f(t)$ at $t = a$, or $f(a)$.

12.5.1 *The Fourier transform of a delta function*

We begin with a delta function at t=0.

$$\mathcal{F}[\delta(t-0)] = \Delta\omega = \int_{-\infty}^{\infty} \delta(t-0)\mathrm{e}^{-\mathrm{i}\omega t}\,\mathrm{d}t. \tag{12}$$

The integration is done using the sifting property of delta functions which gives:

$$\mathcal{F}[\delta(t-0)] = \Delta\omega = 1. \tag{13}$$

since e^0 is one. This result shows that the delta function contains all frequencies having equal magnitude. One way to envisage this is to imagine an infinite series of cosine functions, the only point in time at which they sum together constructively is at the position of the delta function. Elsewhere they sum to zero due to their relative phases. This makes the delta function an ideal test signal and it is the basis of impulse response analysis of a system.

An ideal delta function is infinitely short in time duration and has an infinite extent in frequency. It is a direct consequence of the reciprocal nature of time and frequency.

12.5.2 *The inverse Fourier transform of a delta function in frequency*

The Inverse Fourier transform of a delta function in frequency, $\delta(\omega - \omega_0)$ is given by

$$\mathcal{F}^{-1}[\delta(\omega-\omega_0)] = \frac{1}{2\pi}\int_{-\infty}^{\infty} \delta(\omega-\omega_0)\mathrm{e}^{\mathrm{i}\omega t}\,\mathrm{d}\omega. \tag{14}$$

Using the sifting property again we obtain

$$\mathcal{F}^{-1}[\delta(\omega-\omega_0)] = \frac{\mathrm{e}^{\mathrm{i}\omega_0 t}}{2\pi}. \tag{15}$$

The inverse Fourier Transform of a delta function is a complex exponential. This now allows us to deduce the Fourier transform of sin and cos functions since these may be expressed in complex exponential form:

$$\cos\omega_0 t = \frac{\mathrm{e}^{\mathrm{i}\omega_0 t} + \mathrm{e}^{-\mathrm{i}\omega_0 t}}{2}, \tag{16}$$

$$\sin\omega_0 t = \frac{\mathrm{e}^{\mathrm{i}\omega_0 t} - \mathrm{e}^{-\mathrm{i}\omega_0 t}}{2\mathrm{i}}. \tag{17}$$

Since we know that each complex exponential is the Fourier transform of a delta function, we can see that the Fourier transform of a cosine or sine is a pair of delta functions in frequency.

$$\mathcal{F}[\cos\omega_0 t] = \pi[\Delta(\omega - \omega_0) + \delta(\omega + \omega_0)], \tag{18}$$

$$\mathcal{F}[\sin\omega_0 t] = -\mathrm{i}\pi[\Delta(\omega - \omega_0) - \delta(\omega + \omega_0)]. \tag{19}$$

The pair of delta functions occur at $+\omega_0$ and $-\omega_0$. To be consistent in our analysis we must include negative frequencies. These have no physical significance, but must be included to give a complete description.

12.6 Laplace Transform Analysis

The time behaviour of LTI systems can be analyzed by solving the differential equations that describe the system using the Laplace transform

$$\mathcal{L}[f(t)] = \int_0^\infty f(t)\mathrm{e}^{-st}\,\mathrm{d}t, \tag{20}$$

where the Laplace variable s may be complex. Clearly, if the real part of s is negative, the integral will not converge. The real part of s therefore represents growth/damping in a physical system and the imaginary part represents oscillation. The coordinates of s in the complex plane determine the system behaviour.

The general Laplace procedure is:

(1) Write down the differential equation for the system (mass/spring, electrical circuit etc.)
(2) Convert using the Laplace variable s.
(3) Re-arrange to obtain rational polynomials in s.
(4) Inverse Laplace transform to obtain function in time.

Alternatively, the magnitude frequency behaviour of the system may be obtained by factorising the s equation into a rational polynomial to obtain:

$$H(s) = \frac{(s - a_1)(s - a_2)(s - a_3)\cdots}{(s - b_1)(s - b_2)(s - b_3)\cdots}. \tag{21}$$

Since s is a complex number, the a and b coefficients define the positions of **poles** and **zeros** in the complex plane which determine the system response. The Laplace transform applies to continuous variables. However, we are generally restricted to discrete variables in DSP, it is useful to derive a discrete transform, the Z-transform for analysis of digital systems and signals.

12.6.1 *The Z-transform*

We now consider the discrete variable $x(n)$ which is the nth term of a sequence having the value $x(nT)$. Suppose that $x(n)$ consists of N values, generated by integrating the continuous variable with a sampling function consisting of a sequence of delta functions at equal time intervals T

$$x(n) = \sum_{n=0}^{N} x(t)\delta(t - nT). \tag{22}$$

The Laplace transform of this is given by

$$\mathcal{L}[X(n)] = \int_0^\infty \sum_{n=0}^{N} x(t)\delta(t - nT)\mathrm{e}^{-st}\,\mathrm{d}t. \tag{23}$$

This may be integrated using

$$\mathcal{L}[\delta(t - nT)] = \mathrm{e}^{-snT}, \tag{24}$$

and the sifting property of delta functions to obtain

$$\mathcal{L}[X(n)] = \sum_{n=0}^{N} x(nT)\mathrm{e}^{-snT}. \tag{25}$$

Now make the substitution

$$Z = \mathrm{e}^{-snT}, \tag{26}$$

to obtain

$$X(Z) = \sum_{n=0}^{N} x(n)z^{-n}. \tag{27}$$

The sample time T can be taken out as it is a constant, and we are left with a time independent discrete transformation to complex space. The Z-transform will be developed further in §12.7.2. Essentially, Z is a complex frequency variable, and $X(Z)$ can be expressed as a factorised rational polynomial in Z to give the complex coordinates of poles and zeros which determine the system response in a similar way to s-plane poles and zeros. The essential difference is that the discrete variable is strictly limited in frequency to f_N, the Nyquist frequency.

12.6.2 *The discrete Fourier transform*

Recall that for a continuous variable, the Fourier transform is obtained by integrating the function in time with an infinite series of complex exponentials.

$$\mathcal{F}[x(t)] = X(\omega) = \int_{-\infty}^{\infty} x(t)\mathrm{e}^{-\mathrm{i}\omega t}\,\mathrm{d}t. \tag{28}$$

This returns a two-sided distribution in complex frequency space, the positive part of which can be regarded as the frequency spectrum of the function $x(t)$. Now, comparing this equation with the Laplace transform for a discrete series $x(n)$

$$\mathcal{L}[x(n)] = \sum_{n=0}^{N} x(nT)\mathrm{e}^{-snT}, \tag{29}$$

we can see that they are very similar. The continuous variable t is replaced by the discrete variable nT (or simply n since T, the sampling time, is a constant) Recall that the term s is a complex number $s = \sigma + \mathrm{i}\omega$ so that if we make s entirely imaginary, and carry out the integration over all time we have

$$\mathcal{L}[x(n)] = \sum_{n=-\infty}^{\infty} x(nT)\mathrm{e}^{-\mathrm{i}\omega nT}, \tag{30}$$

which is equivalent to the expression for the Fourier Transform given above. In real practical terms, we are generally dealing with finite length data sets so the summation must be carried out over the length N of the data set. We must also introduce a discrete frequency variable $2\pi f$ (where f is an integer) to replace ω, and rewrite the equation to obtain

$$\mathcal{F}[x(n)] = X(f) = \sum_{n=-N/2}^{N/2} x(n)\mathrm{e}^{-\mathrm{i}2\pi fn/N}. \tag{31}$$

This equation represents a matrix operation on the discrete variable $x(n)$ which returns a discrete complex quantity $X(k)$ which directly represents frequency. This is the basis of the Discrete Fourier Transform, one of the most powerful and widely applied principles in DSP. It is usually implemented in the form of the Fast Fourier Transform or FFT which exploits symmetries in the matrix operation to reduce the number of calculations required. There is a proviso with the FFT that the length N of the data set to be transformed must be a power of two.

For a data set of N discrete values, the frequency resolution in Hz is given by

$$\Delta f = \frac{1}{NT}, \tag{32}$$

where T is the sample time. The DFT bandwidth is 0 to $T/2$ Hz as one would expect from the Nyquist criteria.

12.7 Digital Filters

12.7.1 *Introduction*

The digital filters described in this section operate in the time domain by implementing a discrete convolution of the filter time domain impulse response and the input data. The impulse response is either coded as a set of discrete coefficients in the case of Finite Impulse Response (FIR) filters, or implemented recursively in the case of Infinite Impulse Response (IIR) filters.

12.7.2 *FIR filters*

The general expression for an FIR filter is

$$y_n = \sum_{k=0}^{N} b_k x_{n-k}, \tag{33}$$

which represents a discrete convolution operation. Here y_n is the filter output, N the number of coefficients (filter order), b_k the filter coefficients and x_{n-k} the current and previous inputs.

Recall that in a linear system, the frequency response of the system is the Fourier transform of the impulse response. Since the coefficients b_k of our digital filter represent the discrete time impulse response we can obtain the frequency response by using the Z-transform which operates on a digitized sequence and returns a complex function in frequency. There are standard Z-transforms which can be applied to the terms in the filter difference equation.

Standard Z-transforms are:

$$x_n \Rightarrow X(z), \tag{34}$$

$$x_{n-1} \Rightarrow X(z)z^{-1}, \tag{35}$$

$$x_{n-2} \Rightarrow X(z)z^{-2}, \tag{36}$$

$$\beta x_{n-2} \Rightarrow \beta X(z)z^{-2}. \tag{37}$$

The general method is to apply these standard transforms to each term of the filter difference equation and form a new equation in Z from which the frequency response of the filter may be obtained.

12.7.3 *Three point moving average FIR filter*

Consider the difference equation

$$y_n = \sum_{k=0}^{2} \tfrac{1}{3} x_{n-k} = \tfrac{1}{3}(x_n + x_{n-1} + x_{n-2}). \tag{38}$$

To obtain an expression for the transfer function each term is Z-transformed according to the rules given above,

$$Y(Z) = \tfrac{1}{3}(X(Z) + X(Z)Z^{-1} + X(Z)Z^{-2}), \tag{39}$$

and rearranged to obtain the transfer function which is given by the filter output $Y(Z)$ divided by the input $X(Z)$

$$\frac{Y(Z)}{X(Z)} = \tfrac{1}{3}(1 + Z^{-1} + Z^{-2}), \tag{40}$$

and the final step is to divide through by Z^2 to obtain a rational polynomial in Z

$$\frac{Y(Z)}{X(Z)} = \frac{1}{3}\left(\frac{Z^2 + Z + 1}{Z^2}\right). \tag{41}$$

The filter response depends entirely on the right hand side of the equation. The terms on the top govern the filter response minima (zeros) and the terms on the bottom govern the filter maxima (poles).

We are now ready to describe the filter response using a pole–zero diagram. Here Z is a unit vector which describes a circle in the complex plane called the unit circle, shown in Fig. 12.6.

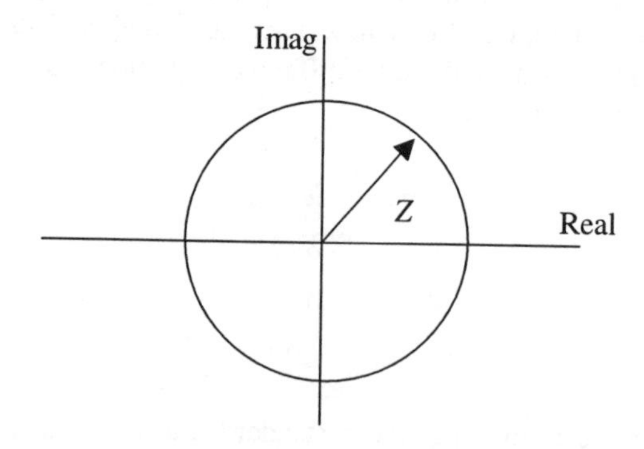

Fig. 12.6 The unit circle in the Z-plane.

The distance around the circle represents increasing frequency from d.c. at 1 on the real axis and the Nyquist frequency at -1 on the real axis. The frequency at any point on the unit circle is given by $\omega = \theta F_n/\pi$.

Just as an analogue filter has poles and zeros in the complex plane that govern the filter response, so does the digital filter. The response for a particular frequency is given by the ratio of the distance to the zero to the distance to the pole from the point on the unit circle representing that frequency. For the three point moving average, we obtain the equation:

$$\frac{Y(Z)}{X(Z)} = \frac{1}{3}\left(\frac{Z^2 + Z + 1}{Z^2}\right). \tag{42}$$

The general rule is that the roots of the numerator determine the position of the zeros (at these points, the filter response is zero) and the roots of the denominator determine the position of the poles where the filter response is maximal.

In this example, the numerator has roots at

$$\frac{-1 \pm \sqrt{1-4}}{2} = -\frac{1}{2} \pm \mathrm{i}\frac{\sqrt{3}}{2} = -0.5 + 0.866\mathrm{i} \quad \text{and} \quad -0.5 - 0.866\mathrm{i}. \tag{43}$$

The denominator has two roots at the origin. The pole–zero plot is therefore as shown in Fig. 12.7 and the filter response will be as shown in Fig. 12.8. This is the response for a low-pass filter as expected. Notice that the form of the filter response is reflected about f_n, and since we are not interested in frequencies above f_n we can disregard this. The form of the filter response is repeated to $f = \infty$.

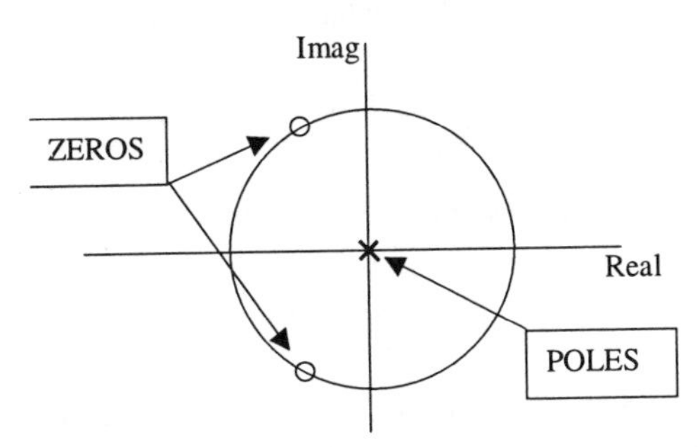

Fig. 12.7 Pole–zero plot for the system defined by Eq. (38).

12.7.4 *High-pass FIR filter*

Suppose that instead of summing the discrete values of a sequence we subtract the previous value from the current value. This is a numerical differentiation opera-

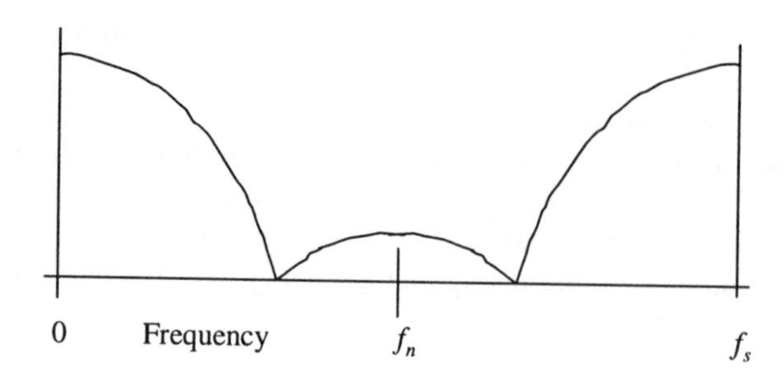

Fig. 12.8 Frequency response of the system defined by Eq. (38).

tion, and the outcome will be a sequence whose values represent the difference between subsequent values of the digitized signal. This will be greater for high frequencies than for low frequencies and we will have the basis of a simple high-pass filter.

The difference equation is now:

$$y_n = x_n - x_{n-1}, \tag{44}$$

applying the Z-transform as before we obtain

$$Y(Z) = X(Z) - X(Z)Z^{-1}, \tag{45}$$

and multiplying through by $X(Z)$ to obtain the transfer function

$$\frac{Y(Z)}{X(Z)} = 1 - Z^{-1}. \tag{46}$$

Multiplying through by Z to obtain a rational polynomial gives

$$\frac{Y(Z)}{X(Z)} = \frac{Z-1}{Z}. \tag{47}$$

The top of the equation has one root at $Z = 1$ on the real axis and a pole at $Z = 0$. The pole–zero plot is shown in Fig. 12.9 and the frequency response is shown in Fig. 12.10.

These two examples are the simplest form of high and low-pass FIR filter. If we want to design more sophisticated filters we can define a frequency response and Fourier transform this to obtain an impulse response which can be sampled and implemented as a set of filter coefficients in the general FIR difference equation. Alternatively we might use a known impulse response from an analogue filter. The main disadvantages of FIR filters are that the number of coefficients required for accurate filtering can be large, and this increases the computation time.

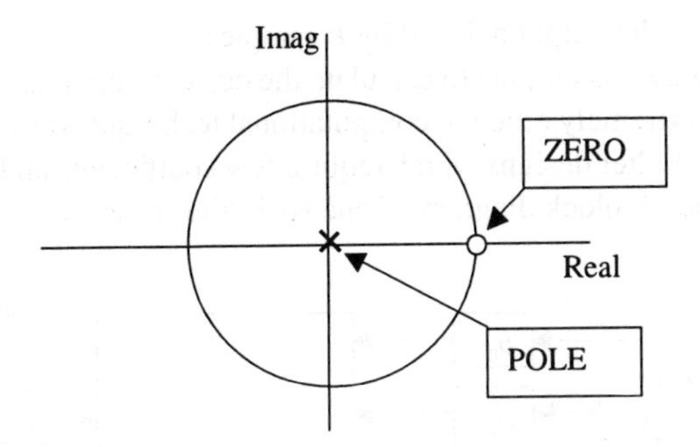

Fig. 12.9 Pole–zero plot for a high-pass FIR filter.

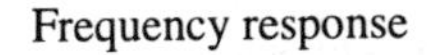

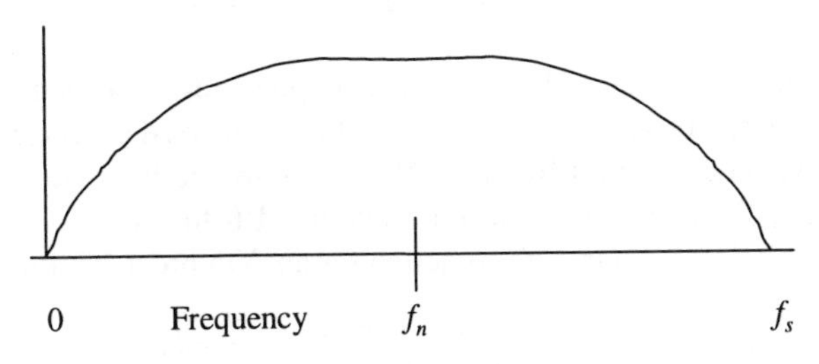

Fig. 12.10 Frequency response of a high-pass FIR filter.

Shortening the impulse response to speed up the filter introduces 'truncation' errors in the filter response.

12.7.5 *Infinite impulse response (IIR) filters*

The shortcomings of FIR filters can be overcome by considering what happens if we use previous outputs as well as previous inputs. We can then write a general difference equation for the filter as follows:

$$y_n = \sum_{k=0}^{L} b_k x_{n-k} - \sum_{k=1}^{M} a_k y_{n-k}. \tag{48}$$

If we consider each of the summations in turn, the first one is simply an FIR filter and y_n is the filter output and x_{n-k} the input delayed by k samples.

The second summation is almost identical. However, instead of using the current and previous inputs, it involves *previous* outputs. Here a_k are filter coefficients and y_{n-k} is the output delayed by k samples.

This use of previous outputs to calculate the current output is known as **recursion** and it is an extremely efficient computational technique, resulting in this case in versatile digital filter designs which require few coefficients and hence minimal computation time. A block diagram of one such filter is shown in Fig. 12.11. IIR

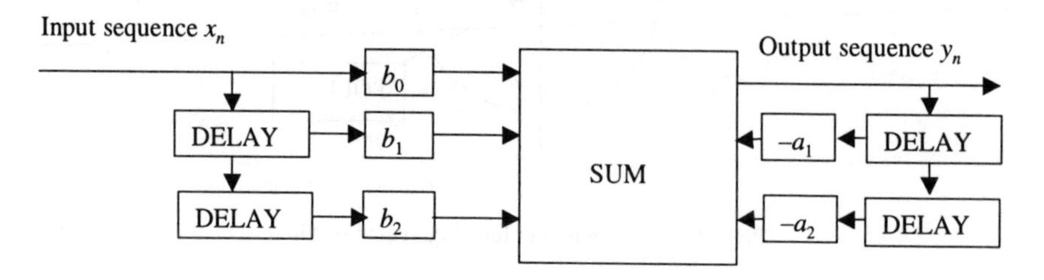

Fig. 12.11 Block diagram of an IIR filter.

filters can be thought of as 'calculating their impulse response as they go', rather than calculating the product of the signal and a set of fixed coefficients to obtain each output value as in the FIR filter. We will now see how we can use the Z transform and pole–zero plots to design some real IIR filters.

The general Z domain transfer function for an Nth order IIR filter is

$$H(z) = \frac{b_0 + b_1 Z^{-1} + b_2 Z^{-2} + \cdots + b_L Z^{-L}}{1 + a_1 Z^{-1} + a_2 Z^{-2} + \cdots + a_M Z^{-M}}. \tag{49}$$

See if you can derive this equation by taking the Z-transform of Eq. (48) and rearranging. This equation allows the filter poles to be placed anywhere in the Z-plane, and as such it can be used to develop a variety of digital filters. We will look at two types, bilinear and biquadratic.

12.7.6 *Bilinear IIR filters (bilins)*

As the name implies, this design method involves using a pair of linear equations in Z to derive the filter response. We begin with a difference equation of the form

$$y_n = x_n + b_1 x_{n-1} - a_1 y_{n-1} \tag{50}$$

and carry out a Z-transform as before and re-arrange

$$Y(Z) = X(Z) + b_1 X(Z) Z^{-1} - a_1 Y(Z) Z^{-1} \tag{51}$$

$$Y(Z) + a_1 Y(Z) Z^{-1} = X(Z) + b_1 X(Z) Z^{-1} \tag{52}$$

$$Y(Z)(1 + a_1 Z^{-1}) = X(Z)(1 + b_1 Z^{-1}), \tag{53}$$

and finally

$$\frac{Y(Z)}{X(Z)} = \frac{(1 + b_1 Z^{-1})}{(1 + a_1 Z^{-1})}, \tag{54}$$

an expression in Z for the transfer function as before. To eliminate the Z^{-1} terms we multiply through by Z to obtain

$$\frac{Y(Z)}{X(Z)} = \frac{(Z + b_1)}{(Z + a_1)}, \tag{55}$$

which has poles and zeros in the complex plane as before. The positions of the poles and zeros is governed by the values of a_1 and b_1 as before.

12.7.7 *Bilinear high-pass filter example*

We can implement a high-pass filter with a zero at DC by making $b_1 = -1$ and $a_1 = 0$. This will give the high-pass filter that we obtained previously. In this case we can modify the filter characteristic by assigning a value to b_1 which has the effect of moving the pole away from the origin where it can have an effect on the filter behaviour, specifically, the pole radius controls the 'corner frequency', as shown in Fig. 12.12. The three curves show the filter response for varying a_1. Notice that the filter response becomes steeper as a_1 approaches 1. In general, the pole radius must be less than 1 or the filter will be unstable.

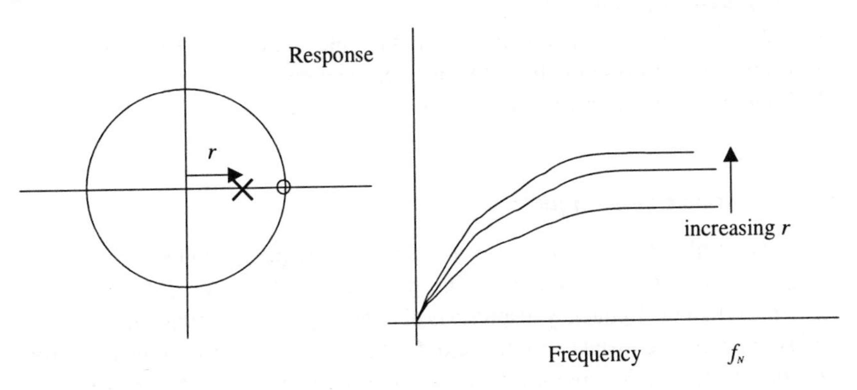

Fig. 12.12 A high-pass filter.

The bilinear design method allows us to move the pole away from the origin, however the pole is confined to the real axis. To place poles and zeros anywhere within the unit circle we need to consider a higher order filter and use the biquad method.

12.7.8 *Biquadratic IIR filter design (biquads)*

In this case, we derive a quadratic Z equation.

$$\frac{Y(Z)}{X(Z)} = \frac{(Z^2 + b_1 Z + b_2)}{(Z^2 + a_1 Z + a_2)}. \tag{56}$$

Just as before, the roots of the top of the equation give the zero positions and the roots of the bottom give the pole positions. The general design method is to decide where we want to put poles and zeros to obtain the required filter, and to calculate the a and b coefficients from the pole/zero coordinates in the Z-plane. Poles and zeros in the Z-plane always exist in pairs (complex conjugates) since they arise from the solution of a quadratic equation.

If we have a pole pair in the Z-plane with coordinates $R_p \pm \mathrm{i}I_p$ the a coefficients are given by $a_1 = -2R_p$ and $a_2 = (R_p^2 + I_p^2)$. Similarly if we have a pair of zeros, the b coefficients are given by $b_1 = -2R_z$ and $b_2 = (R_z^2 + I_z^2)$. Having decided on the pole and zero positions, and calculated the a and b coefficients we then inverse transform the Z equation to get back to a difference equation in time which can be coded into a digital filter. The design procedure is

(1) Decide on pole and zero positions for the filter required. This takes a certain amount of judgment and experience, we will use simplified examples in our filter design examples.
(2) Calculate the a and b coefficients from the pole and zero coordinates.
(3) Inverse Z-transform to obtain the difference equation.
(4) Implement the difference equation as a digital filter.

12.7.9 *Variable Q notch filter*

This type of filter has zero response at a particular frequency and passes all other frequencies within the base-band. The pole–zero plot is shown in Fig. 12.13. Remember that the frequency response is obtained by moving round the unit circle and is given by the distance to the zero divided by the distance to the pole from the point on the circle representing a particular frequency. The Q (or sharpness) of the filter is determined by the pole radius.

The difference equation is

$$y_n = x_n - 2\cos\Omega x_{n-1} + x_{n-2} + 2r\cos\Omega y_{n-1} - r^2 y_{n-2}, \tag{57}$$

where $\Omega = \pi f/f_N$ and f is the frequency to be filtered out.

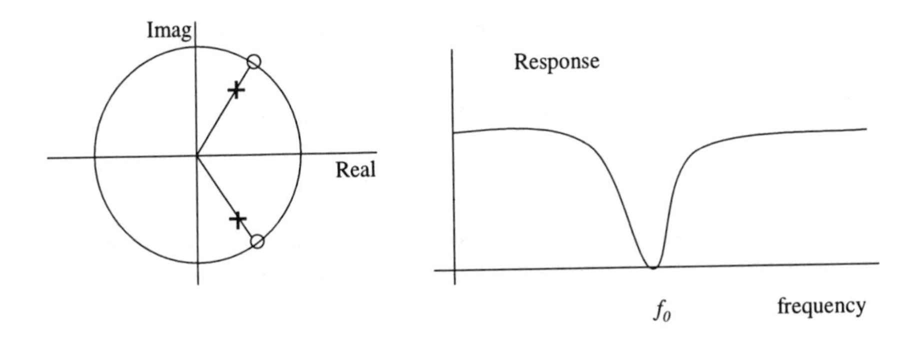

Fig. 12.13 Pole–zero plot for a notch filter.

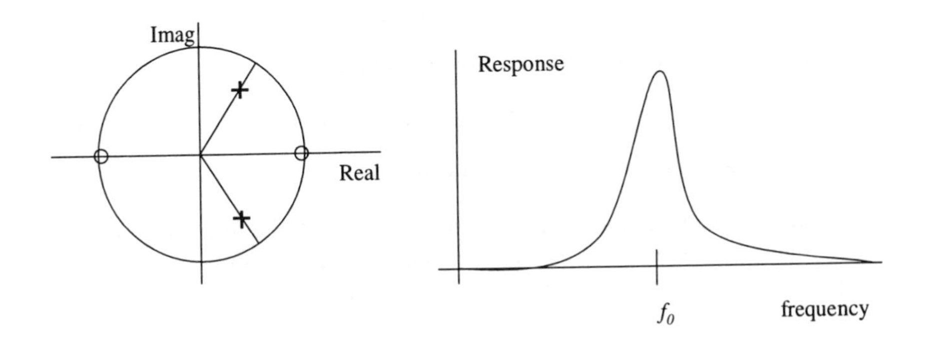

Fig. 12.14 A notch-pass filter.

12.7.10 *Notch-pass filter*

The difference equation in this case is

$$y_n = x_n + x_{n-2} + 2rCos\Omega y_{n-1} - r^2 y_{n-2}. \tag{58}$$

The response is shown in Fig. 12.14. Using these design methods we can implement a range of useful filters to suit our requirements.

12.7.11 *The bilinear transform*

The bilinear transform maps the continuous transfer function $H(s)$ to a discrete equivalent $H(Z)$ using the relationship

$$s = \frac{2(1 - Z^{-1})}{T(1 + Z^{-1})}. \tag{59}$$

Every s term in the Laplace $H(s)$ transfer function expression is replaced by the above expression.

Example 12.1 A high-pass filter has the s-domain expression

$$H(s) = \frac{s}{s + 1000}. \tag{60}$$

To obtain the Bilinear transform, we need the sample time T, in this case $T = 10^{-4}$. We replace all occurrences of s with the expression given above.

$$H(z) = \frac{\frac{2(1-Z^{-1})}{T(1+Z^{-1})}}{\frac{2(1-Z^{-1})}{T(1+Z^{-1})} + 1000} \tag{61}$$

$$= \frac{2(1 - Z^{-1})}{2(1 - Z^{-1}) + 1000 \times T \times 2(1 + Z^{-1})}. \tag{62}$$

Then inverse transform to obtain the difference equation

$$y_n = 0.952x_n - 0.952x_{n-1} + 0.9048y_{n-1}, \tag{63}$$

and the discrete equivalent of the filter can be realized.

What the Bilinear transform does is to 'wrap' the infinite frequency range of the continuous s variable into the finite frequency range of the discrete Z variable, such that the circumference of the unit circle can be regarded as a frequency 'axis' in the Z-plane.

12.8 Summary

In all cases, digital filtering is a discrete convolution operation, and in practice this amounts to a series of shift, multiply and accumulate operations which can be easily implemented in hardware or software to realize extremely efficient, compact and versatile real time filters. Since the filter frequency response is determined entirely by the values of the coefficients, it is possible to create 'adaptive' filters where the coefficient values are incorporated into a feedback loop and to optimize the filter performance. Some ways in which this can be done are discussed in the next chapter.

Further Reading

Oppenheim, A. V. and Schafer, R. (1975) *Digital Signal Processing*. Prentice Hall.

Lynn, P. A. and Fuerst, W. (1989) *Introductory Digital Signal Processing with Computer Applications.* Wiley.
Smith, S. W. (1997) *The Scientist and Engineer's Guide to Digital Signal Processing.* California Technical Publishing
http://www.jhu.edu/ signals/ (Highly recommended)
http://www.bores.com/courses/intro/
http://www.hr/josip/DSP/sigproc.html
http://www.dsptutor.freeuk.com/

Chapter 13

Measurement of Linear Time-Invariant Systems

T. J. Cox
Salford Acoustics Audio and Video
University of Salford

P. Darlington
Apple Dynamics Ltd

13.1 Introduction

It is common to have to classify an acoustic system in terms of its transfer function, whether that is a transducer, a filter, a room acoustic etc. Such a measurement can be done in many ways, but an efficient and popular method for most linear time-invariant (LTI) systems is to use maximum length sequences (MLS). Commercial implementations of these have been hugely successful (e.g. http://www.nvo.com/winmls/door/, http://mlssa.com/). Along the way we shall meet some important mathematical concepts such as such signal and system statistics (auto and cross spectra/covariance/correlation, coherence), MLS generation etc.

In measuring the LTI system we shall monitor both the input and output signals, these will then be processed to calculate the transfer function and impulse response of the LTI system. As we are using noise signals, we are essentially estimating the statistics of the signals and systems.

13.2 Estimating Statistics Using Fourier Methods

You may already be familiar with estimating the power spectrum of an acoustic signal using the Fourier transform method. We can also use the same methods to estimate the important time domain statistics such as the auto and cross-covariance function and important frequency domain statistics such as auto and cross power spectra.

It is possible to make a decomposition of the statistics of *signals* in amplitude, time or frequency (mean, variance, PDF, power spectrum, etc.) In this chapter, we shall turn our attention to descriptions of the performance of *systems*. The systems will be assumed to be linear, and to have a performance which does not change with time—linear time-invariant (LTI) systems. Fourier analysis allows us to represent any signal in terms of the amplitude and phase of the signal's constituent simple harmonic components. An LTI system will respond to each of the Fourier components presented at its input *independently*; each Fourier component at the input will generate a component at the output *at the same frequency*. The system simply causes the Fourier components of the input to be scaled in amplitude and shifted in time before passing them to the output—the system has a **gain** and **phase** associated with each frequency.

We conventionally combine the gain and phase into a single complex valued function of frequency, called the frequency response, $H(\omega)$, in which gain is given by $|H(\omega)|$ and phase angle is given by $\angle H(\omega)$.

At any frequency the action of the system in forming its output Y to input X is expressed by

$$Y(\omega) = X(\omega)H(\omega), \tag{1}$$

in which X and Y are the Fourier transforms of the input and output signals, respectively. Unfortunately, the Fourier transform is impossible to calculate for practical signals, so, although the input–output relationship above for an LTI system is perfectly correct, it has little practical value. Practically speaking, we are forced to make best estimates of transfer functions using estimated statistics of the input and output signals. This estimation (mainly applied in the frequency domain) is the subject of the next few sections.

13.2.1 *Autocovariance function*

The autocovariance function is a lag statistic that describes how a signal varies over time—the similarity between a signal and a delayed version of itself. The **autocovariance function** (ACVF), is defined by the averaged value of the product of a signal and a time-shifted version of itself. For example, the autocovariance of the signal x is

$$s_{xx}(\tau) = \lim_{T\to\infty} \frac{1}{T} \int_{-T/2}^{T/2} x(t)x(t+\tau)\,\mathrm{d}t. \tag{2}$$

We are only going to concern ourselves with digital signals that have been sampled. Assuming some ensemble averaging

$$s_{xx}(\tau) = \lim_{N\to\infty} \frac{1}{N^2} \sum_{j,k}^{N} x_j(t)x_k(t+\tau). \tag{3}$$

The autocovariance function is calculated using a very similar approach to the calculation of the mean square value, in fact

$$s_{xx}(0) = \left\langle x^2 \right\rangle. \tag{4}$$

Some other important properties of the autocovariance function are that $s_{xx}(\tau)$ is an even function of τ,

$$s_{xx}(\tau) = s_{xx}(-\tau), \tag{5}$$

and $s_{xx}(\tau)$ is bounded by $s_{xx}(0)$

$$|s_{xx}(\tau)| \leq s_{xx}(0). \tag{6}$$

If $x(t)$ contains a periodic component then $s_{xx}(\tau)$ contains a periodic component of equal period. If $\lim_{\tau\to\infty} s_{xx}(\tau) \neq 0$ then $\lim_{\tau\to\infty} s_{xx}(\tau) = \mu_x^2$.

The autocovariance is related to (and very frequently confused with) the autocorrelation function. The autocorrelation function is simply a normalized version of the ACVF:

$$R_{xx}(\tau) = \frac{s_{xx}(\tau)}{s_{xx}(0)}. \tag{7}$$

The autocorrelation function has unit value for zero lag, and is bounded by unity.

13.2.2 *Auto power spectral density function*

We formally define the auto power spectral density function (PSD) as

$$S_{xx}(\omega) = \lim_{T\to\infty} \frac{1}{T} \left| \int_{-T/2}^{T/2} x(t)\mathrm{e}^{-\mathrm{i}\omega t}\,\mathrm{d}t \right|^2, \tag{8}$$

which is what people are referring to if they are talking about the power spectrum of a signal.

13.2.3 *Wiener–Khintchine theorem*

The power spectral density function attempts to describe the distribution of a signal's power (i.e. mean square value) over different frequencies. This is related to the information carried by the ACVF. In fact the PSD and the ACVF are a Fourier transform pair. This is a statement of the Wiener–Khintchine theorem:

$$S_{xx}(\omega) = \int_{-\infty}^{\infty} s_{xx}(\tau)\mathrm{e}^{-\mathrm{i}\omega\tau}\,\mathrm{d}\tau. \tag{9}$$

13.2.4 *Cross power spectral density function*

Just as the power spectral density gives a frequency domain decomposition of the autocovariance (through the Wiener–Khintchine theorem), it is possible to define a similar decomposition of the cross-covariance

$$S_{xy}(\omega) = \int_{-\infty}^{\infty} s_{xy}(\tau)\mathrm{e}^{-\mathrm{i}\omega\tau}\,\mathrm{d}\tau, \tag{10}$$

where the function S_{xy} is the **cross power spectral density function** (CPSD) between the signals x and y. The CPSD is a frequency domain expression of the degree of *linear relationship* between x and y. Therefore, the CPSD has major application in the measurement (or, more properly, estimation) of linear transfer functions when x and y are the input and output signals through a system.

13.2.5 *Estimating CPSDs*

We will have problems calculating power spectral density functions for anything but simple deterministic signals, so we need some technique for *estimating* auto and cross power spectra.

Estimating CPSDs uses the Fourier transform method, which may be familiar from estimating auto power spectra, and is a process by which 'raw estimates' formed by the product of truncated Fourier transforms, are averaged

$$S_{xy}(\omega) = \frac{1}{N}\sum_{i=1}^{N} \frac{X_i(\omega)Y_i^*(\omega)}{T}, \tag{11}$$

where the truncated Fourier transforms are calculated over the same time intervals to preserve phase synchrony between the signals. To simplify the notation we shall introduce the idea of the **expected value operator**, $\mathrm{E}[\cdot]$, which returns the averaged value of the operand inside the brackets

$$\mathrm{E}\left[X_T(\omega)Y_T^*(\omega)\right] = \frac{1}{N}\sum_{i=1}^{N} X_i(\omega)Y_i^*(\omega), \tag{12}$$

therefore

$$S_{xy}(\omega) \approx \mathrm{E}\left[\frac{X_T(\omega)Y_T^*(\omega)}{T}\right]. \tag{13}$$

In order to begin to understand the application of the CPSD function, consider the cross power spectral density function between the input, x and output, y of a linear system, H.

$$\begin{aligned} S_{xy}(\omega) &\approx \mathrm{E}\left[\frac{X_T(\omega)Y_T^*(\omega)}{T}\right] \\ &= \mathrm{E}\left[\frac{X_T(\omega)(X_T(\omega)H(\omega))^*}{T}\right]. \end{aligned} \tag{14}$$

Since the transfer function H is a fixed property of the system, we may remove it as a fixed factor from the expected value operator, leaving

$$S_{xy}(\omega) = H(\omega)S_{xx}(\omega). \tag{15}$$

It appears from this equation that the CPSD function should have application in transfer function measurement—the CPSD is actually more powerful than is suggested by our example above, as demonstrated in the following example. A summary of statistical measures is given in Table 13.1.

Table 13.1 Summary of statistical measures, here $X_T = \mathcal{F}\{x_T\}$ etc.

Autospectrum	$S_{xx}(\omega) \approx \mathrm{E}[X_T(\omega)X_T^*(\omega)/T]$
Autocovariance	$s_{xx}(\tau) = \mathcal{F}^{-1}\{S_{xx}(\omega)\}$
Autocorrelation	$R_{xx}(\tau) = s_{xx}(\tau)/s_{xx}(0)$
Cross-spectrum	$S_{xy}(\omega) \approx \mathrm{E}[X_T(\omega)Y_T^*(\omega)/T]$
Cross-covariance	$s_{xy}(\tau) = \mathcal{F}^{-1}\{S_{xy}(\omega)\}$
Cross-correlation	$R_{xy}(\tau) = s_{xy}(\tau)/s_{xy}(0)$

13.2.6 *Transfer function measurement in noise*

Consider Fig. 13.1, which is typical of real world observation of the linear system H. When an input x is applied we cannot observe the response associated with that input—we are always faced with some amount of contaminating noise, n. If we calculate the cross power spectral density function between input and observed

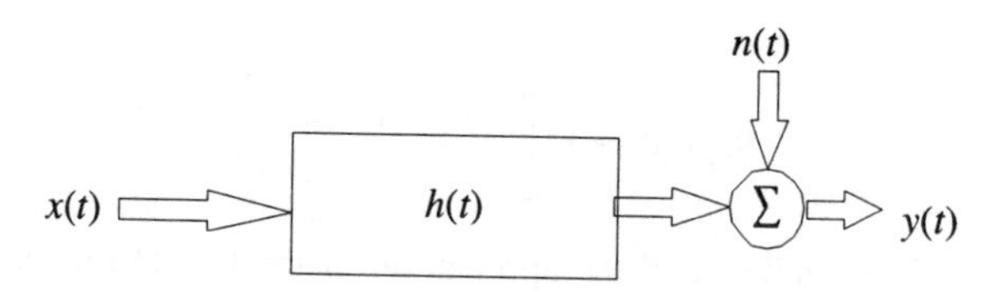

Fig. 13.1 A typical linear system.

response (which includes the inevitable noise) we get

$$S_{xy}(\omega) \approx \mathrm{E}\left[\frac{X_T Y_T^*}{T}\right]$$
$$= \mathrm{E}\left[\frac{X_T(X_T H + N_T)^*}{T}\right]. \tag{16}$$

If we assume that the unwanted noise is independent of X, the CPSD simplifies as X and N produce zero correlation

$$S_{xy}(\omega) = \mathrm{E}\left[\frac{X_T(X_T H)^* + X_T N_T^*}{T}\right] = \mathrm{E}\left[\frac{X_T(X_T H)^*}{T}\right] = S_{xx}(\omega)H(\omega). \tag{17}$$

This important result shows that CPSD functions allow us to estimate linear transfer functions even in noise. In summary:

- When two signals are uncorrelated (as in the example above when $H = 0$), their CPSD function has zero value.
- When two signals are perfectly correlated (as in the example above when $n = 0$), the cross power spectral density's magnitude takes its highest value.
- When two signals are only partially correlated (as in the example above when H and n are non-zero) the CPSD function has an intermediate value.

13.2.7 *The ordinary coherence function*

We have seen that when there is no correlation between two signals, the CPSD function between them has zero value. Conversely, when two signals x and y are perfectly correlated

$$Y(\omega) = X(\omega)H(\omega), \tag{18}$$

and their CPSD takes value

$$S_{xy}(\omega) = H(\omega)S_{xx}(\omega). \tag{19}$$

The magnitude squared CPSD function for such perfectly correlated signals is

$$\begin{aligned} |S_{xy}(\omega)|^2 &= |H(\omega)|^2 |S_{xx}(\omega)|^2 \\ &= S_{xx}(\omega) S_{yy}(\omega). \end{aligned} \tag{20}$$

These two cases represent the limiting values which the magnitude squared CPSD function can assume, and these limits are expressed by **Schwartz's inequality**:

$$0 \leq |S_{xy}(\omega)|^2 \leq S_{xx}(\omega) S_{yy}(\omega). \tag{21}$$

We can quantify how close to perfectly correlated or perfectly uncorrelated two signals are, as a function of frequency, using a statistic derived from Schwartz's inequality. This statistic is called the ordinary coherence function (usually simply called coherence) and is defined as

$$\gamma_{xy}^2(\omega) = \frac{|S_{xy}(\omega)|^2}{S_{xx}(\omega) S_{yy}(\omega)}. \tag{22}$$

It is evident from Schwartz's inequality that the coherence function can range between zero and unity. The coherence function is a measure of the proportion of the *power* in y that is due to linear operations on the signal x. When estimating transfer functions, the coherence function is a useful check on the quality of the data used. In our example of estimating a transfer function in the presence of additive noise, the coherence has unit value when the noise is zero and gets smaller as the noise increases. In order to get the best possible estimate of the transfer function we should try to work in the absence of noise, so transfer function measurements should be made with good coherence.

Frequency analysers capable of measuring transfer functions will be able to display coherence—you should *always* check the coherence associated with 'twin channel' measurements—practically, keeping $\gamma^2 > 0.9$ will ensure good results.

In summary: low coherence implies statistical independence, which could be the result of:

(1) Nonlinear relationship (perhaps the relationship is nonlinear or perhaps there is no relationship at all).
(2) Presence of dependent noise (perhaps another signal path connecting x and y).

13.2.8 *Response of LTI systems to random input*

We are now in a position to describe the response of an LTI system to nondeterministic input. Because the input can only be described statistically, we should anticipate that it will only be possible to make a statistical description of the response.

We are familiar with the fact that an LTI system can be characterized by a **frequency response function**, describing the response of the system to a complex exponential input of any frequency. The frequency response function, $H(\omega)$, must report the gain and phase of the relationship between input and output at each frequency and is, therefore, generally a *complex* function of frequency.

Since we can now relate the frequency domain statistics of input and output to a LTI system, we can begin to estimate the transfer function of an unknown system. Specifically,

$$|H(\omega)|^2 = \frac{S_{yy}(\omega)}{S_{xx}(\omega)}. \tag{23}$$

Unfortunately, this only gives us information about the gain of an LTI system—we are not yet able to observe the phase component. To do this we need to consider cross statistics between input and output,

$$H(\omega) = \frac{S_{xy}(\omega)}{S_{xx}(\omega)}. \tag{24}$$

13.3 Maximum Length Sequences

13.3.1 *Transfer function measurement using maximum length sequences*

Transfer measurement using noise was good, but became extraordinarily useful when it was carried out with maximum length sequences (pseudorandom noise) instead of random noise. Maximum length measurement systems enable the fast and accurate measurement of LTI systems, including acoustics systems. It has much greater immunity to noise than other transfer function measurement systems.

MLS measurement systems use a very similar theory to that used by dual channel FFT analysers; MLS is a particular ruthless and efficient implementation of the method. Consider a noise signal $x(t)$ be applied to an LTI acoustic system with an impulse response $h(t)$. The output from the measurement is a signal $y(t)$.

The time and frequency input/output relationships are

$$y(t) = x(t) * h(t), \tag{25}$$

$$Y(\omega) = X(\omega)H(\omega), \tag{26}$$

where $*$ denotes convolution and capital letters denote the frequency spectrum. Any reasonable dual channel FFT analyser will carry out measurement of H and h using cross-spectra techniques. While this has some noise immunity, it is not a quick process as averaging has to be undertaken. This is because the noise input,

$x(t)$, should ideally be measured over an infinite amount of time to account for the natural fluctuations that any random noise signal has. The measurement has to be undertaken for an infinite time to gain population statistics rather than best estimates. When measuring over a finite time we get inaccuracies; typically, 1024 averages gives an accuracy of about 3%. Using a maximum length sequence noise source removes the need for averaging in many cases.

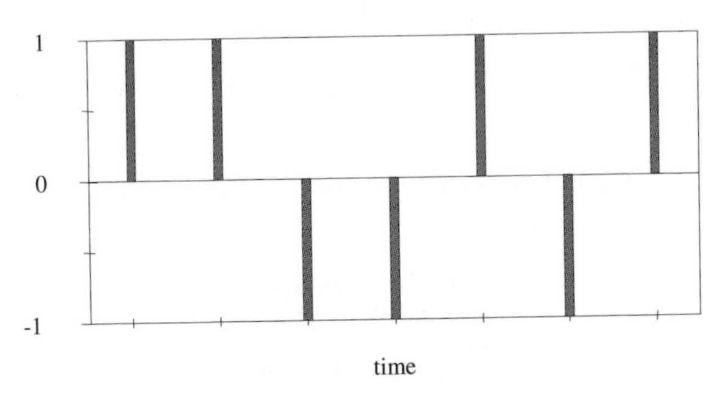

Fig. 13.2 Maximum length sequence for $N = 7$.

The maximum length sequence (MLS) has many properties similar to white noise so the cross-correlation techniques presented above can be used. MLSs are binary sequences; a string of $+1$s and -1s. The sequence is $2^n - 1$ bits long, e.g. the $n = 12$ sequence is 4095 bits long. A length which isn't 2^n is awkward because we can't apply standard FFT algorithms. This problem has been overcome, however, by using a fast Hadamard Transform (see references for more details). With modern computing power, a DFT (discrete Fourier Transform) is usually fast enough where evaluation time is not critical.

MLSs are generated from feedback shift registers constructed from a primitive polynomial. Consider constructing an MLS of length $N = 2^4 - 1 = 15$. A primitive polynomial of length $n = 4$ is:

$$b(x) = x^4 + x^3 + 1. \tag{27}$$

Primitive polynomials of other lengths can be found from tables in books (see references) or the World Wide Web. The corresponding shift register to generate the MLS is shown in Fig. 13.3. Some starting conditions are given to the gates a_n, and then an MLS is generated (note: XOR is equivalent to addition modulo 2). The results are shown in Table 13.2.

In software, it is easier to generate this using a modulo 2 sum

$$a_n = \left(\sum_{i=1}^{n} b_i a_{n-i} \right) \bmod 2. \tag{28}$$

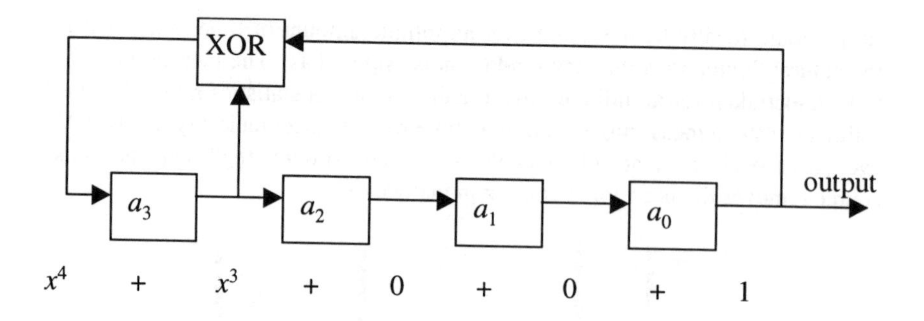

Fig. 13.3 Feedback shift register to generate a MLS.

Table 13.2 Shift register generation of an MLS.

a_3	a_2	a_1	a_0
1	0	0	0
1	1	0	0
1	1	1	0
1	1	1	1
0	1	1	1
1	0	1	1
0	1	0	1
1	0	1	0
1	1	0	1
0	1	1	0
0	0	1	1
1	0	0	1
0	1	0	0
0	0	1	0
0	0	0	1
1	0	0	0
(Starts to repeat)			

This is what is implemented in the script `mlsgenerator.m`. A list of primitive polynomials is given in Table 13.3.

13.3.2 *Measure once*

It is only necessary to measure one complete period of the MLS sequence to determine the transfer function. As the MLS sequence is periodic, one complete period gives population statistics. This is in contrast to the white noise case where to get population statistics we had to average for an infinite amount of time. Conse-

Table 13.3 A selection of primitive polynomials that can be used to generate MLSs.

$x^2 + x + 1$
$x^3 + x + 1$
$x^4 + x + 1$
$x^5 + x^2 + 1$
$x^6 + x + 1$
$x^7 + x + 1$
$x^8 + x^4 + x^3 + x^2 + 1$
$x^9 + x^4 + 1$
$x^{10} + x^3 + 1$
$x^{11} + x^2 + 1$
$x^{12} + x^6 + x^4 + x + 1$
$x^{13} + x^4 + x^3 + x + 1$
$x^{14} + x^5 + x^3 + x + 1$
$x^{15} + x + 1$
$x^{16} + x^5 + x^3 + x^2 + 1$
$x^{17} + x^3 + 1$
$x^{18} + x^7 + 1$
$x^{19} + x^5 + x^2 + x + 1$
$x^{20} + x^3 + 1$

quently, MLS measurement is quicker than using white noise. Note that the MLS only has perfect autocorrelation properties when it is periodic, so in a measurement system it is necessary to play two sequences, one to excite the filter being measured, and the second to actually make the calculations from.

13.3.3 *No truncation errors*

FFT analysers will calculate the cross-correlation via Fourier techniques. This leads to the need for windowing the signals to prevent truncation errors. The MLS is periodic, so provided the calculation is carried out over exactly one period then no truncation errors result.

13.3.4 *Crest factor*

The crest factor (peak/rms) is 1 for an MLS signal, which means that the signal uses the maximum available headroom in making measurements. This provides for a very high signal-to-noise ratio.

Imagine you're trying to record onto tape a piece of classical music and a piece of modern dance music. The dance music is highly compressed and so has a small crest factor. The signal spends most of its time near its maximum level. Consequently, when recording onto tape, the music is nearly always exploiting the maximum signal to noise ratio of the system. With most classical music, however, there is a wide dynamic range. You set the maximum level for the record, but then the quiet bits slip into the noise floor of the tape. Then modern music has a small crest factor, and it is like MLS, whereas random noise is more like the classical music case with a high crest factor.

Note, however, that there are suspicions that this small crest factor can cause problems in some transducers, leading to non-linearities and increased noise. The differences between MLS and random noise for measuring transfer functions is summarised in Table 13.4.

Table 13.4 Comparison of random noise and MLS transfer function measurement systems.

MLS	Dual channel FFT
Used for LTI systems.	Used for LTI systems.
Cross-correlation provides noise immunity.	Cross-correlation provides noise immunity.
Averaging often not needed.	Averaging always needed.
High crest factor: good SNR.	Lower crest factor: lower SNR.
No windowing used.	Windowing and overlap required.
Output correlated with internal input.	Two transducers can be used.
Distortion distorts transfer function.	Distortion appears as discrete impulses in the impulse response.

13.4 Practical

These exercises involve running MATLAB scripts, which can be downloaded from http://www.acoustics.salford.ac.uk/research/mathsbook/.

13.4.1 *FFTs*

This first exercise generates a sine wave, and estimates the power spectrum using a Fast Fourier Transform. If you have never calculated an FFT within MATLAB, you might wish to quickly run through this exercise, otherwise skip to §13.4.2. Load up MATLAB, you'll need the script file `fourierdemo.m`.

The script `fourierdemo.m` creates a sinusoidal wave, takes the FFT and displays the power spectrum. Features to note:

- The application of a window to the time signal before applying a Fourier transform. Why is this done?
- That only the first $N/2 + 1$ points of the spectrum are used. The other points in the spectrum are for negative frequencies and can be ignored for the work here.

13.4.2 *LTIs*

13.4.2.1 `impulse.m`

This script shows the relationship between the input and output signals and the transfer function for an LTI system. The impulse response of the LTI system is also shown. The LTI system is a simple low-pass filter. There is no noise present.

13.4.2.2 `impulse_noise.m`

Noise is now added to the system and the situation becomes more complex. The output power spectrum is shown with and without noise; if the noisy version is used in the calculation of the transfer function using Eq. (23) we would not get a very good estimate for the transfer function. This should give you a sense of why cross spectral techniques are useful.

13.4.3 *Transfer function measurement using noise*

13.4.3.1 `impulse_averages.m`

This script roughly simulates the type of measurement that a dual channel FFT analyser uses to measure transfer functions using noise. The script compares estimates for transfer functions using Eqs. (23) and (24) more closely. We must remember that we are dealing with *estimates* which naturally mean averaging over time. Consequently it is not possible to examine Eqs. (23) and (24) properly unless some averaging is done. The script is set up with 128 averages, and demonstrates the noise immunity of Eq. (24) through `figure(1)`.

Calculate the coherence function for the measurement with and without noise present. (The end of the script lists the defined variables that you will need.) Notice that in the pass band of the filter, the coherence function is OK in the presence of noise, but measurements in the reject band are less certain. Why is the coherence < 1 in the reject band even though there is no noise?

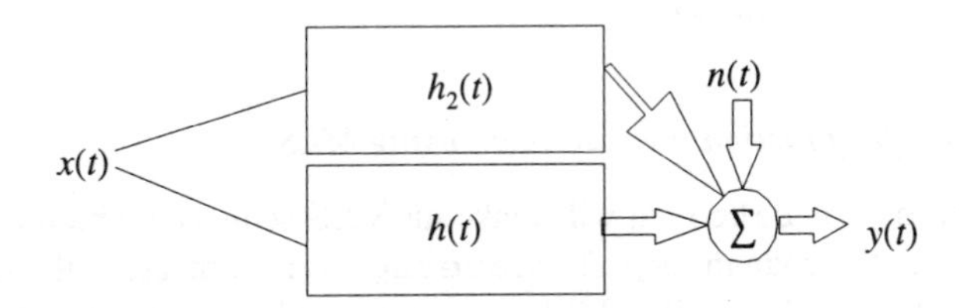

Fig. 13.4 A linear system with dependent noise.

Introduce some dependent noise into the system—this can be done by adding another signal path as shown in Fig. 13.4. Investigate how this corrupts the measurement, the effect this has on measurement accuracy and the coherence function.

Investigate how averaging affects the accuracy of the measurement system. How much improvement in measurement accuracy do you get (in decibels) for a doubling of the number of averages?

If n samples are taken of an autospectrum, then the variance, σ^2 associated with the measurement is

$$\sigma^2\left(|S_{xx}|\right) = \frac{|S_{xx}|^2}{n}. \tag{29}$$

If n samples are taken of a complex cross spectrum $S_{xy} = C_{xy} + \mathrm{i}Q_{xy}$, then the variance, σ^2, associated with the real and imaginary parts of the measurement is

$$\sigma^2\left(C_{xy}\right) = \frac{S_{xx}S_{yy} + C_{xy}^2 - Q_{xy}^2}{n}, \tag{30}$$

$$\sigma^2\left(Q_{xy}\right) = \frac{S_{xx}S_{yy} + Q_{xy}^2 - C_{xy}^2}{n}. \tag{31}$$

This can be used to construct an error in the final transfer function measured. The variance of the measured transfer function (if all the errors are assumed independent) is

$$\sigma^2\left(|H|\right) = \left(\frac{\partial |H|}{\partial Q_{xy}}\right)^2 \sigma^2\left(Q_{xy}\right) + \left(\frac{\partial |H|}{\partial C_{xy}}\right)^2 \sigma^2\left(C_{xy}\right) + \left(\frac{\partial |H|}{\partial S_{xz}}\right)^2 \sigma^2\left(S_{xx}\right), \tag{32}$$

where $|H|$ is given by

$$|H| = \frac{\sqrt{\partial C_{xy}^2 + \partial Q_{xy}^2}}{S_{xx}}. \tag{33}$$

Calculate the expected variance given the formulation of Eqs. (29–33). Compare this to the actual variance measured using the signals from MATLAB. How successful are the formulations?

13.4.4 *Transfer function measurement using MLS*

The code `mlsgenerator.m` will make an MLS sequence which can then be combined with the code in `impulse_averages.m` to measure the LTI system using MLS. Use an $N = 1023$ MLS sequence, to allow ready comparison with the noise results previously generated. Investigate four LTI systems: one perfect, one with noise on the output, one with noise on the input and one with distortion.

By using cross and auto spectral methods, demonstrate the use of MLS in transfer function measurement.

Important notes. To exploit MLS's advantages you *must* use a 1023 length Fourier transform in the cross and auto spectra estimations. MLS only has its wonderful properties if periodicity is assumed. This means MATLAB function `xcorr()` won't be useful here because it uses zero padding. The MATLAB function `fft(x,1023)` will do a non-2^n transform for you. As the system is assumed to be undergoing periodic excitation, you must also play two sequences of the MLS, one to prime the system, calculations are then done off the second period. You should not use windowing with MLS.

Compare the MLS method to using uniform probability noise generated using the MATLAB `rand()` function. The script `impulse_averages.m` might be a useful resource.

Using MATLAB you should investigate the following claims and comments regarding MLS measurement:

(1) When compared to random noise, using MLS removes the need for averaging transfer function estimates in a noise-free environment. Consequently, MLS is a faster measurement method.
(2) MLS is a better test signal than random noise because it equally excites all frequencies (except d.c.) and because the signal is ideally decorrelated from delayed versions of itself.
(3) With noise present, it is often necessary to carry out averaging to get an accurate estimation of the transfer function even when using an MLS signal.
(4) MLS has more immunity to noise interference than random noise measurement because of its low crest factor.
(5) Distortion appears as discrete impulses in the impulse response, this can appear noise-like depending on the type of distortion noise; this cannot be removed by averaging in a time-invariant system. Examine the impulse response of the LTI system with and without distortion.

The distortion added in the example script uses clipping to produce distortion. One of the key issues with MLS is its immunity to time variance. Simulate an LTI system where its characteristics vary over time (you will have to alter the MATLAB function `filter()`), and investigate what effect this has on the transfer function and impulse response. Recent evidence has shown that a swept sine wave (chirp) test signal gives better immunity to problems of time variance. Implement a measurement system using a swept sine wave, and test this claim.

Further Reading

Bendat, J. S. and Piersol, A. G. (1986) *Random Data: Analysis and Measurement Procedures.* John Wiley.

A very useful general text on transfer function measurements, not that accessible, but the most authoritative.

Piersol, A. G. and Bendat, J. S. (1993) *Engineering Applications of Correlation and Spectral Analysis.* John Wiley.

A more applied text book from the authors of the authoritative text.

Bradley, J. S. (1996) Optimizing the decay range in room acoustics measurements using maximum length sequence techniques, *Journal of the Audio Engineering Society* **44**, 4, pp. 266–273.

Interesting sections on distortion artifacts and their influence of RT measurement.

Vanderkooy, J. (1994) Aspects of MLS measuring systems, *Journal of the Audio Engineering Society* **42**, 4, pp. 219–231.

Rife, D. D. and Vanderkooy, J. (1989) Transfer function measurement with maximum length sequences, *Journal of the Audio Engineering Society* **37**, 6, pp. 419–444.

The original paper setting out the basis for the commercial MLS system MLSSA.

Vorlander, M. and Kob, M. (1997) Practical aspects of MLS measurements in building acoustics, *Applied Acoustics* **52**, 3–4, pp. 239–258.

Svensson, U. P. and Nielsen, J. L. (1999) Errors in MLS measurements caused by time variance in acoustic systems, *Journal of the Audio Engineering Society* **47**, 11, pp. 907–927.

Press, W. H., Teukolsky, S. A. Vetterling, W. T. and Flannery, B. P. (1993) *Numerical Recipes in C: The Art of Scientific Computing.* Cambridge University Press.

The numerical techniques book with a nice explanation of FFTs; FORTRAN and PASCAL editions are also available.

Farina, A. (2000) Simultaneous measurement of impulse response and distortion with a swept-sine technique, *Proc. 108th AES Convention, Paris.*

Stan, G. B., Embrechts, J. J. and Archambeau, D. (2002) Comparison of different impulse response measurement techniques, *Journal of the Audio Engineering Society* **50**, 4, pp. 249–262.

There has been a renaissance in the use of swept sine measurement because it is better at dealing with time variance than MLS.

Fan, P. and Darnell, M. (1996) *Sequence Design for Communications Applications.* Wiley.

The best 'cookbook' for making number sequences.

Reed, M. J. and Hawksford, M. O. J. (1996) Identification of discrete Volterra series using maximum length sequences, *IEE Proceedings Circuits Devices & Systems* **143**, 5, pp. 241–248.

How to measure distortion artefacts using MLS.

Chapter 14

Numerical Optimisation

T. J. Cox
Salford Acoustics Audio and Video
University of Salford

P. Darlington
Apple Dynamics Ltd

14.1 Introduction

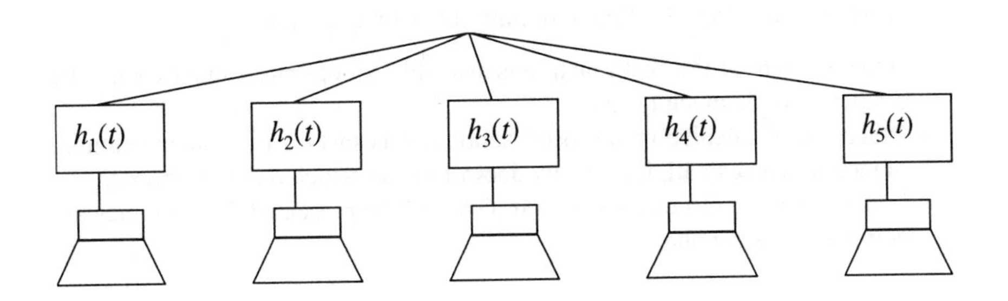

Fig. 14.1 A loudspeaker array optimisation problem.

This chapter will examine some numerical techniques that are commonly used to iteratively solve mathematical problems: optimisation algorithms and adaptive filtering. Numerical optimisation is used in design work, for example consider the system shown in Fig. 14.1. A signal is fed to five loudspeakers via a bank of five digital filters. By altering the digital filter coefficients, we can change the radiation characteristics of the array. What would be the most appropriate filter coefficients

a to achieve omnidirectional radiation? This is the type of task that we can ask a numerical optimiser to solve. But before reaching for such a tool, you need to consider:

- Is there an analytical solution to the problem (in this case there is for an omnidirectional characteristic)?
- What physical understanding of the problem do we have, and how can we capitalise on this to maximise the success of the algorithm (omnidirectional array from highly directional loudspeakers is going to be difficult)?
- Do I have a theoretical (analytical or empirical) model which correctly models the real life situation?
- Will the theoretical model run fast enough?

Essentially the problem is to find a set of numbers **a** (filter coefficients) that minimises some cost function ε (figure of merit, error). In the case above, the cost function would measure the closeness of the radiated polar response to the desired radiated polar response. This could be the reciprocal of a mean square parameter

$$\varepsilon(\mathbf{a}) = \left(\sum_{m=1}^{M} (L_{d,m} - L_{\mathbf{a},m})^2 \right)^{-1}, \tag{1}$$

where there are M measurements on the polar response, $L_{d,m}$ are the desired levels in the polar response and $L_{\mathbf{a},m}$ are the actual levels for the particular set of parameters **a**. Part of using an optimisation problem is determining the most appropriate cost function.

There are a number of different optimisers that we can use:

- If the gradient of the cost function is available, an optimiser that exploits the gradient will be much faster.
- Otherwise, if calculating the cost function is very fast (< 1 s), and the number of parameters **a** small, it probably doesn't matter which one you choose.
- In this chapter a Genetic Algorithm (GA) will be presented, because they are currently very popular.

14.2 Genetic Algorithms

A genetic algorithm mimics the process of evolution that occurs in biology. A population of individuals is randomly formed. Each individual is determined by their genes; in this case the genes are simply the set of numbers a_n which describe the filter coefficients. Each individual (or set of coefficients) has a fitness value (figure of merit) that indicates how good they are at radiating in an omnidirectional manner. Over time new populations are produced by combining (breeding) previous shapes, and the old population dies off. Offspring are produced by pairs

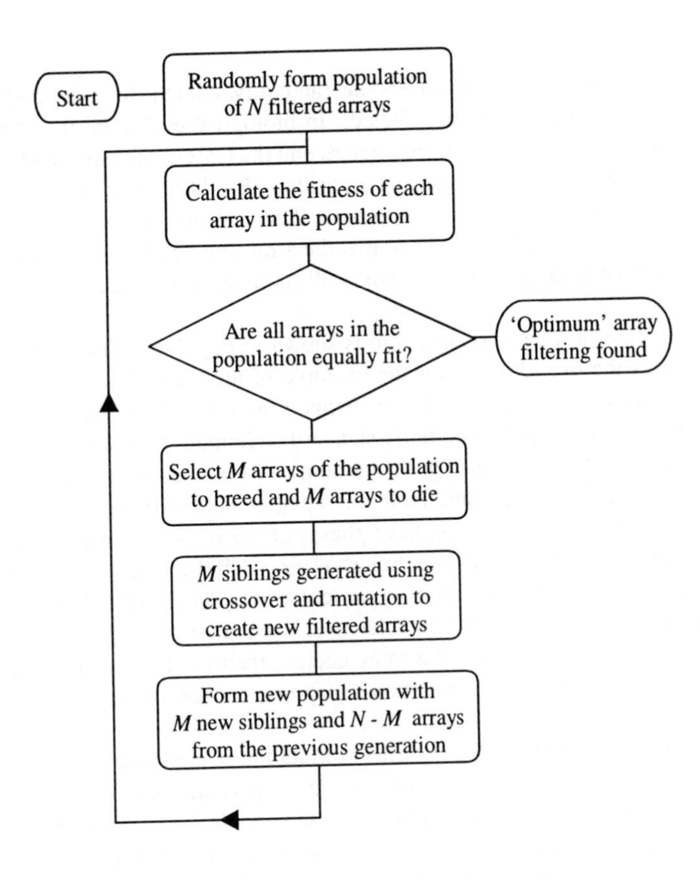

Fig. 14.2 Genetic algorithm flowchart.

of parents breeding, and the offspring has genes that are a composite of the genes from the parents. The offspring shape will then have features drawn from the parent shapes, in the same way that facial features of a child can often be seen in the parents. A common method for mixing the genes is called multipoint crossover. For each gene, there is a 50% chance of the child's gene coming from parent A, and a 50% chance of the gene being from parent B (assuming that the genetic code is held in a binary format.)

If all that happened was a combination of the parent genes, then the system would never look outside the parent population for better solutions. The fish array would never get lungs and walk about on the land. As with biological populations, to enable dramatic changes in the population of array filters, mutation is needed. This is a random procedure whereby there is a small probability of any gene in the child sequence being randomly changed, rather than directly coming from the parents.

Selecting shapes to die off can be done randomly, with the least fit (the poorest arrays) being most likely to be selected. In biological evolution, the fittest are most likely to breed and pass on their genes, and the least fit the most likely to die, this is also true in an artificial genetic algorithm. By these principles, the fitness of successive populations should improve. This process is continued until the population becomes sufficiently fit so that the array can be classified as optimum. A common termination criterion is when all the members of the population are identical.

The process of a genetic algorithm is shown in Fig. 14.2. The script `GA.m`[1] demonstrates the use of a GA for the example of an array of 13 point sources. There are many variations on a GA, in terms of how breeding, mutation, etc are implemented. The script gives a fairly standard technique. It may not be the most efficient coding, but it demonstrates some important points about GAs.

The task is to make an omnidirectional radiation pattern from a set of 13 point sources. These point sources can have phases of $+1$ or -1. Consequently, each individual has 13 genes which are either $+1$ or -1 indicating the phases of the loudspeaker drivers. The fitness of a particular array is calculated via the standard deviations of the radiated energies over a polar response. If all receivers on the polar response received the same energy, then the fitness would be zero. Any deviations from the omnidirectional response causes the standard deviation to increase.

The algorithm needs tuning in terms of mutation rate, population size, rate of dying etc. This is annoying if you are using the algorithm for a one-off test—not so bad for repeat runs. In the example given, the population consists of 50 arrays. A sixth of the population dies in each generation, and the fittest in the generation always survives. For convenience, the number breeding and the number dying are set equal. The choice of the particular individuals to breed and die is done via the cumulative probability distribution. Consider choosing an individual to die. For the nth individual, the cumulative probability distribution c_n is given by:

$$c_n = \frac{\sum_{m=n}^{N} f_m - \min \mathbf{F}}{\max \mathbf{F} - \min \mathbf{F}} \tag{2}$$

where $\mathbf{F}$ is the matrix of fitnesses, with f_n being the fitness of the nth array (individual). There are assumed to be N individuals in the population. A dice is then rolled from 0 to 1 to determine which individual should die. As individuals with poor fitness, in this case large standard deviations, will be associated with large steps in the cumulative distribution, they are most likely to be chosen. The fittest individual, with the smallest fitness value cannot be chosen in this scheme. There

[1] A MATLAB program that can be downloaded from http://www.acoustics.salford.ac.uk/research/mathsbook/.

are other ways of choosing who should breed and die, but this method is fast and efficient.

When breeding, for each gene a dice is rolled from 0 to 1. If the dice value is > 0.5 then the gene comes from parent A, if the dice value is < 0.5 then the gene comes from parent B. The mutation rate is set so there is a 2% chance for each gene mutating. Whether a gene mutates is determined by rolling a dice, and if the gene mutates the state is flipped from -1 to $+1$ or *vice versa*. This is quite a high mutation rate, but is fine because there are a small number of genes in this example.

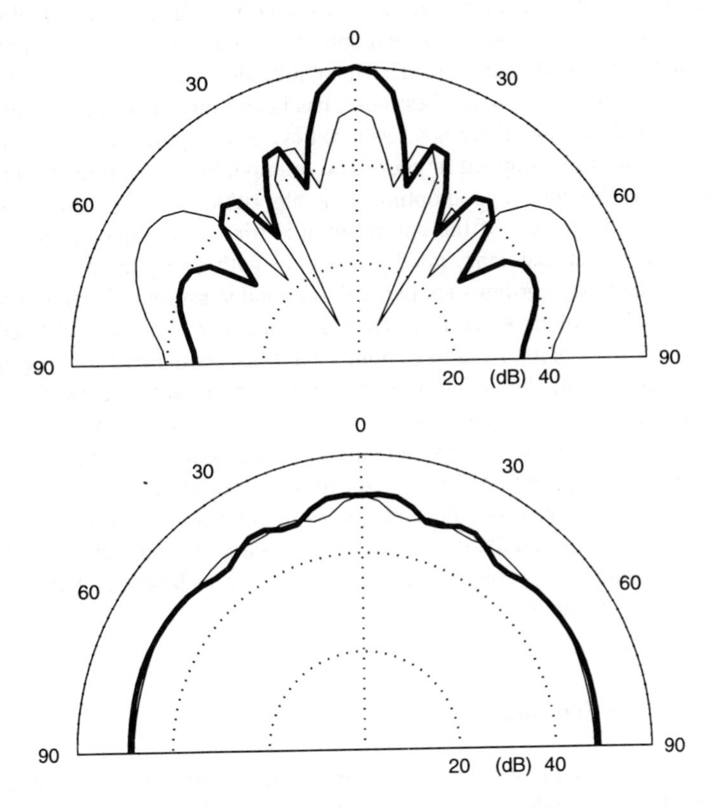

Fig. 14.3 The radiation from four arrays. Top: plane surface (bold) and linear array (normal). Bottom: GA solution (bold) and Barker array (normal).

Figure 14.3 shows the polar responses from various arrays. It compares a line array, where all the point sources radiate in phase with the optimised solution produced by the GA. It shows that the GA produces more arrays that radiate more evenly in all directions. Also shown is an array when the phases are set completely randomly. Most interesting is the line labelled Barker. This is using a $N = 13$ length Barker sequence (Fan and Darnell, 1996) to determine the filter coefficients. This is a number sequence with good autocorrelation properties, i.e. one whose Fourier Transform is flattest possible for a binary sequence. The significance here is that it would have been possible to guess this solution (labelled Barker) and get the coefficients from a text with a lot less effort than involved encoding a GA algorithm. If a possible good solution is known, it is a good idea to include one of these individuals in the starting population.

In any optimisation problem there will be a large number of local minima, but somewhere there will be the numerically lowest point, called the global minimum. A usual analogy for a two-dimensional optimisation is finding the lowest point on a hilly landscape (while blindfolded). The blindfolded person might find the nearest valley bottom and think the best point has been found, not realising that over the next mountain ridge there is a lower valley. The key to a good optimisation algorithm is not to be trapped in poor local minima, but to continue to find deep local minima. Provided a good optimisation algorithm is chosen, this should not be a problem. Especially if the optimisation is tried many times from different starting points as is customary good practice, or with a very large population. It is claimed that GA algorithms are particularly good at avoiding local minima, but they still can be fooled. When there are a large number of degrees of freedom in an optimisation problem, i.e. a large number of filter coefficients to be optimised, the surface describing the variation of the figure of merit with the coefficients becomes very complex. There will be a very large number of minima. It is virtually impossible to find the global minimum unless a large amount of time is used with the optimisation algorithm being started over and over again from a wide variety of places on the error space. Fortunately, as the number of degrees of freedom increase, the need to find the numerical global minimum becomes less important for many optimisation problems, because there will be a large number of equally good solutions available.

14.3 Adaptive Filtering

The LTI systems that have been dealt with previously have been 'constant'. By 'constant' I mean that they have not changed with the input signal. But there is no reason why a filter cannot be designed to adapt to the input signal, and this then opens up some advanced processing techniques—adaptive filtering. By adapting, I mean that the filter coefficients are allowed to change in time.

Consider the problem shown in Fig. 14.4. Noise is fed to the room and collected by a microphone. The noise is also fed to the adaptive filter which is expected to process the signal in exactly the same way as the room system. The difference between the microphone signal and the output of the adaptive filter is

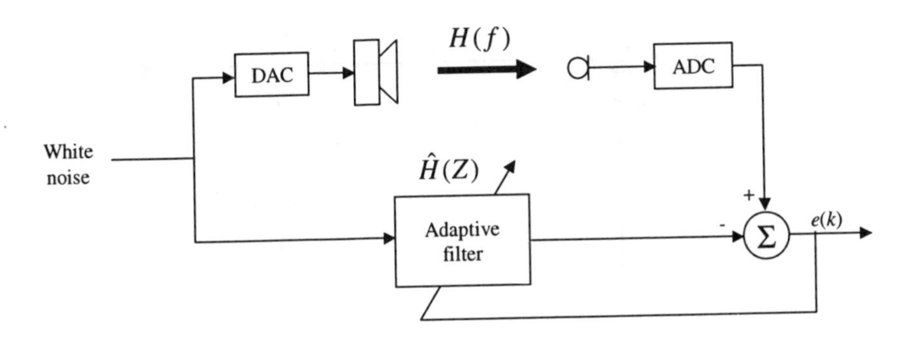

Fig. 14.4 An optimisation problem.

the error. Ideally, the error should be zero. If it isn't zero, this error is used to update the adaptive filter, so it makes a better estimation in the future. The adaptive filter is simply an FIR or IIR filter, what changes in the adaptive filter are the filter coefficients. So in many ways this is like a numerical optimisation, which might be tackled by a GA, but in adaptive filtering (such as active noise control) the Least Mean Square (LMS) approach is more common. The above is an example of system identification.

So why is this useful? Take a speech communication system as an example. There are a number of points within a speech communication system in which the intelligibility of the spoken message is threatened by noise. This noise may be acoustically or electrically generated, or it may be the reverberant effects of a long system impulse response generating a signal which appears to be noise e.g. room reverberation. Fortunately, as a consequence of linearity (which applies at small signal levels in the acoustic domain and, given appropriate circuit design, in the electronic domain) it is possible to remove some of the additive noise using signal processing techniques.

14.3.1 *Basics*

Consider the system described by Fig. 14.5. The speech signal s is corrupted by the addition of the noise signal n at the first summing node, generating the observable signal d. At the second summing node, a signal y is subtracted from d. The result of this subtraction is the error signal, e.

- If the signal y is a copy of signal n, then the noise corruption on the signal s is removed, $e = s$.
- If y is a reasonable approximation of n, then some of the noise contamination is removed, $e \approx s$.
- If y is largely uncorrelated with n then the second summing node represents

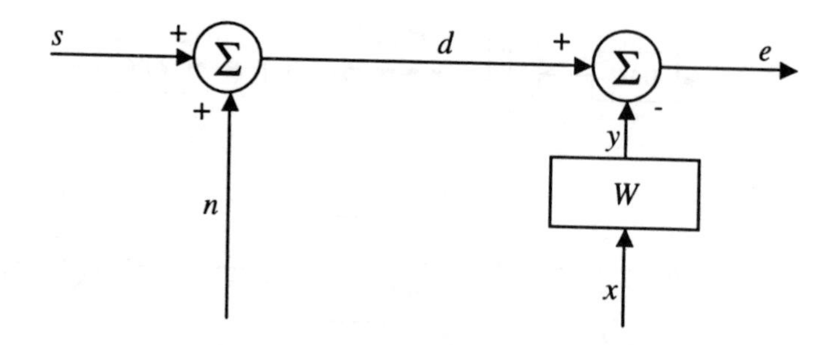

Fig. 14.5 The fundamental structure of a noise cancelling system.

an additional source of noise, further corrupting the speech component in e.

The cancelling signal y is derived by filtering operations (through the filter block W which is an adaptive filter) on the reference signal x. We now consider the necessary relationship between the signals and the optimal configuration of the filter which result in attenuation of the noise component of d.

The error signal, e, can be written as

$$e = s + n - Wx, \tag{3}$$

in which all the signal variables and the filter response are complex functions of frequency. The magnitude squared error is then

$$|e|^2 = |s+n|^2 - |W|^2|x|^2 - 2\,\mathrm{Re}[(s+n)(Wx)^*], \tag{4}$$

in which * denotes the complex conjugate and $\mathrm{Re}[\cdot]$ denotes real part. In order that the analysis can deal with the non-deterministic signal types, we introduce some averaging of the mean squared error signal, using the expected value operator

$$\begin{aligned}\mathrm{E}[|e|^2] = \mathrm{E}[|s|^2] + \mathrm{E}[|n|^2] + 2\,\mathrm{E}[sn^*] + |W|^2\,\mathrm{E}[|x|^2] \\ \text{-}2\,\mathrm{Re}[W\,\mathrm{E}[sx^*]] - 2\,\mathrm{Re}[W\,\mathrm{E}[nx^*]].\end{aligned} \tag{5}$$

We now make some assumptions about the statistical relationship between the signals:

(1) The noise n is uncorrelated with the speech s.
(2) The reference x is correlated with the noise n (and so, by 1), is uncorrelated with the speech s.

Under these assumptions, Eq. (5) simplifies (since the averaged product of uncorrelated signals is zero) to

$$\mathrm{E}[|e|^2] = \mathrm{E}[|s|^2] + \mathrm{E}[|n|^2] + |W|^2\,\mathrm{E}[|x|^2] - 2\,\mathrm{Re}[W\,\mathrm{E}[nx^*]]. \tag{6}$$

Notice that Eq. (6) is a positive definite quadratic function of W—the mean square error will have a unique minimum value when the (complex) filter W is correctly adjusted to W_{opt} which minimises the error, which we shall now identify.

If we differentiate Eq. (6) with respect to **w**

$$\frac{\partial \, \mathrm{E}[|e|^2]}{\partial W} = 2|W| \, \mathrm{E}[|x|^2] - 2 \, \mathrm{E}[nx^*], \tag{7}$$

then the derivative will be zero at the minimum value of the mean square error. Equating Eq. (7) to zero allows us to solve for the optimal filter W_{opt}

$$\frac{\partial \, \mathrm{E}[|e|^2]}{\partial W} = 2|W| \, \mathrm{E}[|x|^2] - 2 \, \mathrm{E}[nx^*] = 0,$$

so

$$W_{opt} = \frac{\mathrm{E}[nx^*]}{\mathrm{E}[|x|^2]}. \tag{8}$$

The expected value of nx^* is proportional to the cross power spectral density function between n and x and the expected value of the magnitude square of x is proportional to the auto power spectral density of x (see previous chapter). So Eq. (8) may be rewritten as

$$W_{opt} = \frac{S^*_{nx}(\omega)}{S_{xx}(\omega)}, \tag{9}$$

which is (by definition) the transfer function between the reference signal x (interpreted as input) and the noise signal n (interpreted as output). In other words, the optimal configuration of the cancelling filter W_{opt}, is the inverse of the filter relating n (input) and x (output).

This can be seen in Fig. 14.6, in which the transfer function e/s is seen by inspection to be trivially 1; the noise added at the first summing node is perfectly cancelled at the second summing node. The noise is attenuated by ∞ dB. Such

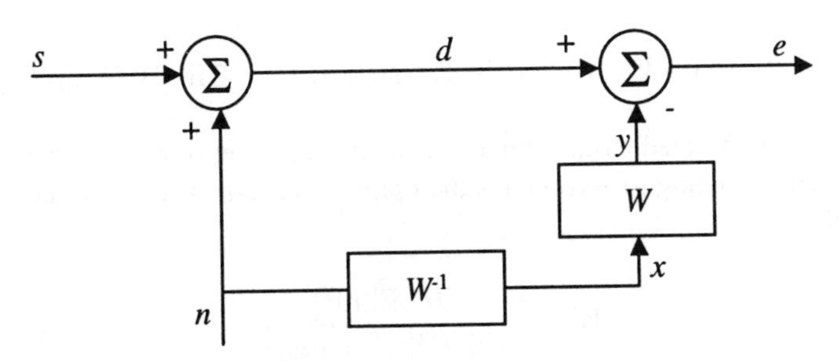

Fig. 14.6 Operation of the noise canceller in the idealised case of perfect correlation between n and x.

perfect performance is never achieved in practice for several reasons, most important among which are:

(1) Imperfect implementation of the cancelling filter.
(2) Imperfect correlation between the noise n and the reference x.

We shall examine the second cause in detail. Imperfect correlation between n and x can be modelled by the system depicted in Fig. 14.7, in which the additional noise n_2 represents those components of x which are not correlated with n (we further assume that n_2 is independent of s). The LTI transfer function relating the correlated components of x to n is relabelled H for generality. The error signal in

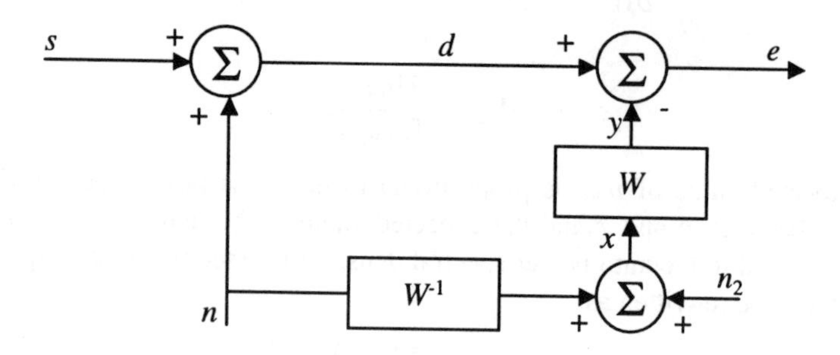

Fig. 14.7 Noise Cancelling with imperfect correlation between noise and reference (second uncorrelated noise signal, n_2, models uncorrelated component of x.)

Fig. 14.7 is now

$$E = s + n - W(Hn + n_2), \tag{10}$$

and the mean square error (which is still a positive definite quadratic function of W) is now

$$\mathrm{E}[|e|^2] = \mathrm{E}[|s|^2] + \mathrm{E}[|n|^2] + |W|^2\,\mathrm{E}[|Hn + n_2|^2] - 2\,\mathrm{Re}[WH\,\mathrm{E}[|n|]^2], \tag{11}$$

where the uncorrelated terms which vanish have been set to zero. Differentiating Eq. (11) and equating to zero yields the optimal cancelling filter for the configuration of Fig. 14.4

$$W_{\text{opt}} = \frac{H^*\,\mathrm{E}[|n|^2]}{\mathrm{E}[|H|^2|n|^2 + |n_2|^2]}. \tag{12}$$

Again we shall introduce the auto and cross spectra. But in addition, we shall also bring in the ordinary coherence γ function that we met when dealing with transfer function measurement in noise:

$$\gamma_{n,x}^2 = \frac{|S_{nx}|^2}{S_{nn}S_{xx}} = \frac{|HS_{nn}|^2}{S_{nn}(|H|^2 S_{nn} + S_{n_2 n_2})}. \tag{13}$$

This then gives the optimum cancellation condition as

$$W_{\text{opt}} = \frac{\gamma_{n,x}^2(\omega)}{H(\omega)}. \tag{14}$$

Notice that the introduction of the uncorrelated noise does not change the phase of the optimal filter (defined by the inverse of H), it simply scales the optimal canceller gain defined in Eq. (10) by the coherence (remember that the coherence function is purely real.)

The noise is attenuated by (consider Fig. 14.4)

$$10\log_{10}\left[\frac{S_{nn}(\omega)}{S_{nn}(\omega)|1 - WH|^2 + |W|^2 S_{n_2 n_2}(\omega)}\right], \tag{15}$$

which, assuming the optimal filter W_{opt}, is used, can be written as

$$10\log_{10}\left[\frac{S_{nn}(\omega)}{S_{nn}(\omega)(1-\gamma_{n,x}^2)^2 + S_{n_2 n_2}(\omega)\gamma_{n,x}^2/|H|^2}\right]$$
$$= 10\log_{10}\left[\frac{1}{(1-\gamma_{n,x}^2)^2 + (1-\gamma_{n,x}^2)\gamma_{n,x}^2/|H|^2}\right]. \tag{16}$$

This shows that the attenuation of the noise component on d is a function of the coherence between noise and reference signal (and the magnitude of the transfer function H). The attenuation increases as the coherence increases (although, of course, the coherence function is bounded in magnitude between zero and 1.) The relationship between attenuation and coherence is plotted in Fig. 14.5, in which the magnitude of H has been normalised to unity.

Note that useful levels of noise attenuation can only be achieved with high coherence between the reference and the noise signal to be cancelled—this poses a particular problem in electroacoustic applications. Given the expression Eq. (16) for the amount of noise attenuation expected from the cancelling system, the expected improvement in speech to noise ratio is numerically equal to the amount of noise attenuation achieved (the speech being unaffected by the system when the noise and speech are truly uncorrelated).

14.3.2 *Discrete-time formulation of the noise cancelling problem*

Implementing the optimal cancelling filter is a non-trivial task and noise cancellation only really became feasible when electronic components allowed the filter to be implemented digitally. This not only allowed a precise, repeatable filter to

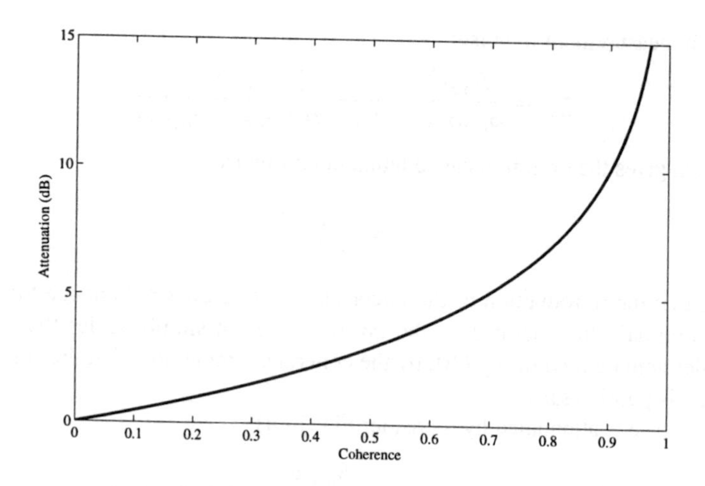

Fig. 14.8 The attenuation possible for various degrees of coherence between noise and reference.

be constructed but also allowed the system to become self-tuning, using adaptive filtering techniques (see next section). In order to understand the contemporary adaptive filtering approaches to noise cancelling it is necessary to recast the (frequency domain) analysis of the previous section into a (discrete) time analysis—this is the purpose of the present section.

In discrete time form, the error sequence in Fig. 14.5 is given by

$$e_k = s_k + n_k - \mathbf{w}_k^T \mathbf{x}_k, \tag{17}$$

in which k is the iteration index i.e. time. The superscript T denotes transposition, such that $\mathbf{w}^T\mathbf{x}$ is the scalar product of the impulse response vector of the filter $\mathbf{w}$ and the vector of present and past reference inputs, $\mathbf{x}$. The length of $\mathbf{w}$ and $\mathbf{x}$ is L, such that Eq. (17) could be written as

$$e_k = s_k + n_k - \sum_{j=0}^{L-1} w_j x_{k-j}. \tag{18}$$

So the last term in Eq. (17) is a normal filtering process, and this writing with transposition is essentially a vector representation of a filtering process you should now be fairly familiar with. We write in this form because it is standard practice in adaptive filtering, not just to confuse! The squared instantaneous error is

$$e_k^2 = s_k^2 + n_k^2 + 2s_k n_k - 2s_k \mathbf{w}_k^T \mathbf{x}_k - 2n_k \mathbf{w}_k^T \mathbf{x}_k + \mathbf{w}_k^T \mathbf{x}_k \mathbf{x}_k^T \mathbf{w}_k. \tag{19}$$

If we make the same assumptions about the statistical relationship between the signals as in the previous section and apply averaging, the expected squared error is

$$\mathrm{E}[e_k^2] = \mathrm{E}[s_k^2] + \mathrm{E}[n_k^2] - 2\mathbf{w}^T\,\mathrm{E}[n_k\mathbf{x}_k] + \mathbf{w}^T\,\mathrm{E}[\mathbf{x}_k\mathbf{x}_k^T]\mathbf{w}. \tag{20}$$

Notice that the expected squared error is a positive real quadratic function of the L-dimensional space spanned by $\mathbf{w}$—it has a minimum value at a unique position, $\mathbf{w}_{\text{opt}}$ which we identify by the same process of differentiation and equating to zero as before. The derivative is

$$\frac{\partial\,\mathrm{E}[e_k^2]}{\partial\mathbf{w}} = -2\,\mathrm{E}[n_k\mathbf{x}_k] + 2\,\mathrm{E}[\mathbf{x}_k\mathbf{x}_k^T]\mathbf{w}, \tag{21}$$

which leads to an optimal impulse response vector for the transversal (FIR) filter $\mathbf{w}$ of

$$\mathbf{w}_{\text{opt}} = \frac{\mathrm{E}[n_k\mathbf{x}_k]}{\mathrm{E}[\mathbf{x}_k\mathbf{x}_k^T]}, \tag{22}$$

in which $\mathrm{E}[n\mathbf{x}]$ is the $1 \times L$ cross covariance vector between n and x and $\mathrm{E}[\mathbf{x}\mathbf{x}^T]$ is the $L \times L$ (auto) covariance matrix of x. (Note the subscripts k disappear if the signals are stationary). Equation (22) is the discrete-time equivalent of Eq. (8) (and has been written in similar notation to emphasise the similarity.)

The computation of the coefficients of the optimal filter, defined by the impulse response vector $\mathbf{w}_{\text{opt}}$, is a non-trivial computational task, as a result of:

- The averaging process (implied by the expectation operators) in the computation of the covariances.
- The matrix inversion.

Fortunately, a computationally efficient iterative approach to the identification of $\mathbf{w}_{\text{opt}}$ exists—an adaptive filter. This is studied in the next section.

14.3.3 *Adaptive methods for noise cancellation*

We have noted that the expected squared error Eq. (20) associated with the noise cancelling system of Fig. 14.1 is a simple quadratic function of $\mathbf{w}$. Although it is possible to identify the optimal $\mathbf{w}$, associated with the minimum expected squared error, in a one-step process: Eq. (22), it is possible to find the minimum value using an iterative technique, using gradient searching methods.

The 'surface' defined by function in Eq. (20) has a value of expected squared error associated with every filter $\mathbf{w}$. If the current filter is $\mathbf{w}_j$, then we may refine the filter design by moving in the direction of the negative gradient ('falling downhill' towards the minimum at $\mathbf{w}_{\text{opt}}$)

$$\mathbf{w}_{j+1} = \mathbf{w}_j - \alpha \left.\frac{\partial\,\mathrm{E}[e^2]}{\partial\mathbf{w}}\right|_{\mathbf{w}=\mathbf{w}_j}, \tag{23}$$

in which α is a positive scalar which determines the step size or update rate and the subscript j is an iteration index for the update process (the averaging processes invoked in Eq. (23) may not allow the update to occur at each sample instant, i.e. $j \neq k$).

Since we have an expression for the gradient, Eq. (21), we may substitute this into Eq. (23) to give

$$\mathbf{w}_{j+1} = \mathbf{w}_j + \alpha[2\,\mathrm{E}[n\mathbf{x}] - 2\,\mathrm{E}[\mathbf{x}\mathbf{x}^T]\mathbf{w}_j], \tag{24}$$

which will converge towards $\mathbf{w}_{\text{opt}}$, as long as α is sufficiently small to make the process stable. The stability of the search process can (in this case) be examined analytically.

The gradient searching method of Eq. (24) has the advantage of simplicity and assured convergence, but it is computationally expensive if either the system is changing or the signals are not stationary on the timescale of the adaptation of the filter, in which case the covariances have to be continually estimated. As both of these conditions are likely to be met in a speech communication system an alternative approach may be more useful. Such an approach can be generated by attempting to minimise the instantaneous squared error Eq. (19) rather than the expected squared error. The inevitable noise on the instantaneous signals (when they are interpreted as estimates of the population statistics) gives the resulting gradient search technique its name.

14.3.4 *Stochastic gradient search methods*

A search strategy which uses

$$\mathbf{w}_{k+1} = \mathbf{w}_k - \alpha \left.\frac{\partial e_k^2}{\partial \mathbf{w}}\right|_{\mathbf{w}=\mathbf{w}_k}, \tag{25}$$

has been found to converge to $\mathbf{w}_{\text{opt}}$, provided the step size parameter α is suitably chosen. Substituting for the derivative gives

$$\begin{aligned}\mathbf{w}_{k+1} &= \mathbf{w}_k - \alpha[-2n_k\mathbf{x}_k + 2\mathbf{x}_k\mathbf{x}_k^T\mathbf{w}_k] \\ &= \mathbf{w}_k - 2\alpha[(y_k - n_k)\mathbf{x}_k],\end{aligned} \tag{26}$$

(y is defined in Fig. 14.1). Noting that $y - n$ is the error when $s = 0$ (and is equivalently the 'correlated' error even in the presence of the non-zero but uncorrelated s) allows Eq. (26) to be re-written as

$$\mathbf{w}_{k+1} = \mathbf{w}_k + 2\alpha e_k\mathbf{x}_k. \tag{27}$$

This is the **Least-Mean-Square (LMS) algorithm** discovered by Widrow and Hoff. It has been found to be robustly stable in many practical applications and is clearly a simple and computationally efficient approach to identifying $\mathbf{w}_{opt}$. It forms the basis of most contemporary adaptive noise cancelling systems.

The performance of the LMS algorithm is illustrated below by example and also in the MATLAB script `simple_lms.m`.

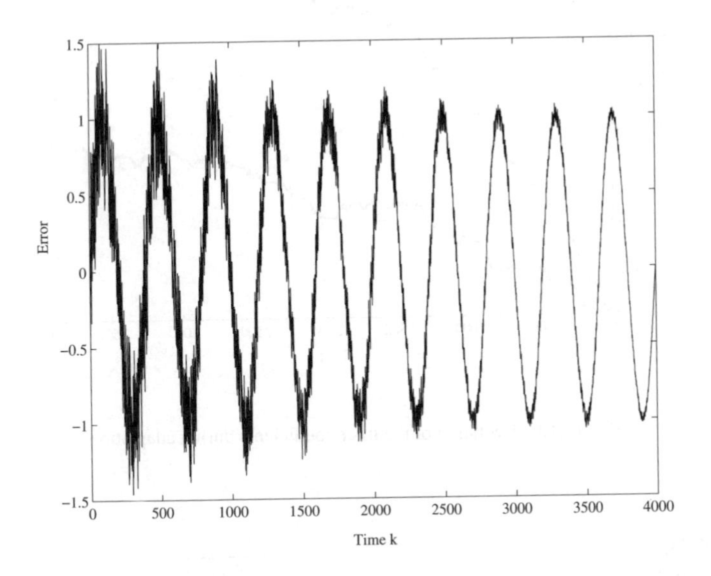

Fig. 14.9 Evolution of the error signal during adaptation.

A simulation of a discrete time implementation of Fig. 14.6 was coded, in which

$$n_k = 0.5x_k + 0.2x_{k-1}, \tag{28}$$

and an $L = 2$ adaptive filter W was updated using the LMS algorithm Eq. (27) in a signal environment in which s was a simple sinusoid and x a random process. The error signal is shown in Fig. 14.9.

The initial noise is seen to be quickly cancelled, leaving a pure sinusoid—the signal s. Notice that the decay of the noise follows a roughly exponential form. This is due to the fact that the convergence behaviour of the LMS algorithm approximates the first order convergence of the true steepest descent algorithm. However, precise stable bounds for the LMS algorithm can only be determined in the context of certain simple deterministic reference signals.

The convergence of the two elements of the $L = 2$ impulse response vector (the weights of the adaptive filter $\mathbf{w}$) are shown in Fig. 14.10. The weights are seen to approach the optimal values implied by Eq. (28). This is very similar to

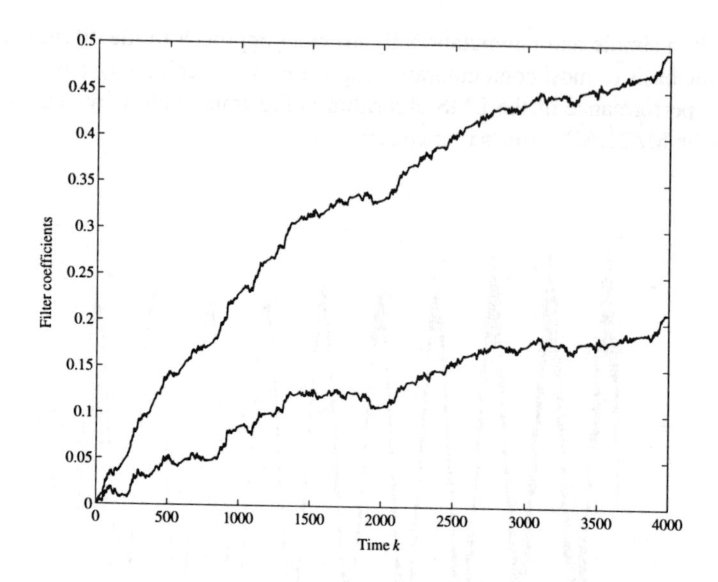

Fig. 14.10 Evolution of the filter coefficients during adaptation.

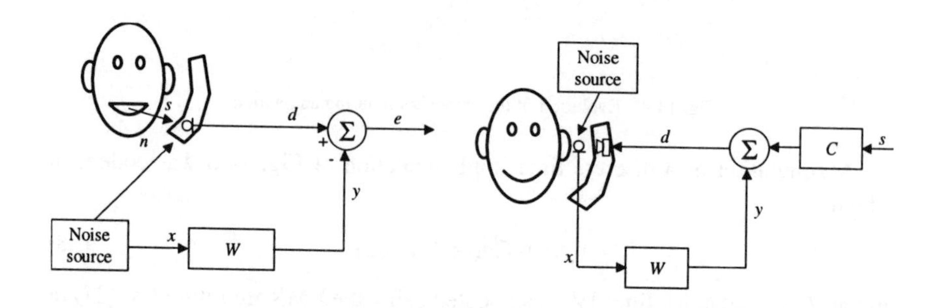

Fig. 14.11 Noise cancelling on a telephone microphone (left); active noise reduction at the earpiece of a telephone (right).

the system shown in Fig. 14.6, except that the white noise signal is fed direct to x and then filtered to get the signal n. These algorithms are used in a variety of systems, two examples are given in Fig. 14.11, others are given by Nelson and Elliott (1992).

References

Fan, P. and Darnell, M. (1996) *Sequence Design for Communication Applications.* Wiley.

Nelson, P. A. and Elliott, S. J. (1992) *Active Control of Sound.* Academic Press.

Widrow, B. and Stearns, S. D. (1985) *Adaptive Signal Processing.* Prentice Hall.

Index